Y0-AAA-540

Communicating Mathematics

Brett Groehler/University of Minnesota Duluth

Joseph A. Gallian

Contemporary Mathematics

479

Communicating Mathematics

A Conference in Honor of Joseph A. Gallian's 65th Birthday
July 16–19, 2007
University of Minnesota
Duluth, Minnesota

Timothy Y. Chow
Daniel C. Isaksen
Editors

American Mathematical Society
Providence, Rhode Island

2000 *Mathematics Subject Classification.* Primary 00B20, 00B30.

Library of Congress Cataloging-in-Publication Data

Communicating mathematics : proceedings of a conference in honor of Joseph A. Gallian's 65th birthday, July 16–19, 2007, Duluth, Minnesota / Timothy Y. Chow, Daniel C. Isaksen, editors.
p. cm. — (Contemporary mathematics ; v. 479)
Includes bibliographical references.
ISBN 978-0-8218-4345-1 (alk. paper)
1. Communication in mathematics—United States—Congresses. 2. Mathematics—Research—United States—Congresses. 3. Mathematics—Study and teaching—United States—Congresses. 4. Mathematics—United States—Data processing—Congresses. I. Gallian, Joseph A. II. Chow, Timothy Y., 1969– III. Isaksen, Daniel C., 1972–

QA41.4.C66 2009
510—dc22 2008036377

♾ The paper used in this book is acid-free and falls within the guidelines
established to ensure permanence and durability.
Visit the AMS home page at http://www.ams.org/

10 9 8 7 6 5 4 3 2 1 14 13 12 11 10 09

Contents

Preface

As anyone who has met him knows, Joseph Gallian is unique. It is not just his achievements that make him unique, even though there is a long list of them: his highly successful summer research experience for undergraduates (REU), started in the days when most people believed undergraduates incapable of doing mathematics research; his textbooks *Contemporary Abstract Algebra* and *For All Practical Purposes: Mathematical Literacy in Today's World* (co-author); his co-directorship of Project NExT; his numerous research articles; his countless awards and honors for teaching and service; and his legendary mathematics lectures that are accessible to broad audiences. Rather, Gallian is unique because of his dynamic personality, his infectious enthusiasm, and his love for mathematics, all of which have combined to inspire research, encourage networking among mathematicians, and communicate mathematics effectively.

The value of Gallian's contribution to the mathematical community cannot be overstated. The star researcher may contribute to mathematics by writing seminal papers, and the gifted expositor may contribute to mathematics by writing classic textbooks, but the one who *enables* dozens of young mathematicians to become star researchers and gifted expositors contributes far more. Gallian is the enabler *par excellence.* He has an egalitarian philosophy that mathematics is not an exclusive, elitist club, but is open to all. It is no accident that many alumni of the Duluth REU are not only outstanding researchers but also excellent communicators, because clear exposition is critical to breaking down the barriers that stand in the way of mathematical progress.

To celebrate the Gallian vision of mathematics, the conference *Communicating Mathematics* was held at the University of Minnesota Duluth in July 2007. This conference marked the 30th anniversary of the Duluth REU, as well as Gallian's 65th birthday. The overarching objective of the gathering was to inspire research productivity and enthusiasm among mathematicians at all stages of their careers. The conference program consisted of colloquium-style lectures delivered by experts to communicate current research ideas in a wide variety of mathematical fields, together with shorter contributed talks, and many opportunities for participants to develop and renew meaningful collaborations.

In keeping with Gallian's uniqueness, we have tried to make the volume you are now holding unique, and more than just another conference proceedings. Some articles are research articles and others are purely expository, but in all cases we have pushed the authors to produce works of high expository quality. The papers are intended to be accessible to a general mathematics audience, including first-year or second-year graduate students. We hope that this volume will inspire research productivity and celebrate fine exposition, and will be especially useful for junior researchers as well as mathematicians considering a change in research area. We also expect that it will be used more generally by mathematicians looking to enrich themselves and their research programs by looking at the problems and techniques used in other areas. We would especially like to draw the reader's attention to the biography of Gallian written by Aparna Higgins, one of Gallian's close collaborators.

In short, if the reader of this volume catches some of the excitement of the Gallian legacy and is inspired to create and communicate mathematics effectively, then we will have succeeded in our aims. Enjoy!

Acknowledgements. The editors appreciate the valuable contributions of many people in preparing and assembling this volume. In particular, they thank Brett Groehler of the University of Minnesota Duluth for the photograph of Gallian; Geir Helleloid for editorial assistance; and Christine Thivierge of the AMS for assistance and guidance throughout the publication process.

Contemporary Mathematics
Volume **479**, 2009

A Journey of Discovery: Orthogonal Matrices and Wireless Communications

Sarah Spence Adams

Dedicated to Joseph Gallian on his 65th birthday and the 30th anniversary of his Duluth REU

ABSTRACT. Real orthogonal designs were first introduced in the 1970's, followed shortly by the introduction of complex orthogonal designs. These designs can be described simply as square matrices whose columns are formally orthogonal, and their existence criteria depend on number theoretic results from the turn of the century. In 1999, generalizations of these designs were applied in the development of successful wireless communication systems, renewing interest in the theory of these orthogonal designs. This area of study represents a beautiful marriage of classical mathematics and modern engineering.

This paper has two main goals. First, we provide a brief and accessible introduction to orthogonal design theory and related wireless communication systems. We include neither the mathematical proofs of the relevant results nor the technical implementation details, rather hoping that this gentle introduction will whet the reader's appetite for further study of the relevant mathematics, the relevant engineering implementations, or, in the best case scenario, both. Second, in light of the dedication of this paper to Joe Gallian, who was an extraordinary undergraduate research advisor to so many of us and who inspired so many of us to become undergraduate research advisors ourselves, we describe the involvement of undergraduates in research on these orthogonal designs and related communications systems.

1. Introduction and Motivation

Three accepted facts regarding wireless communications systems are that bandwidth is scarce, that multipath fading must be combated, and that wireless communications systems have grown into extremely complex systems made of many interacting components [**25**]. *Space-time block codes* (defined formally in Section 2) address each of these issues by elegantly combining two forms of diversity: space diversity and time diversity. This combination of diversity was a major step in moving the capacity of wireless communications systems towards the theoretical limits.

1991 *Mathematics Subject Classification.* Primary 05B20, Secondary 94A05, 62K05, 62K10.
The author was supported in part by NSA Grant H98230-07-1-0022.

A certain class of space-time block codes, known as complex orthogonal space-time block codes, are built using mathematical constructs known as generalized complex orthogonal designs. Orthogonal designs were first introduced in the mid-1970's [**10, 12, 13, 14**], however, mathematical results in other contexts dating from the 1890's laid the foundation for these combinatorial structures [**1, 2, 3, 16, 17, 24**]. The application of orthogonal designs and their generalizations as as complex orthogonal space-time block codes has been successful. For example, these codes are used in the Third Generation Mobile Standard and the proposed standard for wireless LANs IEEE 802.11n.

Our introductory overview of complex orthogonal space-time block codes does not attempt to match the technical rigor or completeness of Calderbank and Naguib's survey on this topic [**7**]. We refer the reader to their survey, as well as to the other articles in the bibliography, for a more advanced treatment of the relevant mathematics and implementation details. The books by Jafarkhani [**18**] and Larsson and Stoica [**20**] provide particularly comprehensive treatments of the topic.

In Section 2, we provide the necessary definitions and background information. In Section 3, we discuss the two main research problems in this area, including a narrative journey of undergraduate involvement in the solution of one of these main problems. Section 4 provides a brief overview of other related research problems, and the paper is concluded in Section 5.

2. Definitions and Preliminaries

In modern wireless communication systems, data are sent from one or more transmit antennas to one or more receive antennas. In this paper, we consider one model for sending data over fading channels using multiple transmit antennas and one or more receive antennas. The motivation for using multiple transmit antennas is that multiple copies of data sent from different physical positions may be corrupted in different ways during transmission, which may better allow the receiver to recover the intended data. In other words, through appropriate signal processing at the receiver, the independent paths provided by the multiple transmit antennas can be considered as one channel that is more reliable than any of the single independent paths. This idea is known as space diversity. The model we consider also sends multiple copies of data at different timesteps, again increasing the likelihood that the receiver can recover the intended data. This idea is known as time diversity. Space and time diversity are explained in more depth in several of our references, e.g., [**7, 18, 20**]. In the model that we consider, the data (or signals) sent between antennas are modelled using complex variables.

In the most general form, a *space-time block code* for n transmit antennas is a mapping from k complex variables $\{z_1, \ldots, z_k\}$ onto an $r \times n$ matrix $\mathbf{G}$, wherein each of the n columns of $\mathbf{G}$ represents the transmissions of a distinct antenna and each of the rows represents the transmissions at a given timestep. The entries in the matrix determine which antenna should send which signal at which time, so that an entry of z_l in position (i, j) indicates that the signal corresponding to the complex variable z_l should be transmitted by the jth antenna in the ith timestep. An entry of 0 indicates that the corresponding antenna does not transmit during the corresponding timestep. As the number of rows r represents the number of timesteps required to send the given data, and as this model requires us to receive all of the data before we can decode, r is referred to as the *decoding delay* of the code. The ratio k/r of number of distinct variables to the number of rows is a measure of the code's efficiency, and this is referred to as the *rate* of the code.

Complex orthogonal space-time block codes (COSTBCs) require that the matrix $\mathbf{G}$ is a *generalized complex orthogonal design*, the formal definition of which we will develop below. Qualitatively, COSTBCs require that the columns of the matrix $\mathbf{G}$ are orthogonal, and it is this orthogonality (in part) that makes COSTBCs attractive in practice: The orthogonality permits a simple maximum-likelihood decoding algorithm which, through only linear combining at the receiver, decouples the signals transmitted from the multiple antennas [**30**]. These codes also achieve full diversity [**30**].

Real orthogonal designs were first defined and studied by Geramita, Geramita, and Seberry Wallis [**10, 12, 13, 14**]:

DEFINITION 2.1. A *real orthogonal design* of order n and type $(s_1, s_2, \ldots, s_k)$ $(s_l > 0)$ in real commuting variables $x_1, x_2, \ldots, x_k$, is an $n \times n$ matrix $\mathbf{A}$ with entries from the set $\{0, \pm x_1, \pm x_2, \ldots, \pm x_k\}$ satisfying

$$\mathbf{A}\mathbf{A}^T = \mathbf{A}^T\mathbf{A} = \sum_{l=1}^{k} s_l x_l^2 \mathbf{I}_n,$$

where $\mathbf{I}_n$ is the $n \times n$ identity matrix.

The generalization to the complex domain followed shortly thereafter [**9**]:

DEFINITION 2.2. A *complex orthogonal design* of order n and type $(s_1, s_2, \ldots, s_k)$ $(s_l > 0)$ in real commuting variables $x_1, x_2, \ldots, x_k$, is an $n \times n$ matrix $\mathbf{C}$ with entries in the set $\{0, \pm x_1, \pm x_2, \ldots, \pm x_k, \pm ix_1, \pm ix_2, \ldots, \pm ix_k\}$ satisfying

$$\mathbf{C}^H\mathbf{C} = \sum_{l=1}^{k} s_l x_l^2 \mathbf{I}_n,$$

where H is the Hermitian transpose (the transpose complex conjugate) and $i^2 = -1$.

An alternative definition for a complex orthogonal design $\mathbf{C}$ has entries from the set $\{0, \pm z_1, \ldots, \pm z_k, \pm z_1^*, \ldots, \pm z_k^*\}$, where the z_l are complex commuting variables and z_l^* denotes the complex conjugate of z_l, such that

$$\mathbf{C}^H\mathbf{C} = \sum_{l=1}^{k} s_l |z_l|^2 \mathbf{I}_n.$$

This latter definition is commonly used in the signal processing literature for its application to space-time block codes [**21, 22, 30**].

EXAMPLE 2.3. The following matrix $\mathbf{C}$ is a complex orthogonal design on complex variables z_1 and z_2, using the latter definition above:

$$\mathbf{C} = \begin{pmatrix} z_1 & z_2 \\ -z_2^* & z_1^* \end{pmatrix}$$

In fact, this simple complex orthogonal design was the first to be employed as a space-time block code: The application was presented by Alamouti in 1998 [**5**]. (Alamouti himself did not refer to his code as a space-time block code, as that term was coined later by Tarokh, Jafarkhani, and Calderbank [**30**].) Alamouti [**5**] presented a very simple maximum-likelihood decoding scheme for this code, later reviewed and generalized by Tarokh et al. [**30**], that hinges on the orthogonality of the columns of $\mathbf{C}$.

Classical results from the turn of the century [**1, 2, 3, 16, 17, 24**] have been shown to imply that nontrivial real and complex orthogonal designs exist only for limited values of n. The mathematics literature refers to the existence problem for orthogonal designs as the Hurwitz-Radon problem [**11**], and the results utilize the Hurwitz-Radon function [**27**], also referred to in the mathematics literature as the Radon-Hurwitz numbers [**2, 6**] or the Hurwitz-Radon-Eckmann formula [**8, 26**]. We recall the following definitions:

DEFINITION 2.4. A set of $n \times n$ real matrices $\{\mathbf{B}_1, \mathbf{B}_2, \ldots, \mathbf{B}_m\}$ is called a size m *Hurwitz-Radon family of matrices* if $\mathbf{B}_i^T \mathbf{B}_i = \mathbf{I}_n$ and $\mathbf{B}_i^T = -\mathbf{B}_i$ for all $1 \leq i \leq m$ and if $\mathbf{B}_i \mathbf{B}_j = -\mathbf{B}_j \mathbf{B}_i$ for all $1 \leq i < j \leq m$.

DEFINITION 2.5. Let $n = 2^a b$, b odd, and write $a = 4c + d$ where $0 \leq d < 4$. *Radon's function* is the arithmetic function $\rho(n) = 8c + 2^d$.

Radon showed that a Hurwitz-Radon family of $n \times n$ matrices contains strictly less than $\rho(n)$ matrices, with the bound of $n-1$ matrices being achieved if and only if n = 2, 4, or 8 [**24**]. Tarokh et al. provide the details showing the equivalence of Radon's result to the fact that real orthogonal designs of order n exist only for n = 2, 4, or 8 [**30**], assuming all entries in the design are nonzero. An extension of this result shows that complex orthogonal designs of order n exist only for $n = 2$, assuming all entries in the design are nonzero [**30**]. Liang also provides a detailed summary of Hurwitz-Radon theory in the context of his treatment of real orthogonal designs [**21**].

Another connection between real orthogonal designs and Hurwitz-Radon theory was provided by Geramita, Geramita, and Seberry Wallis in their answer to the fundamental question concerning the maximum number of variables permissable in an $n \times n$ real orthogonal design. In particular, they showed that the number of variables in an $n \times n$ real orthogonal design is less than or equal to $\rho(n)$ [**10**]. Additionally, Calderbank and Naguib summarize certain connections among the rate of real orthogonal designs, Hurwitz-Radon theory, amicable orthogonal designs, and Clifford algebras [**7**].

Tarokh et al. recognized that Alamouti's successful coding scheme based on the matrix $\mathbf{C}$ given in Example 2.3 could be generalized by employing more general matrices with mutually orthogonal columns. Due to the severe limitations on the sizes of orthogonal designs and upon realizing the potential for utilizing orthogonal designs in wireless communications systems, Tarokh et al. [**30**] defined generalized, or rectangular, complex orthogonal designs as follows:

DEFINITION 2.6. A *generalized complex orthogonal design (GCOD)* of order n is an $r \times n$ matrix $\mathbf{G}$ with entries from $\{0, \pm z_1, \ldots, \pm z_k, \pm z_1^*, \ldots, \pm z_k^*\}$, or products of these complex indeterminants with the imaginary unit i, such that

$$\mathbf{G}^H \mathbf{G} = \sum_{l=1}^{k} |z_l|^2 \mathbf{I}_n.$$

If the entries of $\mathbf{G}$ are allowed to be complex linear combinations of the complex variables and their conjugates, then the design $\mathbf{G}$ is called a *generalized complex linear processing orthogonal design.*

Using a well-known normalization result [**30**], it is assumed throughout that each column of a generalized complex orthogonal design includes exactly one position occupied by $\pm z_l$ or $\pm z_l^*$, for each $1 \leq l \leq k$ and each row includes at most one position occupied by $\pm z_l$ or $\pm z_l^*$, for each $1 \leq l \leq k$. Geramita and Seberry provide a comprehensive reference on orthogonal designs [**11**], and Liang reviews and defines all of the relevant generalizations [**22**].

By extending the definition of a complex orthogonal design from the square case to the rectangular case, Tarokh et al. were able to consider nontrivial orthogonal matrices for any numbers of columns. In turn, this allowed the development of a theory for complex orthogonal space-time block codes for any number of antennas [**30**]. As with the Alamouti code [**5**], the columns of these rectangular designs are orthogonal, and therefore the associated codes enjoy a simple decoding algorithm [**7, 30**]. Furthermore, these rectangular codes provide full transmit diversity and increase the capacity of wireless channels [**7, 30**].

3. Important Parameters: Maximum Rate and the Road to Minimum Delay

As with any combinatorial structure, there are natural optimization questions to ask concerning orthogonal designs. We mentioned above the question of how many variables can be included in an $n \times n$ real orthogonal design. This question is generalized as follows: What is the *maximum rate* (ratio of number of variables to number of rows) achievable by a GCOD with n columns? A complementary question, known as the "fundamental problem for generalized complex orthogonal designs [**30**]," is as follows: What is the *minimum decoding delay* achievable by a maximum rate generalized complex orthogonal design with n columns?

Taken together, we see that the overarching goal for any number of columns is to arrange as many variables as possible into as few rows as possible. In other words, for any number of transmit antennas, we would like to send as many distinct signals as possible in the shortest possible amount of time. Of course, this optimization is subject to the orthogonality constraint on the columns.

Liang answered the first main research question by proving that GCODs with $2m-1$ or $2m$ columns have the same maximum rate of $\frac{m+1}{2m}$, where m is any natural number [**21**]. Furthermore, Liang provided an algorithm for constructing maximum rate GCODs for any number of columns [**21**]. Several other authors also worked towards determining the maximum rate and developing algorithms to produce high-rate GCODs [**23, 29, 32, 33**].

Liang's result was hot off the press when I became interested in orthogonal designs and their associated codes, thus it was clear that I should tackle the question on minimum decoding delay. At that time, the only cases for which the minimum delay was known were the trivial cases of 2, 3, and 4 columns, the cases of 5 and 6 columns addressed by Liang [**21**], and the cases of 7 and 8 columns addressed by Kan and Shen [**19**]. The arguments for these cases were specialized to the number of columns, and thus could not be elegantly generalized.

My initial strategy was to focus on the combinatorial structure of maximum rate codes, hoping to exploit any found patterns and/or substructures. As I can tend to be old-fashioned, I began with a by-hand analysis of all known maximum rate GCODs. I looked for hidden patterns by printing out examples, literally cutting and taping rows and columns in different orders, and using colored highlighters to track various variables. Indeed, this "arts and crafts project" led to the identification of beautiful patterns and substructures within the maximum rate GCODs. The next challenge, of course, was to determine which patterns and substructures were significant and what they might imply about the minimum delay. This challenge was compounded by the size of the matrices under consideration: Even a "small" example has 8 columns and 112 rows, and every possible permutation of rows and columns ought to have been considered. It was clearly time to turn to computational power, a perfect project for undergraduate research students.

As luck would have it, just around this time, two Olin sophomores Nathan Karst and Jon Pollack came into my office looking for some advice on choosing a topic for their course project in my Discrete Mathematics class. (For the record, I believe their project proposal was due the next day.) These students were talented programmers, and I suggested that they implement Liang's new algorithm for maximum rate GCODs and create software to manipulate the generated designs. I showed them what I had been doing by hand: The avid programmers were horrified that I would do such work manually, and they left my office motivated to create software to automate all of my pattern-searching actions. Soon, they were hooked on the excitement of working on unsolved problems, and they became my research partners for the rest of their undergraduate careers (and beyond, in the case of Karst).

The new (and constantly evolving) software allowed us to generate examples larger than those previously available, which allowed us to test our hypotheses on larger sets of data. Even more importantly, the software allowed us to quickly reconfigure particular examples to identify additional patterns and substructures. We ultimately focused on the zero patterns of the rows in the maximum rate designs. A key observation was that every possible pattern of $m-1$ zeros in a row of length $2m$ seemed to appear in at least one row of any maximum rate GCOD with $2m$ columns. Linear algebraic and combinatorial arguments were employed to prove that this observation holds. This then led naturally to the result that a lower bound on the decoding delay for a maximum rate GCOD with $2m$ columns is $\binom{2m}{m-1}$ [**4**]. We later extended this result to hold for the case of $2m-1$ columns [**4**].

Naturally, we then needed to determine if our bound was tight. For the cases of $2m-1$ columns and $2m$ columns with m even, an algorithm already existed to generate maximum rate codes that happened to achieve the lower bound on delay [**23**]. For the case of $2m$ columns with m odd, however, no such algorithms existed. In fact, all known algorithms for maximum rate GCODs generated examples that achieved twice our lower bound on delay when the number of columns was $2m$ for m odd [**21, 23, 29**]. Also, Liang had proven that for the case of 6 columns, the minimum decoding delay was 30, or twice our lower bound [**21**].

This case of $2m$ columns with m odd was elusive for quite a while, although we did have an early success in proving that if such a GCOD does not meet the lower bound, then it can at best meet twice the lower bound. Since algorithms did exist that happened to generate maximum rate codes that meet twice the lower bound in this case [**21, 23, 29**], we were left to prove that GCODs in this case cannot achieve the lower bound.

After months of analysis, it became clear that we needed some new tools to attack this final case of $2m$ columns for m odd. Several new undergraduate research students had rotated through the group during this time, two of whom, Alex Dorsk and Andy Kalcic, helped developed a theory connecting GCODs to signed graphs, which are graphs whose edges are assigned weights of -1 or 1. Unfortunately, the restatement in terms of signed graphs turned out to be an unsolved problem studied (and known to be difficult) by graph theorists. However, this new paradigm allowed for new methods of attack, and it pleased me that my students were learning to initiate creative approaches to the problem.

In the end, the beast was tackled using a combination of combinatorial analysis, the signed graph paradigm, and a linear algebraic technique that was developed primarily by another undergraduate, Mathav Kishore Murugan, who was visiting me as a summer research student from the Indian Institute of Technology in Kharagpur, India. This was truly a collaborative effort, involving contributions from several undergraduates with complementary strengths and interests. The final result was as expected: For a maximum rate GCOD with $2m$ columns for m odd, the best achievable decoding delay is $2\binom{2m}{m-1}$, or twice the lower bound.

Related results came as corollaries to our main work on minimum decoding delay. For example, we showed that a maximum rate GCOD with $2m-1$ or $2m$ columns must utilize at least $\frac{1}{2}\binom{2m}{m}$ variables, with this bound being tight for $2m-1$ columns and $2m$ columns for m even. Similar to the result on minimum decoding delay, for the case of $2m$ columns with m odd, the minimum required number of variables is twice this bound, or $\binom{2m}{m}$ [4].

4. Additional Research Topics

4.1. Rate 1/2 Complex Orthogonal Designs. The lower bound of $\binom{2m}{m-1}$ on the decoding delay of a maximum rate GCOD with $2m-1$ or $2m$ columns grows quickly with respect to the number of columns. This suggests that non-rate-optimal GCODs may be preferable in practice (or more interesting in theory), as the decoding delay is prohibitively large in maximum rate GCODs with large numbers of columns. It is known that the maximum rate of GCODs with $2m-1$ or $2m$ columns is $\frac{m+1}{2m}$ [21], which approaches 1/2 as the number of columns increases. Hence there are two natural questions to ask: What is the minimum decoding delay for rate 1/2 codes? Would the sacrifice in rate be worth the trade-off for a smaller decoding delay? While these two questions are motivated by the application of designs to coding systems, the first question in particular reveals some interesting mathematics.

Several undergraduates have been involved in the effort to determine the minimum decoding delay of rate 1/2 GCODs, including Matthew Crawford, Caitlin Greeley, Nathan Karst, Mathav Kishore Murugan, and Bryce Lee. Our current work indicates that the minimum delay for a rate 1/2 GCOD with $2m-1$ or $2m$ columns is 2^{m-1} or 2^m, depending on the number of columns modulo 8. Although this delay is significantly smaller than the delay for maximum rate codes, it still grows exponentially.

It remains to examine the performance of these rate 1/2 codes, and it remains to answer whether we can slightly decrease the rate below 1/2 in order to gain a significant savings on decoding delay.

4.2. Conjugation Patterns and Transceiver Signal Linearization. As is often the case, a question that seems to have only theoretical importance turns out to have practical significance as well. During our work determining the minimum decoding delay for maximum rate GCODs, we began to notice a connection between whether a given design met the bound on decoding delay and whether it was possible to arrange the matrix using equivalence operations so that every nonzero entry in a given row was either conjugated or non-conjugated. We studied various patterns of conjugation mostly out of mathematical curiosity, however this property is also useful in the application of these designs as COSTBCs.

Certain COSTBCs enjoy a property known as *transceiver signal linearization*, which can facilitate decoding: This linearization allows the code to be backward compatible with existing signal processing techniques and standards, and it allows for the design of low complexity channel equalizers and interference suppressing filters [**28**]. It has been shown that a COSTBC can achieve transceiver signal linearization if each row in the code has either all conjugated entries or all non-conjugated entries [**28**], which is precisely the property we analyzed while looking for mathematical patterns in the underlying designs.

We have been able to show that if a maximum rate COSTBC has an odd number of columns $2m-1$ and achieves the minimum delay of $\binom{2m}{m-1}$ or has an even number of columns $2m$ and achieves twice the minimum delay $2\binom{2m}{m-1}$, then transceiver signal linearization is achievable. If a maximum rate COSTBC has an even number of columns $2m$ and achieves the minimum delay, then transceiver signal linearization is not possible.

This work highlights the fact that trade-offs are omnipresent when designing a communication system. For example, there are maximum rate COSTBCs with $2m \equiv 0 \pmod 4$ columns that achieve transceiver signal linearization, and there are such codes that achieve the lower bound on decoding delay, but no such code can achieve both properties simultaneously.

Our current work examines which rate 1/2 COSTBCs can achieve transceiver signal linearization, and again the work indicates that the answer depends on the number of columns modulo 8.

5. Conclusions

This paper has provided a gentle introduction to orthogonal designs and their application as complex orthogonal space-time block codes. We have reviewed the major results of both theoretical and practical importance regarding these designs/codes, presenting the maximum rate as determined by Liang and providing a narrative of the determination of the minimum decoding delay by this author working with an army of undergraduates. We have also briefly described certain related research topics, including some open problems.

Recent breakthroughs in antenna technology [**15, 31**] and a growing interest in distributed systems contribute to a growing interest in determining the optimal trade-off between rate and delay for COSTBCs with arbitrary numbers of antennas. However, we must note that it is only for two antennas that we can simultaneously achieve full rate, square size (meaning lowest possible delay), and transceiver linearization. As the number of antennas increases, there is a sacrifice in some or all of these parameters. The analysis of such trade-offs is at the core of the challenge of designing wireless communication systems.

Acknowledgements

I would like to thank all of the undergraduates who have done research with me on complex orthogonal space-time block codes, as well as those who have conducted preliminary research on orthogonal designs defined over the quaternion domain in a quest to develop a new class of space-time-polarization block codes: Herbert Chang, Matthew Crawford, Alex Dorsk, Caitlin Greeley, Andy Kalcic, Nathan Karst, Mathav Kishore Murugan, Bryce Lee, David Nelson, Aaron Peterson, Jon Pollack, and Russell Torres. Special thanks go to my first two undergraduate research students, Nathan Karst and Jon Pollack, who worked with me for three years until I had to let them graduate. I expect that many of these students will continue in the legacy of Joe Gallian and become undergraduate research advisors themselves.

I would also like to thank Jennifer Seberry for sparking my initial interest in combinatorial designs and their related codes.

References

[1] J. F. Adams. Vector fields on spheres. *Ann. of Math. (2)*, 75:603–632, 1962.

[2] J. F. Adams, P. D. Lax, and R. S. Phillips. On matrices whose real linear combinations are non-singular. *Proc. Amer. Math. Soc.*, 16:318–322, 1965.

[3] J. F. Adams, P. D. Lax, and R. S. Phillips. Correction to "On matrices whose real linear combinations are nonsingular". *Proc. Amer. Math. Soc.*, 17:945–947, 1966.

[4] S. Adams, N. Karst, and J. Pollack. The minimum decoding delay of maximum rate complex orthogonal space-time block codes. *IEEE Trans. Inform. Theory*, 53(8):2677–2684, 2007.

[5] S. M. Alamouti. A simple transmit diversity technique for wireless communications. *IEEE Journal on Selected Areas in Communications*, 16(8):1451–1458, 1998.

[6] M. A. Berger and S. Friedland. The generalized Radon-Hurwitz numbers. *Compositio Math.*, 59(1):113–146, 1986.

[7] A. R. Calderbank and A. F. Naguib. Orthogonal designs and third generation wireless communication. In *Surveys in combinatorics, 2001 (Sussex)*, volume 288 of *London Math. Soc. Lecture Note Ser.*, pages 75–107. Cambridge Univ. Press, Cambridge, 2001.

[8] B. Eckmann. Gruppentheoretischer Beweis des Satzes von Hurwitz-Radon über die Komposition quadratischer Formen. *Comment. Math. Helv.*, 15:358–366, 1943.

[9] A. V. Geramita and J. M. Geramita. Complex orthogonal designs. *J. Combin. Theory Ser. A*, 25(3):211–225, 1978.

[10] A. V. Geramita, J. M. Geramita, and J. S. Wallis. Orthogonal designs. *Linear and Multilinear Algebra*, 3(4):281–306, 1975/76.

[11] A. V. Geramita and J. Seberry. *Orthogonal designs*, volume 45 of *Lecture Notes in Pure and Applied Mathematics*. Marcel Dekker Inc., New York, 1979. Quadratic forms and Hadamard matrices.

[12] A. V. Geramita and J. S. Wallis. Orthogonal designs. III. Weighing matrices. *Utilitas Math.*, 6:209–236, 1974.

[13] A. V. Geramita and J. S. Wallis. Orthogonal designs. II. *Aequationes Math.*, 13(3):299–313, 1975.

[14] A. V. Geramita and J. S. Wallis. Orthogonal Designs. IV. Existence questions. *J. Combinatorial Theory Ser. A*, 19:66–83, 1975.

[15] B. N. Getu and J. B. Andersen. The MIMO cube: A compact MIMO antenna. *IEEE Trans. Wireless Commun.*, 4:1136–1141, 2005.

[16] A. Hurwitz. Uber die komposition der quadratischen formen von beliebig vielen variablem. *Nachr. Gesell. d. Wiss. Gottingen*, pages 309–316, 1898.

[17] A. Hurwitz. Uber die komposition der quadratischen formen. *Math. Ann.*, 88(5):1–25, 1923.

[18] H. Jakarhami. *Space-time coding theory and practice*. Cambridge University Press, Cambridge, 2005.

[19] H. Kan and H. Shen. A counterexample for the open problem on the minimal delays of orthogonal designs with maximal rates. *IEEE Trans. Inform. Theory*, 51(1):355 – 359, 2005.

[20] E. G. Larsson and P. Stoica. *Space-time block coding for wireless communications*. Cambridge University Press, Cambridge, 2003.

[21] X.-B. Liang. Orthogonal designs with maximal rates. *IEEE Trans. Inform. Theory*, 49(10):2468–2503, 2003. Special issue on space-time transmission, reception, coding and signal processing.

[22] X.-B. Liang and X.-G. Xia. On the nonexistence of rate-one generalized complex orthogonal designs. *IEEE Trans. Inform. Theory*, 49(11):2984–2989, 2003.

[23] K. Lu, S. Fu, and X.-G. Xia. Closed-form designs of complex orthogonal space-time block codes of rates (k+1)/2k for 2k-1 or 2k transmit antennas. *IEEE Trans. Inform. Theory*, 51(12):4340–4347, 2005.

[24] J. Radon. Lineare scharen orthogonaler matrizen. *Abhandlungen aus dem Mathematischen Seminar der Hamburgischen Universitat*, pages 1–14, 1922.

[25] T. Rappaport. *Wireless communications: Principles and Practice, Second Edition*. Prentice Hall, New York, 1993, 2001.

[26] J. T. Schwartz. *Differential Geometry and Topology*. Gordon and Breach, New York, 1968.

[27] D. B. Shapiro. *Compositions of quadratic forms*, volume 33 of *de Gruyter Expositions in Mathematics*. Walter de Gruyter & Co., Berlin, 2000.

[28] W. Su, S. N. Batalama, and D. A. Pados. On orthogonal space-time block codes and transceiver signal linearization. *IEEE Communications Letters*, 8(60):458–460, 2004.

[29] W. Su, X.-G. Xia, and K. J. R. Lui. A systematic design of high-rate complex orthogonal space-time block codes. *IEEE Communications Letters*, 8(6):380 – 382, 2004.

[30] V. Tarokh, H. Jafarkhani, and A. R. Calderbank. Space-time block codes from orthogonal designs. *IEEE Trans. Inform. Theory*, 45(5):1456–1467, 1999.

[31] C. Waldschmidt and W. Wiesbeck. Compact wide-band multimode antennas for MIMO and diversity. *IEEE Trans. Antennas Propagat.*, 52:1963–1969, 2004.

[32] H. Wang and X.-G. Xia. Upper bounds of rates of complex orthogonal space-time block codes. *IEEE Trans. Inform. Theory*, 49(10):2788 – 2796, 2003.

[33] C. Xu, Y. Gong, and K.B. Letaief. High-rate complex orthogonal space-time block codes for high number of transmit antennas. *2004 IEEE International Conference on Communications*, 2:823 – 826, 2004.

Franklin W. Olin College of Engineering, Needham, MA 02492, USA
E-mail address: sarah.adams@olin.edu

Contemporary Mathematics
Volume **479**, 2009

Probabilistic Expectations on Unstructured Spaces

John Beam

ABSTRACT. We introduce the reader to a conceptually powerful, but still widely unknown, way of defining probabilities and their expectations, pioneered by Bruno de Finetti circa 1930. His model requires no structure on the underlying collection of events. Such a probability can be thought of as a fair pricing allocation, and is an accessible tool for a wide variety of problems.

1. Introduction to Coherent Probabilities

Hilbert's sixth problem [**17**] was "To treat in the same manner, by means of axioms, those physical sciences in which mathematics plays an important part; in the first rank are the theory of probabilities and mechanics." Responses to this call varied, but in 1933, Kolmogorov [**20**] laid a set of axioms that has been widely accepted as the foundation of modern probability theory. His probability measure is essentially Lebesgue's measure, but normalized and defined over abstract spaces. In particular, one begins with a nonempty set Ω of possible outcomes of some experiment, chooses a Boolean algebra $\mathcal{A}$ of subsets, called *events*, of Ω — that is, a collection which includes Ω, the complements of its members, and unions and intersections of finite collections of its members — and then defines a countably-additive measure P on $\mathcal{A}$ as follows:

DEFINITION 1.1. A map $P : \mathcal{A} \to [0,1]$ is said to be a *countably-additive probability* if:

i) $P(\Omega) = 1$;

ii) $P(\bigcup_{n\geq 1} A_n) = \sum_{n\geq 1} P(A_n)$ for any disjoint countable collection of sets $A_n \in \mathcal{A}$ whose union is in $\mathcal{A}$.

Bruno de Finetti proposed an alternative way to axiomatize the subject. He developed his theory between 1926 and 1931 (see [**14**], reprinted in [**21**]), gradually publishing the results in article form, and then in book form with [**11**], [**12**], and [**13**] in the 1970s. His "coherent" probability (defined below) generalizes Kolmogorov's probability measure, both in that $\mathcal{A}$ need not be an algebra ($\mathcal{A}$ can be any collection of subsets of Ω), and also in that P need be only finitely additive if $\mathcal{A}$ is an algebra.

2000 *Mathematics Subject Classification.* Primary: 60-06, 60-02, 60A05, 60A10, 28A25.

Key words and phrases. de Finetti, subjective probability, coherence, finite additivity, integration, expectation.

A *finitely-additive probability* is defined by replacing the word *countable* with *finite* in condition *ii*) of Definition 1.1. To illustrate the distinction, take Ω to be the set of positive integers and $\mathcal{A}$ to be the algebra of finite or cofinite sets on Ω. (A cofinite set is one whose complement is finite.) Define P to be 0 on each finite set and 1 on each cofinite set. Then P is finitely but not countably additive.

DEFINITION 1.2. Let $\mathcal{A}$ be any collection of subsets of a non-empty set Ω. A map $P : \mathcal{A} \to [0,1]$ is said to be a *coherent probability* if there does not exist any finite collection of sets $A_1, \ldots, A_N \in \mathcal{A}$ and values $a_1, \ldots, a_N \in \mathbb{R}$ such that $\sum_{n=1}^{N} a_n \big(1_{A_n} - P(A_n) \big) > 0$ everywhere on Ω. (By 1_{A_n} we denote the indicator function on the set A_n — the function which takes the value 1 on A_n and 0 elsewhere.)

P can be thought of as a pricing assignment: $P(A)$ is the price of a bet whose payoff is 1 dollar if the event $A \in \mathcal{A}$ occurs. Two gamblers trade bets with each other on collections $A_1, \ldots, A_N \in \mathcal{A}$, one paying $P(A_n)$ dollars to the other for each one-dollar-payoff bet on A_n. The coefficient a_n represents the number of one-dollar-payoff bets placed on A_n; it can be positive or negative, depending on whether the bet is bought or sold. The assignment P is coherent if the odds are fair — if it is impossible for one gambler to arrange some sure-win finite combination against the other. In Proposition 1.5 we will establish the connection between coherence and finite additivity.

EXAMPLE 1.3. (of a non-coherent assignment). Two coins (not specified to be fair or independent) are to be flipped. A one-dollar payoff bet on "at least one head" costs 60 cents, while a one-dollar payoff bet on "at least one tail" costs 25 cents. This pricing assignment is not coherent, because by recognizing the collection as underpriced and buying one of each of these bets, you assure yourself of netting either \$0.15 or \$1.15. Formally, write

$$\Omega = \{HH, HT, TH, TT\},$$

$$\mathcal{A} = \big\{\{HH, HT, TH\}, \{HT, TH, TT\}\big\},$$

$$P(\{HH, HT, TH\}) = 0.6, \;\; P(\{HT, TH, TT\}) = 0.25.$$

P is not coherent because

$$\big(1_{\{HH,HT,TH\}} - 0.6\big) + \big(1_{\{HT,TH,TT\}} - 0.25\big)$$

is strictly positive, equaling 0.15 on $\{HH, TT\}$ and 1.15 on $\{HT, TH\}$. This example could be modified to yield a coherent probability by changing the price on (at least one tail) to 50 cents. □

EXAMPLE 1.4. (of a coherent probability on a collection which is not an algebra). Let Ω be the set of positive integers. Define the *density* P of a subset A to be the limit as n approaches ∞, if such limit exists, of the proportion of integers in $\{1, 2, \ldots, n\}$ that belong to A. The collection of sets on which P is defined is easily verified not to be an algebra, and P is finitely additive (more accurately, coherent) but not countably additive on this collection; both of these problems appear as an exercise in (2.15 of [**3**]). □

The following proposition and its proof appear in ([**11**], sec. 5.9 and [**12**], sec. 3.8). We present the proof here in part because the references are not as widely available as they should be, and in part because de Finetti's style of writing is often difficult to follow; his terminology and notation are often nonstandard, and his proofs tend to be informal.

PROPOSITION 1.5. *Let $\mathcal{A}$ be an algebra on a non-empty set Ω and let $P : \mathcal{A} \to [0,1]$. Then P is a coherent probability if and only if it is a finitely-additive probability.*

PROOF. Suppose P is finitely additive on $\mathcal{A}$. By way of contradiction, suppose there exist constants $a_1, \ldots, a_n \in \mathbb{R}$ and sets $A_1, \ldots, A_n \in \mathcal{A}$ such that

$$\sum_{n=1}^{N} a_n\big(1_{A_n} - P(A_n)\big) > 0 \text{ on } \Omega. \tag{1.1}$$

We may assume, without loss of generality, that $\{A_1, \ldots, A_N\}$ is an $\mathcal{A}$-partition, all of whose elements are non-empty.Now for each k, the function on the left of (1.1) equals $a_k - \sum_{n=1}^{N} a_n P(A_n)$ on A_k, so that for all k, we get $a_k > \sum_{n=1}^{N} a_n P(A_n)$. But this is not possible, since the additivity of P implies $\sum_{n=1}^{N} P(A_n) = 1$.

Conversely, if P is coherent on $\mathcal{A}$, then $P(\Omega) = 1$, since if $P(\Omega)$ were less than 1, then $1_\Omega - P(\Omega) > 0$ everywhere on Ω. To establish additivity, if A_1 and A_2 are disjoint sets in $\mathcal{A}$, then $1_{A_1} + 1_{A_2} = 1_{A_1 \cup A_2}$, so that

$$\big(1_{A_1} - P(A_1)\big) + \big(1_{A_2} - P(A_2)\big) - \big(1_{A_1 \cup A_2} - P(A_1 \cup A_2)\big)$$

equals

$$P(A_1 \cup A_2) - \big(P(A_1) + P(A_2)\big),$$

a constant value on Ω. By the coherence of P, this value cannot be greater than 0 (and, since we could have begun by multiplying each term by -1, it also cannot be less than 0); hence, it equals 0, so that $P(A_1 \cup A_2) = P(A_1) + P(A_2)$. □

Clearly, if $\mathcal{A} \subseteq \mathcal{B}$ are collections of subsets of a non-empty set Ω, then the restriction to $\mathcal{A}$ of any coherent probability on $\mathcal{B}$ must also be coherent. Further, a coherent probability on any $\mathcal{A}$ can be extended to to a coherent probability on the power set of Ω (proofs can be found in [**1**] and [**16**], or, using the language of "partial measures," in [**18**]). Thus, a coherent probability can be characterized as a function on $\mathcal{A}$ that can be extended to a finitely-additive probability on some algebra containing $\mathcal{A}$. In practice, one usually verifies coherence of P on $\mathcal{A}$ by demonstrating that it can be extended to a finitely-additive probability on an algebra containing $\mathcal{A}$. Why, then, don't we just dispense with the notion of coherence and go straight to finitely-additive probability measures? One answer is that by extending P onto a collection larger than $\mathcal{A}$, one has to choose among various possible values on the additional sets in the collection, thus incorporating more information than was originally present, and potentially leading to false inferences. Another answer is that the coherency conditions provide a useful framework for interpreting a number of problems. These ideas will be explored in Sections 3 and 4.

If a probability is finitely additive, then finite combinations of bets are fair, so if a probability is countably additive, does it follow that countable combinations of bets are fair? In his book ([**23**], sec. 1.5), David Pollard rightly expresses skepticism about such an extension of the argument. Confirming his suspicions, it is, in fact, possible to take a countably-additive probability P and construct a bet $\sum_{n\geq 1} a_n\big(1_{A_n} - P(A_n)\big)$ which is everywhere greater than any desired value – that is, to combine a countable collection of fair bets to yield an arbitrarily unfair one. Such an example is constructed for the Lebesgue measure on the unit interval in [**2**], by taking advantage of the conditional nature of the convergence of the alternating harmonic series. By imposing additional regularity conditions in the description of coherence, however, one can incorporate the axiom of countable additivity. Various ways of describing this are presented in [**2**], [**4**], and [**16**].

2. Expectations

"As to the axioms of the theory of probabilities," Hilbert elaborated on his sixth problem, "it seems to me desirable that their logical investigation should be accompanied by a rigorous and satisfactory development of the method of mean values in mathematical physics..." *Mean values* (that is, *expected values*, or *expectations*) are indispensable to our understanding of many probabilistic concepts, not only in physics, but in just about every arena where probability is applied: we calculate expected stock prices, expected casino winnings, expected business costs, expected waiting times, life expectancies, etc...

For a countably-additive probability measure P, Kolmogorov interpreted the expectation as the integral, in the Lebesgue sense, of an $\mathcal{A}$-measurable real-valued function (called a *random variable*) on Ω. For finitely-additive probability measures, one can likewise interpret expectations as integrals, but using the integration theory of Dunford and Schwartz [**9**]. Both of these developments proceed by defining the integral of the indicator functions 1_A, where $A \in \mathcal{A}$, to be $P(A)$, extending this definition linearly to the *simple functions* (finite linear combinations of indicator functions), and then passing through limits to those functions which are approached in some sense by a sequence of simple functions.

De Finetti preferred not to distinguish between the measure and the integral. With a slight modification of the coherency condition, he was able to axiomatize the integral, or expectation (he called it the *prévision*), directly. (Daniell had previously introduced a theory of integration in this manner, in the countably-additive setting, developing the Lebesgue integral as a linear functional; see [**6**].) De Finetti denoted the expectation by P itself, and treated the probability of an event A as the integral of its indicator function; he wrote $P(A)$ for both. In Section 1.4 of [**23**], Pollard makes several convincing arguments for both de Finetti's approach to the integral and his notation. One purpose of the current paper is to present de Finetti's theory from several viewpoints — namely, the integral will be presented as a coherent prévision (hereafter, *coherent expectation*), as a linear functional, and in the traditional style, beginning with a measure — in the hopes that the reader may find some common ground. For clarity, we will denote the probability of an event A by $P(A)$ but the expectation of its indicator function by $P(1_A)$ or by $E_P(1_A)$, depending on the context.

DEFINITION 2.1. Let $\mathcal{F}$ be any collection of real-valued functions on a non-empty set Ω. A map $P : \mathcal{F} \to \mathbb{R}$ is said to be a *coherent expectation* if there does not exist an $\varepsilon > 0$ and a finite collection of functions $X_1, \ldots, X_N \in \mathcal{F}$ and values $a_1, \ldots, a_N \in \mathbb{R}$ such that $\sum\limits_{n=1}^{N} a_n\big(X_n - P(X_n)\big) \geq \varepsilon$ everywhere on Ω.

The gambling context of a coherent probability applies also to a coherent expectation: $P(X)$ is the price of a bet whose payoff is $X(\omega)$ dollars if $\omega \in \Omega$ occurs; the gamblers are trading bets on the functions $X \in \mathcal{F}$. The expectation is coherent if it is impossible for a gambler to arrange some finite combination of bets such that the infimum of his winnings is greater than 0. This fair price allocation was de Finetti's notion of an expectation.

When applied to just the indicator functions on $\mathcal{A}$, Definition 2.1 specializes to Definition 1.2 of a coherent probability. Note that, in the earlier definition, we needn't have specified the codomain of P to be $[0,1]$, but rather did so as a matter of clarity. This restriction is already implied; if P were greater than 1 on some event, then we would just sell a bet on that event, thus winning a sure profit. Likewise, if P were somewhere less than 0, we would buy a bet. We similarly established, in the proof of Proposition 1.5, that $P(\Omega)$ must equal 1 if $\Omega \in \mathcal{A}$; hence, we can, without losing the generality of any argument, adopt the convention that Ω belongs to $\mathcal{A}$.

The following proposition and its proof appear in ([**12**], chap. 3), but again, we include a proof for accessibility.

PROPOSITION 2.2. *Let $\mathcal{F}$ be a vector space of real-valued functions on a non-empty set Ω. A map $P : \mathcal{F} \to \mathbb{R}$ is a coherent expectation if and only if it is a non-negative linear functional satisfying $P(1_\Omega) = 1$.*

PROOF. Suppose P is linear but not coherent; for some $\varepsilon > 0$, functions $X_1, \ldots, X_N \in \mathcal{F}$, and values $a_1, \ldots, a_N \in \mathbb{R}$,

$$\sum_{n=1}^{N} a_n\big(X_n - P(X_n)\big) \geq \varepsilon \tag{2.1}$$

on Ω. Write $X = \sum_{n=1}^{N} a_n X_n$; upon a reorganization of terms, the left side of (2.1) becomes $X - P(X)$, so that

$$X - P(X) - \varepsilon \geq 0.$$

Applying P to the the left side of this inequality results in,

$$P(X) - P(X) \cdot P(1_\Omega) - \varepsilon \cdot P(1_\Omega),$$

and assuming $P(1_\Omega) = 1$, this equals $-\varepsilon$, in which case P is not non-negative.

Conversely, given $X_1, X_2 \in \mathcal{F}$ and $a_1, a_2 \in \mathcal{A}$, observe that the constant

$$P(a_1X_1 + a_2X_2) - a_1P(X_1) - a_2P(X_2)$$

equals

$$a_1\big(X_1 - P(X_1)\big) + a_2\big(X_2 - P(X_2)\big) - \big(a_1X_1 + a_2X_2 - P(a_1X_1 + a_2X_2)\big).$$

By the coherence criterion, this constant cannot be greater than 0 (and also it cannot be less than 0), so it must equal 0; and thus, P is linear. Non-negativity is trivial: if $X \geq 0$ but $P(X) < 0$, then $X - P(X) \geq -P(X) > 0$, violating coherence. Finally, $P(1_\Omega) = 1$ because otherwise either $1 - P(1_\Omega)$ or $-\big(1 - P(1_\Omega)\big)$ would be a positive constant, violating coherence. □

In their treatise [**9**], Dunford and Schwartz developed a theory of the integral for a broad class of functions (not necessarily real valued) with respect to finitely-additive measures (not necessarily probability measures) on algebras. Below, we adapt their procedure for coherent probabilities. In this restricted setting, we generalize their work, in that ours applies to unstructured collections $\mathcal{A}$, and we also simplify their work, in that ours requires only one convergence condition rather than their two. We begin by defining a semi-norm (see Definition 2.4 below) — it is a generalization of the $\mathcal{L}^1$-semi-norm — on the space of real-valued functions on Ω; we consider to be integrable (see Definition 2.5) those functions which are approached under this semi-norm by a sequence of $\mathcal{A}$-simple functions (finite linear combinations of indicator functions 1_A, where $A \in \mathcal{A}$).

DEFINITION 2.3. Let P be a coherent probability on any $(\Omega, \mathcal{A})$. We define the *expectation* of the $\mathcal{A}$-simple function $S := \sum_{n=1}^{N} a_n 1_{A_n}$ by

$$E_P(S) = \sum_{n=1}^{N} a_n P(A_n).$$

DEFINITION 2.4. Let P be a coherent probability on any $(\Omega, \mathcal{A})$. Define an extended-real-valued semi-norm $\|\cdot\|_P$ on the vector space of real-valued functions on Ω as follows. If a function X is bounded, let

$$\|X\|_P = \inf \{E_P(S) \mid S \text{ is } \mathcal{A}\text{-simple and } S \geq |X|\}.$$

For the general function X, let

$$\|X\|_P = \sup \{\|Y\|_P \mid Y \text{ is bounded and } 0 \leq Y \leq |X|\}.$$

DEFINITION 2.5. Let P be a coherent probability on any $(\Omega, \mathcal{A})$. A real-valued function X on Ω is said to be *P-integrable* if there exists a sequence of $\mathcal{A}$-simple functions S_n such that $\|X - S_n\|_P \to 0$. We define the *expectation* of such an X by

$$E_P(X) = \lim_{n \to \infty} E_P(S_n).$$

The collection of P-integrable functions forms a vector space over $\mathbb{R}$, containing the $\mathcal{A}$-simple functions as a subspace, and E_P is a non-negative linear functional (i.e., a coherent expectation) on this space. (It is relatively straightforward to show this, the only tricky part being non-negativity.) We now state and partially establish conditions under which E_P is the *unique* coherent expectation.

PROPOSITION 2.6. *Consider any $(\Omega, \mathcal{A})$. Let P be a coherent expectation on the space of $\mathcal{A}$-simple functions (and thus also a coherent probability on $\mathcal{A}$). A real-valued function X on Ω is bounded and P-integrable if and only if $E_P(X)$ is the unique value assignable to X by any extension of P as a coherent expectation.*

PROOF. We will prove the the sufficiency of the first condition. Let X be a bounded P-integrable real-valued function on Ω. By way of contradiction, suppose first that

$$P(X) > E_P(X) + 3\varepsilon \text{ for some } \varepsilon > 0. \tag{2.2}$$

Let S be an $\mathcal{A}$-simple function such that both $\|X - S\|_P < \varepsilon$ and $|E_P(X) - E_P(S)| < \varepsilon$. By the definition of $\|\cdot\|_P$ there exists an $\mathcal{A}$-simple function $T \geq |X - S|$ such that $E_P(T) < \varepsilon$. Then $S + T$ is an $\mathcal{A}$-simple function greater than or equal to X, satisfying $E_P(S + T) < E_P(X) + 2\varepsilon$. But now the coherency conditions have been violated:

$$\big((S + T) - P(S + T)\big) - \big(X - P(X)\big) > \varepsilon,$$

since $S + T - X \geq 0$ and

$$P(X) - P(S+T) = P(X) - E_P(S+T) > 3\varepsilon + E_P(X) - E_P(S+T) > 3\varepsilon - 2\varepsilon.$$

If instead of (2.2) we had supposed that

$$P(X) < E_P(X) - 3\varepsilon \text{ for some } \varepsilon > 0,$$

we would have observed that $S - T$ is less than or equal to X and satisfies $E_P(S - T) > E_P(X) - 2\varepsilon$, again resulting in a violation of the coherency conditions:

$$\big(X - P(X)\big) - \big((S-T) - P(S-T)\big) > \varepsilon.$$

The proof of the other direction of the proposition requires too much machinery for this article; details are provided in Chapter 8 of [**1**]. The objective is to show that if either X is bounded but not P-integrable or X is unbounded, then there exists more than one extension of P as a non-negative linear functional onto a vector space which includes the $\mathcal{A}$-simple functions and X. □

In sections 5.9, 5.33, and 6.5 of [**11**], de Finetti observes something equivalent to Proposition 2.6, but through upper and lower expectations. (For the connection between these ideas, see chapters 4, 6, and 8 of [**1**].) De Finetti was reluctant to work with unbounded functions, because of the arbitrary nature of the values of their expectations. His reluctance appears in some respects to have been unfounded. If one wanted to extend E_P as a coherent expectation from the $\mathcal{A}$-simple functions onto a collection including more than one unbounded function, additional constraints may be imposed upon the freedom to choose the extension. Lester Dubins [**8**] has shown that if $\mathcal{A}$ is a σ-algebra, then any two non-negative linear functionals which are defined on the space of all measurable functions must be identical, provided they agree on the space of bounded measurable functions. The following conjecture is an open problem; the author has a preliminary proof for the special case where $\mathcal{A}$ is an algebra.

CONJECTURE 2.7. *Consider any $(\Omega, \mathcal{A})$. Let P be a coherent expectation on the space of $\mathcal{A}$-simple functions (and thus also a coherent probability on $\mathcal{A}$). Then E_P is the unique extension of P as a coherent expectation onto the space of all P-integrable functions (including the unbounded ones).*

3. Examples

EXAMPLE 3.1. (of a non-integrable function taking only finitely-many values). The following information is known about a certain six-sided die: 1 occurs with probability 1/6; a lower value (1, 2, or 3) occurs with probability 1/2; a power of two (1, 2, or 4) occurs with probability 1/2; a multiple of three (3 or 6) occurs with probability 1/3. No other observations were recorded. Consider the question, "With what probability does an even outcome (2, 4, or 6) occur?" Write $\Omega = \{1,2,3,4,5,6\}$ and $\mathcal{A} = \big\{\Omega, \{1\}, \{1,2,3\}, \{1,2,4\}, \{3,6\}\big\}$, with P as described on this collection. P is coherent because it can be extended to a probability measure on the power set of Ω; however, if we proceeded with this extension in order to apply Kolmogorov's model, we would get an answer to our question, but the answer would not be meaningful — anything between 1/3 and 2/3 might result. The fact is, the function $1_{\{2,4,6\}}$ is not P-integrable: one could verify this either by demonstrating that

$$\|1_{\{2,4,6\}} - S\|_P \geq \frac{1}{3}$$

for all $\mathcal{A}$-simple functions S, or by observing that the expectations $E_Q(\mathit{1}_{\{2,4,6\}})$ vary according to which extension Q of P is chosen. On the other hand, we *can* answer the question, "With what probability does a prime outcome (2, 3, or 5) occur?" The function

$$\mathit{1}_{\{2,3,5\}} = \mathit{1}_\Omega + \mathit{1}_{\{1,2,3\}} - \left(\mathit{1}_{\{1,2,4\}} + \mathit{1}_{\{3,6\}} + \mathit{1}_{\{1\}}\right)$$

is $\mathcal{A}$-simple, and is therefore P-integrable, with expectation

$$E_P(\mathit{1}_{\{2,3,5\}}) = 1 + \frac{1}{2} - \left(\frac{1}{2} + \frac{1}{3} + \frac{1}{6}\right) = \frac{1}{2}.$$

De Finetti's model has incorporated a method for distinguishing the good information from the bad. □

In measure theory, many methods rely on a function's "measurability." A real-valued function on $(\Omega, \mathcal{A})$ is said to be $\mathcal{A}$-measurable if its inverse image of every Borel set is in $\mathcal{A}$. This concept is of very little use if $\mathcal{A}$ is not an algebra. The next example is elementary but gets right to the heart of the matter; in particular, it establishes that an integrable function need not be measurable.

Example 3.2. (of a simple function which is not measurable). Let $\Omega = \{1, 2, 3, 4\}$ and let $\mathcal{A}$ contain the sets Ω, $\{1, 2\}$ and $\{2, 3\}$. Define the $\mathcal{A}$-simple function $S = \mathit{1}_{\{1,2\}} - \mathit{1}_{\{2,3\}}$, which equals $\mathit{1}_{\{1\}} - \mathit{1}_{\{3\}}$. None of the sets $S^{-1}(\{1\})$, $S^{-1}(\{0\})$, or $S^{-1}(\{-1\})$ lies in $\mathcal{A}$. □

When $\mathcal{A}$ is an algebra, the collection of P-integrable functions forms a vector lattice; in particular, if X is P-integrable, then so are $X \vee 0$ (the pointwise maximum of the two functions) and $X \wedge 0$ (the pointwise minimum of the two functions). It is common, when calculating the Lebesgue integral, to integrate these two pieces first. If $\mathcal{A}$ is not an algebra, we can see by the previous example that this may not be possible: define $P(\{1, 2\}) = \frac{2}{3}$ and $P(\{2, 3\}) = \frac{1}{3}$; since S is $\mathcal{A}$-simple, it is P-integrable, but neither $S \vee 0$ nor $S \wedge 0$ is P-integrable — so we cannot know exactly where on Ω the expectation's value comes from. We are, on the other hand, able to split S into the difference of two non-negative pieces when evaluating the expectation:

$$E_P(S) = E_P(\mathit{1}_{\{1,2\}}) - E_P(\mathit{1}_{\{2,3\}}) = 1 \cdot \frac{2}{3} - 1 \cdot \frac{1}{3}.$$

We now demonstrate that it is not always possible to find *any* way of splitting an integrable function into such a pair; this inability fundamentally distinguishes the nature of expectations when $\mathcal{A}$ is unstructured as opposed to an algebra.

Example 3.3. (of a function which is integrable but cannot be written as the difference of two non-negative integrable functions). Let $\Omega = \{4, 5, 6, \ldots\}$, and let $\mathcal{A}$ consist of Ω, the empty set, and the sets A_k, B_k and C_k, for $k \geq 1$, where we define

$$A_k = \{4k + 1, 4k + 2\};$$
$$B_k = \{4k + 2, 4k + 3\};$$
$$C_k = \{4k + 4\} \cup \bigcup_{i \geq k} \{4i + 1, 4i + 3\}.$$

Assign the probability P as follows:

$$\begin{aligned} P(A_k) &= P(B_k) = \frac{2}{4}\left(\frac{1}{2}\right)^{k+1} \\ P(C_k) &= \frac{1}{4}\left(\frac{1}{2}\right)^{k+1} + \sum_{i\geq k}\frac{2}{4}\left(\frac{1}{2}\right)^{i+1} \\ &= \frac{1}{4}\left(\frac{1}{2}\right)^{k+1} + \frac{2}{4}\left(\frac{1}{2}\right)^{k}. \end{aligned}$$

Define the function

$$X = \sum_{k\geq 1} k\big(1_{\{4k+1\}} - 1_{\{4k+3\}}\big).$$

To see that P is coherent, it suffices to observe that it can be extended to a finitely-additive probability on an algebra containing $\mathcal{A}$. First write $\mathcal{B}_0 = \big\{\emptyset, \{4\}\big\}$, and for $1 \leq i \leq 4$, let $\mathcal{B}_i$ be the algebra of all finite or cofinite subsets of the set

$$\Omega_i = \bigcup_{k\geq 1}\{4k+i\}.$$

Let

$$\mathcal{B} = \big\{B_0 \cup B_1 \cup B_2 \cup B_3 \cup B_4 \;\big|\; B_i \in \mathcal{B}_i\big\}.$$

Then $\mathcal{B}$ is an algebra containing $\mathcal{A}$. Choose a constant

$$0 < \alpha \leq \frac{1}{2}.$$

Define Q to be the finitely-additive probability on $\mathcal{B}$ satisfying:

$$\begin{aligned} Q\big(\{4\}\big) &= \alpha; \\ Q\big(\{4k+i\}\big) &= \frac{1}{4}\left(\frac{1}{2}\right)^{k+1} \quad \text{for } 1 \leq i \leq 4 \text{ and } k \geq 1; \\ Q(\Omega_i) &= \frac{1}{4}\sum_{k\geq 1}\left(\frac{1}{2}\right)^{k+1} = \left(\frac{1}{4}\right)\left(\frac{1}{2}\right) \quad \text{for } 1 \leq i \leq 3; \\ Q(\Omega_4) &= \frac{1}{4}\sum_{k\geq 1}\left(\frac{1}{2}\right)^{k+1} + \left(\frac{1}{2} - \alpha\right) \\ &= \left(\frac{1}{4}\right)\left(\frac{1}{2}\right) + \left(\frac{1}{2} - \alpha\right). \end{aligned}$$

It is clear that Q agrees with P on $\mathcal{A}$. (It may be of interest to note that if we choose $\alpha = 1/2$, then Q is countably additive.)

Now we show that X is P-integrable. Define the $\mathcal{A}$-simple functions

$$\begin{aligned} S_1 &= 1_{A_1} - 1_{B_1}; \\ S_n &= S_{n-1} + n\big(1_{A_n} - 1_{B_n}\big) \quad \text{for } n \geq 2. \end{aligned}$$

Let M be any positive integer. If $M \leq n+1$, then

$$|X - S_n| \wedge M \leq (n+1)1_{C_{n+1}}$$

and if $M > n+1$, then

$$|X - S_n| \wedge M \leq (n+1)1_{C_{n+1}} + \sum_{k=n+2}^{M} 1_{C_k}.$$

In either case,

$$\begin{aligned}\big\||X - S_n| \wedge M\big\|_P &\leq (n+1)P(C_{n+1}) + \sum_{k\geq n+2} P(C_k) \\ &< (n+1)\left(\frac{3}{4}\right)\left(\frac{1}{2}\right)^{n+1} + \sum_{k\geq n+2}\left(\frac{3}{4}\right)\left(\frac{1}{2}\right)^{k} \\ &= (n+2)\left(\frac{3}{4}\right)\left(\frac{1}{2}\right)^{n+1}.\end{aligned}$$

This expression is independent of M and approaches 0 as n approaches ∞, so $\|X - S_n\|_P$ approaches 0 as well. Since $E_P(S_n) = 0$ for each n, we get that $E_P(X) = 0$.

We conclude the proof by showing that it is not possible to express X as the difference of two non-negative P-integrable functions $X^{\oplus}$ and $X^{\ominus}$. We shall accomplish this by showing that if a function $X^{\oplus}$ is greater than or equal to both 0 and X, then it cannot be P-integrable. By way of contradiction, suppose there exists a function $X^{\oplus} \geq X \vee 0$ and a sequence of $\mathcal{A}$-simple functions S_n such that

$$\|X^{\oplus} - S_n\|_P \to 0. \tag{3.1}$$

We use the following representation for the functions S_n:

$$S_n = \sum_{i\geq 1} a_{n,i} 1_{A_i} + \sum_{i\geq 1} b_{n,i} 1_{B_i} + \sum_{i\geq 1} c_{n,i} 1_{C_i} + d_n.$$

Of course for each n, all but a finite number of the coefficients $a_{n,i}$, $b_{n,i}$ and $c_{n,i}$ must be 0. We observed earlier in this proof that P can be extended to a probability Q taking a strictly positive value on each element of Ω. It is clear that

$$\begin{aligned}\|X^{\oplus} - S_n\|_P &\geq \|X^{\oplus} - S_n\|_Q \\ &\geq \big|X^{\oplus}(\omega) - S_n(\omega)\big| \cdot Q(\{\omega\})\end{aligned}$$

for any $\omega \in \Omega$, and thus the convergence in (3.1) implies the convergence

$$\lim_{n\to\infty} S_n(\omega) = X^{\oplus}(\omega) \tag{3.2}$$

for each ω. Let $k \geq 1$. For all $n \geq 1$,

$$S_n(4k+2) = a_{n,k} + b_{n,k} + d_n;$$

that is,

$$\begin{aligned}a_{n,k} + b_{n,k} &= S_n(4k+2) - d_n \\ &= S_n(4k+2) - S_n(4). \end{aligned} \tag{3.3}$$

Observe also that

$$S_n(4k+1) = a_{n,k} + \sum_{i=1}^{k} c_{n,i} + d_n$$

and

$$S_n(4k+3) = b_{n,k} + \sum_{i=1}^{k} c_{n,i} + d_n.$$

Adding and rearranging terms and replacing d_n by $S_n(4)$, we get

$$\sum_{i=1}^{k} c_{n,i} = \frac{1}{2}\Big(S_n(4k+1) + S_n(4k+3) - \big(a_{n,k} + b_{n,k}\big)\Big) - S_n(4).$$

Applying (3.3) and then (3.2), we obtain
(3.4)

$$\lim_{n\to\infty} \sum_{i=1}^{k} c_{n,i} = \frac{1}{2}\Big(X^{\oplus}(4k+1) + X^{\oplus}(4k+3) - \big(X^{\oplus}(4k+2) - X^{\oplus}(4)\big)\Big) - X^{\oplus}(4).$$

Next, we note that the inequalities $X^{\oplus} \geq X$ and $X^{\oplus} \geq 0$ imply

$$X^{\oplus}(4k+1) + X^{\oplus}(4k+3) \geq k, \tag{3.5}$$

and we claim that

$$\lim_{k\to\infty} X^{\oplus}(4k+2) = X^{\oplus}(4). \tag{3.6}$$

(To verify the claim, suppose otherwise, that there were some $\varepsilon > 0$ such that $\big|X^{\oplus}(4k+2) - X^{\oplus}(4)\big| > \varepsilon$ for infinitely many k. Applying (3.2), there would then exist an N such that, given any $n \geq N$, $\big|X^{\oplus}(4k+2) - S_n(4)\big| > \varepsilon/2$ for each of those values k. Note that any $\mathcal{A}$-simple function must, for all but finitely-many k, assume the same value on $4k+2$ as it does on 4. Thus, it would follow that $\big|X^{\oplus}(4k+2) - S_n(4k+2)\big| > \varepsilon/2$ for some (dependent on n) infinite subcollection of the aforementioned values k, and then further that any $\mathcal{A}$-simple function lying above $\big|X^{\oplus} - S_n\big| \wedge \varepsilon/2$ must also lie above $\varepsilon/2 \cdot \mathit{1}_{\{4\}}$, so that

$$\|X^{\oplus} - S_n\|_P \geq \Big\|\frac{\varepsilon}{2}\mathit{1}_{\{4\}}\Big\|_P.$$

This value is greater than or equal to $(\varepsilon/2)(1/2)$, since, as we observed earlier, there is an extension of P which takes the value $1/2$ on $\{4\}$. This would hold for every $n \geq N$; but then our assumption (3.1) would be violated. Thus we have established (3.6).) Applying (3.5) and (3.6) to (3.4), we obtain that, for sufficiently large k,

$$\lim_{n\to\infty} \sum_{i=1}^{k} c_{n,i} > \frac{1}{2}k - 1 - X^{\oplus}(4). \tag{3.7}$$

We will end by contradicting (3.7). Since any $\mathcal{A}$-simple function must, for all but finitely-many k, assume the same value on $4k+4$ as it does on 4, the same sort of argument used to prove (3.6) yields

$$\lim_{k\to\infty} X^{\oplus}(4k+4) = X^{\oplus}(4).$$

So there exists some $K \geq 0$ such that if $k \geq K$, then

$$\big|X^{\oplus}(4k+4) - X^{\oplus}(4)\big| < \frac{1}{4}. \tag{3.8}$$

For all k and n,

$$S_n(4k+4) = c_{n,k} + d_n.$$

Replacing d_n by $S_n(4)$ and rearranging terms, we have

$$c_{n,k} = S_n(4k+4) - S_n(4).$$

Letting n approach ∞, we get by (3.2) that

$$\lim_{n\to\infty} c_{n,k} = X^{\oplus}(4k+4) - X^{\oplus}(4),$$

which is, by (3.8), less than $1/4$, provided $k \geq K$. Then for all $k > K$,

$$\lim_{n\to\infty} \sum_{i=1}^{k} c_{n,i} = \lim_{n\to\infty} \sum_{i=1}^{K} c_{n,i} + \sum_{i=K+1}^{k} \lim_{n\to\infty} c_{n,i}$$

$$< \frac{1}{3}k, \text{ for sufficiently large } k.$$

(The finitude of the first term on the right of the equality is guaranteed by (3.4).) This last inequality violates our prior result (3.7), thus completing our argument by contradiction. □

4. Commentary and Applications

In the forward to de Finetti's two-volume *Theory of Probability* [**12**], [**13**], published in 1974 and 1975, D. V. Lindley wrote, "I believe that it is a book destined ultimately to be recognized as one of the great books of the world." In a later reference to these works, he stated that there should be "a moratorium on research so that we can all study it." [**15**]. It has often been speculated that de Finetti's work would be much better known but for problems in communication. He wrote in Italian; at that, to cite Cifarelli and Regazzini [**5**], his writing is "full of nuances... It is then difficult to translate... without losing some of its peculiar stylistic aspects or introducing some simplifications which may distort the original meaning." Nevertheless, his work is followed seriously by a number of mathematical philosophers, for the conceptual value of "subjective" probabilities in decision theory (see, for example, [**19**]), and by some circles of economists (see [**7**]) and statisticians, especially regarding the finitely-additive aspect, for its applications to Bayesian inference (see [**5**], [**24**]). Regarding financial markets, Robert Nau [**22**] writes that in the 1970s, "there was an explosion of interest among finance theorists in models of asset pricing by arbitrage. The key discovery of this period was the fundamental theorem of asset pricing [which] states that there are no arbitrage opportunities in a financial market if and only if there exists a probability distribution with respect to which the expected value of every asset's future payoffs, discounted at the risk free rate, lies between its current bid and ask prices; and if the market is complete, the distribution is unique. This is just de Finetti's fundamental theorem of subjective probability [a result closely related to Proposition 2.6 of the current paper], with discounting thrown in, although de Finetti is not usually given credit in the finance literature for having discovered the same result 40 years earlier."

We close with a natural setting for the gambler's role in the subject; the reader is referred to Heath and Sudderth's more explicit presentation in [**16**]. We are at the track, where a horse race is about to begin. The betting window has just closed; a total of d_n dollars has been placed on each horse n to win among the field $1, 2, \ldots, N$. If we assume that the track deducts no fees, then after the race, each dollar bet on the winning horse, k, will pay out $\left(\sum_{n=1}^{N} d_n\right)/d_k$ dollars; thus the investment in dollars for a 1-dollar payoff on that horse is $d_k/\left(\sum_{n=1}^{N} d_n\right)$. This is exactly how de Finetti would evaluate the probability of horse k winning the race, in the context of his pricing assignments (recall the description following Definition 1.2); it is also is consistent as a probability with the way the track posts the "odds." Now suppose that more than three horses are racing, and that people are also betting in a "show" pool, that a given horse will finish among the top three. Outside of North America, money in this pool is divided into equal parts for the top-three finishers, and each of those amounts is apportioned among the bettors on that horse. If a total of D_n dollars has been placed on each horse n in this pool, then the price of a 1-dollar payoff bet on horse k to show is $3D_k/\left(\sum_{n=1}^{N} D_n\right)$; this defines a coherent probability for the available bets in the show pool, provided no more than one third of the money was placed on any one horse. It is easy to verify that this collection of available bets does not constitute an algebra. As a final note of interest, people may not bet consistently from one pool to another, and so if one defines a P that accounts for both the win and the show pools, he may find it to be incoherent, and thus may be able to rig a sure-win bet. In the aforementioned paper, Heath and Sudderth describe the conditions for which this occurs.

References

[1] J. Beam, *Expectations for Coherent Probabilities,* Unpublished Ph. D. Dissertation, University of Miami, 2002.

[2] J. Beam, Unfair gambles in probability, *Statistics and Probability Letters* **77** (2007), 681-686.

[3] P. Billingsley, *Probability and Measure,* John Wiley & Sons, New York, 1986.

[4] V. S. Borkar, V. R. Konda, and S. K. Mitter, On De Finetti coherence and Kolmogorov probability, *Statistics and Probability Letters* **66** (2004), 417-421.

[5] D. M. Cifarelli and E. Regazzini, De Finetti's contribution to probability and statistics, *Statistical Science* **11** (1996), 253-282.

[6] P. J. Daniell, A general form of integral, *Annals of Mathematics* **19** (1918), 279-294.

[7] E. Diecidue, P. Wakker, and M. Zeelenberg, Eliciting decision weights by adapting de Finetti's betting-odds method to prospect theory, *Journal of Risk and Uncertainty* **34** (2007), 179-199.

[8] L. E. Dubins, On everywhere-defined integrals, *Transactions of the American Mathematical Society* **232** (1977), 187-194.

[9] N. Dunford and J. T. Schwartz, *Linear Operators, Part I,* Interscience Publishers, Inc., New York, 1967.

[10] B. de Finetti, La prévision: ses lois logiques, ses sources subjectives, *Annales de l'Institut Henri Poincaré* **7** (1937), 1-68. English translation in [**21**], pp. 93-158.

[11] B. de Finetti, *Probability, Induction and Statistics,* John Wiley & Sons, New York, 1972.

[12] B. de Finetti, *Theory of Probability: A critical introductory treatment, Volume 1,* translated by A. Machí and A. Smith, John Wiley & Sons, New York, 1974.

[13] B. de Finetti, *Theory of Probability: A critical introductory treatment, Volume 2,* translated by A. Machí and A. Smith, John Wiley & Sons, New York, 1975.

[14] B. de Finetti, Probability: beware of falsifications!, *Scientia* **111** (1976), 283-303.

[15] P. K. Goel and A. Zellner (eds.), *Bayesian Inference and Decision Techniques: Essays in Honor of Bruno de Finetti,* North-Holland, New York, 1986.

[16] D. C. Heath and W. D. Sudderth, On a theorem of de Finetti, oddsmaking, and game theory, *The Annals of Mathematical Statistics* **43** (1972), 2072-2077.

[17] D. Hilbert, Mathematical problems, *Bulletin of the American Mathematical Society* **8** (1902), 437-479 (English transl. by M. W. Newson). The original appeared in *Göttinger Nachrichten* (1900), 253-297, and in *Archiv der Mathernatik una Physik,* 3dser., vol. 1 (1901), 44-63 and 213-237.

[18] A. Horn and A. Tarski, Measures in Boolean algebras, *Transactions of the American Mathematical Society* **64** (1948), 467-497.

[19] R. Jeffrey, *Subjective Probability: The Real Thing,* Cambridge University Press, New York, 2004.

[20] A. Kolmogorov, *Grundbegriffe der Wahrscheinlichkeitsrechnung,* Chelsea Publishing Company, New York, 1946 (originally published in 1933). An English translation: A. Kolmogorov, *Foundations of the Theory of Probability,* translated by N. Morrison, Chelsea Publishing Company, New York, 1950.

[21] H. E. Kyburg and H. E. Smokler (eds.), *Studies in Subjective Probability, 2nd ed.,* Robert E. Krieger Publishing Co., Huntington, NY, 1980.

[22] R. F. Nau, De Finetti was right: probability does not exist, *Theory and Decision* **51** (2001), 89-124.

[23] D. Pollard, *A User's Guide to Measure Theoretic Probability,* Cambridge University Press, New York, 2002.

[24] W. D. Sudderth, Finitely additive priors, coherence and the marginalization paradox, *The Journal of the Royal Statistical Society* **42** (1980), 339-341.

UNIVERSITY OF WISCONSIN OSHKOSH, MATHEMATICS DEPARTMENT, 800 ALGOMA BLVD., OSHKOSH, WI 54901-8631

E-mail address: beam@uwosh.edu

Contemporary Mathematics
Volume **479**, 2009

A beginner's guide to forcing

Timothy Y. Chow

Dedicated to Joseph Gallian on his 65th birthday

1. Introduction

In 1963, Paul Cohen stunned the mathematical world with his new technique of *forcing,* which allowed him to solve several outstanding problems in set theory at a single stroke. Perhaps most notably, he proved the independence of the continuum hypothesis (CH) from the Zermelo-Fraenkel-Choice (ZFC) axioms of set theory. The impact of Cohen's ideas on the practice of set theory, as well as on the philosophy of mathematics, has been incalculable.

Curiously, though, despite the importance of Cohen's work and the passage of nearly fifty years, forcing remains totally mysterious to the vast majority of mathematicians, even those who know a little mathematical logic. As an illustration, let us note that Monastyrsky's outstanding book [**11**] gives highly informative and insightful expositions of the work of almost every Fields Medalist—but says almost nothing about forcing. Although there exist numerous textbooks with mathematically correct and complete proofs of the basic theorems of forcing, the subject remains notoriously difficult for beginners to learn.

All mathematicians are familiar with the concept of an *open research problem.* I propose the less familiar concept of an *open exposition problem.* Solving an open exposition problem means explaining a mathematical subject in a way that renders it totally perspicuous. Every step should be motivated and clear; ideally, students should feel that they could have arrived at the results themselves. The proofs should be "natural" in Donald Newman's sense [**13**]:

> This term ... is introduced to mean not having any ad hoc constructions or *brilliancies.* A "natural" proof, then, is one which proves itself, one available to the "common mathematician in the streets."

I believe that it is an open exposition problem to explain forcing. Current treatments allow readers to verify the truth of the basic theorems, and to progress fairly rapidly to the point where they can *use* forcing to prove their own independence results (see [**2**] for a particularly nice explanation of how to use forcing as a

2000 *Mathematics Subject Classification.* Primary 03E35; secondary 03E40, 00-01.

black box to turn independence questions into concrete combinatorial problems). However, in all treatments that I know of, one is left feeling that only a genius with fantastic intuition or technical virtuosity could have found the road to the final result.

This paper does not solve this open exposition problem, but I believe it is a step in the right direction. My goal is to give a rapid overview of the subject, emphasizing the broad outlines and the intuitive motivation while omitting most of the proofs. The reader will not, of course, master forcing by reading this paper in isolation without consulting standard textbooks for the omitted details, but my hope is to provide a map of the forest so that the beginner will not get lost while forging through the trees. Currently, no such bird's-eye overview seems to be available in the published literature; I hope to fill this gap. I also hope that this paper will inspire others to continue the job of making forcing totally transparent.

2. Executive summary

The negation of CH says that there is a cardinal number, $\aleph_1$, between the cardinal numbers $\aleph_0$ and $2^{\aleph_0}$. One might therefore try to build a structure that satisfies the negation of CH by starting with something that *does* satisfy CH (Gödel had in fact constructed such structures) and "inserting" some sets that are missing.

The fundamental theorem of forcing is that, under very general conditions, one can indeed start with a mathematical structure M that satisfies the ZFC axioms, and enlarge it by adjoining a new element U to obtain a new structure $M[U]$ that also satisfies ZFC. Conceptually, this process is analogous to the process of adjoining a new element X to, say, a given ring R to obtain a larger ring $R[X]$. However, the construction of $M[U]$ is a lot more complicated because the axioms of ZFC are more complicated than the axioms for a ring. Cohen's idea was to build the new element U one step at a time, tracking what new properties of $M[U]$ would be "forced" to hold at each step, so that one could control the properties of $M[U]$—in particular, making it satisfy the negation of CH as well as the axioms of ZFC.

The rest of this paper fleshes out the above construction in more detail.

3. Models of ZFC

As mentioned above, Cohen proved the independence of CH from ZFC; more precisely, he proved that if ZFC is consistent, then CH is not a logical consequence of the ZFC axioms. Gödel had already proved that if ZFC is consistent, then $\neg$CH, the negation of CH, is not a logical consequence of ZFC, using his concept of "constructible sets." (Note that the hypothesis that ZFC is consistent cannot be dropped, because if ZFC is inconsistent then *everything* is a logical consequence of ZFC!)

Just how does one go about proving that CH is not a logical consequence of ZFC? At a very high level, the structure of the proof is what you would expect: One writes down a very precise statement of the ZFC axioms and of $\neg$CH, and then one constructs a mathematical structure that satisfies both ZFC and $\neg$CH. This structure is said to be a *model* of the axioms. Although the term "model" is not often seen in mathematics outside of formal logic, it is actually a familiar concept. For example, in group theory, a "model of the group-theoretic axioms" is just a group, i.e., a set G with a binary operation $*$ satisfying axioms such as: "There exists an element e in G such that $x * e = e * x = x$ for all x in G," and so forth.

Analogously, we could invent a term—say, *universe*—to mean "a structure that is a model of ZFC." Then we could begin our study of ZFC with definition such as, "A *universe* is a set M together with a binary relation R satisfying..." followed by a long list of axioms such as the *axiom of extensionality:*

> If x and y are distinct elements of M then either there exists z in M such that zRx but not zRy, or there exists z in M such that zRy but not zRx.

Another axiom of ZFC is the *powerset axiom:*

> For every x in M, there exists y in M with the following property: For every z in M, zRy if and only if $z \subseteq x$.

(Here the expression "$z \subseteq x$" is an abbreviation for "every w in M satisfying wRz also satisfies wRx.") There are other axioms, which can be found in any set theory textbook, but the general idea should be clear from these two examples. Note that the binary relation is usually denoted by the symbol $\in$ since the axioms are inspired by the set membership relation. However, we have deliberately chosen the unfamiliar symbol R to ensure that the reader will not misinterpret the axiom by accidentally reading $\in$ as "is a member of."

As an aside, we should mention that it is not standard to use the term *universe* to mean "model of ZFC." For some reason set theorists tend to give a snappy name like "ZFC" to a *list of axioms,* and then use the term "model of ZFC" to refer to the *structures that satisfy the axioms*, whereas in the rest of mathematics it is the other way around: one gives a snappy name like "group" to the structure, and then uses the term "axioms for a group" to refer to the axioms. Apart from this terminological point, though, the formal setup here is entirely analogous to that of group theory. For example, in group theory, the statement S that "$x * y = y * x$ for all x and y" is not a logical consequence of the axioms of group theory, because there exists a mathematical structure—namely a non-abelian group—that satisfies the group axioms as well as the negation of S.

On the other hand, the definition of a model of ZFC has some curious features, so a few additional remarks are in order.

3.1. Apparent circularity. One common confusion about models of ZFC stems from a tacit expectation that some people have, namely that we are supposed to suspend all our preconceptions about sets when beginning the study of ZFC. For example, it may have been a surprise to some readers to see that a universe is defined to be a *set* together with.... Wait a minute—what is a set? Isn't it circular to define sets in terms of sets?

In fact, we are not defining sets in terms of sets, but *universes* in terms of sets. Once we see that all we are doing is studying a subject called "universe theory" (rather than "set theory"), the apparent circularity disappears.

The reader may still be bothered by the lingering feeling that the point of introducing ZFC is to "make set theory rigorous" or to examine the foundations of mathematics. While it is true that ZFC can be used as a tool for such philosophical investigations, we do not do so in this paper. Instead, we take for granted that ordinary mathematical reasoning—including reasoning about sets—is perfectly valid and does not suddenly become invalid when the object of study is ZFC. That is, we approach the study of ZFC and its models in the same way that one approaches the study of any other mathematical subject. This is the best way to grasp the mathematical content; after this is achieved, one can then try to apply the technical results to philosophical questions if one is so inclined.

Note that in accordance with our attitude that ordinary mathematical reasoning is perfectly valid, we will freely employ reasoning about infinite sets of the kind that is routinely used in mathematics. We reassure readers who harbor philosophical doubts about the validity of infinitary set-theoretic reasoning that Cohen's proof can be turned into a purely finitistic one. We will not delve into such metamathematical niceties here, but see for example the beginning of Chapter VII of Kunen's book [**10**].

3.2. Existence of examples. A course in group theory typically begins with many examples of groups. One then verifies that the examples satisfy all the axioms of group theory. Here we encounter an awkward feature of models of ZFC, which is that exhibiting explicit models of ZFC is difficult. For example, there are no finite models of ZFC. Worse, by a result known as the *completeness theorem,* the statement that ZFC has *any models at all* is equivalent to the statement that ZFC is consistent, which is an assumption that is at least mildly controversial. So how can we even get off the ground?

Fortunately, these difficulties are not as severe as they might seem at first. For example, one entity that is almost a model of ZFC is V, the class of all sets. If we take $M = V$ and we take R to mean "is a member of," then we see that the axiom of extensionality simply says that two sets are equal if and only if they contain the same elements—a manifestly true statement. The rest of the axioms of ZFC are similarly self-evident when $M = V$.[1] The catch is that a model of ZFC has to be a *set,* and V, being "too large" to be a set (Cantor's paradox), is a proper class and therefore, strictly speaking, is disqualified from being a model of ZFC. However, it is close enough to being a model of ZFC to be intuitively helpful.

As for possible controversy over whether ZFC is consistent, we can sidestep the issue simply by treating the consistency of ZFC like any other unproved statement, such as the Riemann hypothesis. That is, we can assume it freely as long as we remember to preface all our theorems with a conditional clause.[2] So from now on we shall assume that ZFC is consistent, and therefore that models of ZFC exist.

3.3. "Standard" models. Even granting the consistency of ZFC, it is not easy to produce models. One can extract an example from the proof of the completeness theorem, but this example is unnatural and is not of much use for tackling CH. Instead of continuing the search for explicit examples, we shall turn our attention to important properties of models of ZFC.

One important insight of Cohen's was that it is useful to consider what he called *standard* models of ZFC. A model M of ZFC is standard if the elements of M are *well-founded sets* and if the relation R is ordinary set membership. Well-founded sets are sets that are built up inductively from the empty set, using operations such as taking unions, subsets, powersets, etc. Thus the empty set $\{\}$ is well-founded, as are $\{\{\}\}$ and the infinite set $\{\{\}, \{\{\}\}, \{\{\{\}\}\}, \ldots\}$. They are called "well-founded" because the nature of their inductive construction precludes any well-founded set from being a member of itself. We emphasize that if M is standard, then the

[1] Except, perhaps, the axiom of regularity, but this is a technical quibble that we shall ignore.

[2] In fact, we already did this when we said that the precise statement of Cohen's result is that *if ZFC is consistent* then CH is not a logical consequence of ZFC.

elements of M are not amorphous "atoms," as some of us envisage the elements of an abstract group to be, but are *sets*. Moreover, well-founded sets are *not* themselves built up from "atoms"; it's "sets all the way down."

While it is fairly clear that if standard models of ZFC exist, then they form a natural class of examples, it is not at all clear that any standard models exist at all, even if ZFC is consistent.[3] (The class of all well-founded sets is a proper class and not a set and hence is disqualified.) Moreover, even if standard models exist, one might think that constructing a model of ZFC satisfying $\neg$CH might require considering "exotic" models in which the binary relation R bears very little resemblance to ordinary set membership. Cohen himself admits on page 108 of [**6**] that a minor leap of faith is involved here:

> Since the negation of CH or AC may appear to be somewhat unnatural one might think it hopeless to look for standard models. However, we make a firm decision at the point to consider only standard models. Although this may seem like a very severe limitation in our approach it will turn out that this very limitation will guide us in suggesting possibilities.

Another property that a model of ZFC can have is *transitivity*. A standard model M of ZFC is *transitive* if every member of an element of M is also an element of M. (The term *transitive* is used because we can write the condition in the suggestive form "$x \in y$ and $y \in M$ implies $x \in M$.") This is a natural condition to impose if we think of M as a universe consisting of "all that there is"; in such a universe, sets "should" be sets of things that already exist in the universe. Cohen's remark about standard models applies equally to transitive models.[4]

Our focus will be primarily on standard transitive models. Of course, this choice of focus is made with the benefit of hindsight, but even without the benefit of hindsight, it makes sense to study models with natural properties before studying exotic models. When M is a standard transitive model, we will often use the symbol $\in$ for the relation R, because in this case R *is* in fact set membership.

4. Powersets and absoluteness

At some point in their education, most mathematicians learn that all familiar mathematical objects can be defined in terms of sets. For example, one can define the number 0 to be the empty set $\{\}$, the number 1 to be the set $\{0\}$, and in general the number n to be the so-called *von Neumann ordinal* $\{0, 1, \ldots, n-1\}$. The set $\mathbb{N}$ of all natural numbers may be defined to be $\aleph_0$, the set of all von Neumann ordinals.[5] Note that with these definitions, the membership relation on $\aleph_0$ corresponds to the usual ordering on the natural numbers (this is why n is defined as $\{0, 1, \ldots, n-1\}$ rather than as $\{n-1\}$). The ordered pair (x, y) may be defined à la Kuratowski as the set $\{\{x\}, \{x, y\}\}$. Functions, relations, bijections, maps, etc., can be defined as certain sets of ordered pairs. More interesting mathematical structures can be defined as ordered pairs (X, S) where X is an underlying set and S is the structure

[3]It turns out that the existence of a standard model of ZFC is indeed a stronger assumption than the consistency of ZFC, but we will ignore this nicety.

[4]We remark in passing that the *Mostowski collapsing theorem* implies that if there exist *any* standard models of ZFC, then there exist standard transitive models.

[5]We elide the distinction between the cardinality $\aleph_0$ of $\mathbb{N}$ and the order type ω of $\mathbb{N}$.

on X. With this understanding, the class V of all sets may be thought of as being the entire mathematical universe.

Models of ZFC, like everything else, live inside V, but they are special because they look a lot like V itself. This is because it turns out that virtually all mathematical proofs of the existence of some object X in V can be mimicked by a proof from the ZFC axioms, thereby proving that any model of ZFC must contain an object that is at least highly analogous to X. It turns out that this "analogue" of X is often *equal* to X, especially when M is a standard transitive model. For example, it turns out that every standard transitive model M of ZFC contains all the von Neumann ordinals as well as $\aleph_0$.

However, the analogue of a mathematical object X is not *always* equal to X. A crucial counterexample is the powerset of $\aleph_0$, denoted by $2^{\aleph_0}$. Naïvely, one might suppose that the powerset axiom of ZFC guarantees that $2^{\aleph_0}$ must be a member of any standard transitive model M. But let us look more closely at the precise statement of the powerset axiom. Given that $\aleph_0$ is in M, the powerset axiom guarantees the existence of y in M with the following property: For every z in M, $z \in y$ if and only if every w in M satisfying $w \in z$ also satisfies $w \in \aleph_0$. Now, does it follow that y is precisely the set of all subsets of $\aleph_0$?

No. First of all, it is not even immediately clear that z is a subset of $\aleph_0$; the axiom does not require that *every* w satisfying $w \in z$ also satisfies $w \in \aleph_0$; it requires only that *every* w *in* M satisfying $w \in z$ satisfies $w \in x$. However, under our assumption that M is *transitive,* every $w \in z$ is in fact in M, so indeed z is a subset of $\aleph_0$.

More importantly, though, y does not contain *every* subset of $\aleph_0$; it contains *only those subsets of* x *that are in* M. So if, for example, M happens to be *countable* (i.e., M contains only countably many elements), then y will be countable, and so a fortiori y cannot be equal to $2^{\aleph_0}$, since $2^{\aleph_0}$ is uncountable. The set y, which we might call the *powerset of* $\aleph_0$ *in* M, is not the same as the "real" powerset of $\aleph_0$, a.k.a. $2^{\aleph_0}$; many subsets of $\aleph_0$ are "missing" from y.

This is a subtle and important point, so let us explore it further. We may ask, is it really possible for a standard transitive model of ZFC to be countable? Can we not mimic (in ZFC) Cantor's famous proof that $2^{\aleph_0}$ is uncountable to show that M must contain an uncountable set, and then conclude by transitivity that M itself must be uncountable?

The answer is no. Cantor's theorem states that there is no bijection between $\aleph_0$ and $2^{\aleph_0}$. If we carefully mimic Cantor's proof with a proof from the ZFC axioms, then we find that Cantor's theorem tells us that there is indeed a set y in M that plays the role of the powerset of $\aleph_0$ in M, and that there is *no bijection in* M between $\aleph_0$ and y. However, this fact does not mean that there is *no bijection at all* between $\aleph_0$ and y. There might be a bijection *in* V between them; we know only that such a bijection *cannot be a member of* M; it is "missing" from M. So Cantor's theorem does not exclude the possibility that y, as well as M, is countable, even though y is necessarily "uncountable in M."[6] It turns out that something stronger can be said: the so-called *Löwenheim-Skolem theorem* says that if there are any models of ZFC at all, then in fact there exist countable models.

More generally, one says that a concept in V is *absolute* if it coincides with its counterpart in M. For example, "the empty set," "is a member of," "is a subset of,"

[6]This curious state of affairs often goes by the name of *Skolem's paradox.*

"is a bijection," and "$\aleph_0$" all turn out to be absolute for standard transitive models. On the other hand, "is the powerset of" and "uncountable" are *not* absolute. For a concept that is not absolute, we must distinguish carefully between the concept "in the real world" (i.e., in V) and the concept in M.

A careful study of ZFC necessarily requires keeping track of exactly which concepts are absolute and which are not. However, since the majority of basic concepts are absolute, except for those associated with taking powersets and cardinalities, in this paper we will adopt the approach of mentioning non-absoluteness only when it is especially relevant.

5. How one might try to build a model satisfying ¬CH

The somewhat counterintuitive fact that ZFC has countable models with many missing subsets provides a hint as to how one might go about constructing a model for ZFC that satisfies ¬CH. Start with a countable standard transitive model M. The elementary theory of cardinal numbers tells us that there is always a smallest cardinal number after any given cardinal number, so let $\aleph_1, \aleph_2, \ldots$ denote the next largest cardinals after $\aleph_0$. As usual we can mimic the proofs of these facts about cardinal numbers with formal proofs from the axioms of ZFC, to conclude that there is a set in M that plays the role of $\aleph_2$ in M. We denote this set by $\aleph_2^M$. Let us now construct a function F from the Cartesian product $\aleph_2^M \times \aleph_0$ into the set $2 = \{0, 1\}$. We may interpret F as a sequence of functions from $\aleph_0$ into 2. Because M is countable and transitive, so is $\aleph_2^M$; thus we can easily arrange for these functions to be pairwise distinct. Now, if F is already in M, then M satisfies ¬CH! The reason is that functions from $\aleph_0$ into 2 can be identified with subsets of $\aleph_0$, and F therefore shows us that the powerset of $\aleph_0$ in M must be at least $\aleph_2$ in M. Done!

But what if F is missing from M? A natural idea is to add F to M to obtain a larger model of ZFC, that we might call $M[F]$.[7] The hope would be that F can be added in a way that does not "disturb" the structure of M too much, so that the argument in the previous paragraph can be carried over into $M[F]$, which would therefore satisfy ¬CH.

Miraculously, this seemingly naïve idea actually works! There are, of course, numerous technical obstacles to be surmounted, but the basic plan as outlined above is on the right track. For those who like to think algebraically, it is quite appealing to learn that forcing is a technique for constructing new models from old ones by adjoining a new element that is missing from the original model. Even without any further details, one can already imagine that the ability to adjoin new elements to an existing model gives us enormous flexibility in our quest to create models with desired properties. And indeed, this is true; it is the reason why forcing is such a powerful idea.

What technical obstacles need to be surmounted? The first thing to note is that one clearly cannot add only the set F to M and expect to obtain a model of ZFC; one must also add, at minimum, every set that is "constructible" from F together with elements of M, just as when we create an extension of an algebraic

[7]Later on we will use the notation $M[U]$ rather than $M[F]$ because it will turn out to be more natural to think of the larger model as being obtained by adjoining another set U that is closely related to F, rather than by adjoining F itself. For our purposes, $M[U]$ and $M[F]$ can just be thought of as two different names for the same object.

object by adjoining x, we must also adjoin everything that is *generated* by x. We will not define "constructible" precisely here, but it is the same concept that Gödel used to prove that CH is consistent with ZFC, and in particular it was already a familiar concept before Cohen came onto the scene.

A more serious obstacle is that it turns out that we cannot, for example, simply take an arbitrary subset a of $\aleph_0$ that is missing from M and adjoin a, along with everything constructible from a together with elements of M, to M; the result will not necessarily be a model of ZFC. A full explanation of this result would take us too far afield—the interested reader should see page 111 of Cohen's book [**6**]—but the rough idea is that we could perversely choose a to be a set that encodes explicit information about the size of M, so that adjoining a would create a kind of self-referential paradox. Cohen goes on to say:

> Thus a must have certain special properties.... Rather than describe a directly, it is better to examine the various properties of a and determine which are desirable and which are not. The chief point is that we do not wish a to contain "special" information about M, which can only be seen from the outside.... The a which we construct will be referred to as a "generic" set relative to M. The idea is that all the properties of a must be "forced" to hold merely on the basis that a behaves like a "generic" set in M. This concept of deciding when a statement about a is "forced" to hold is the key point of the construction.

Cohen then proceeds to explain the forcing concept, but at this point we will diverge from Cohen's account and pursue instead the concept of a *Boolean-valued model* of ZFC. This approach was developed by Scott, Solovay, and Vopěnka starting in 1965, and in my opinion is the most intuitive way to proceed at this juncture. We will return to Cohen's approach later.

6. Boolean-valued models

To recap, we have reached the point where we see that if we want to construct a model of $\neg$CH, it would be nice to have a method of starting with an arbitrary standard transitive model M of ZFC, and building a new structure by adjoining some subsets that are missing from M. We explain next how this can be done, but instead of giving the construction right away, we will work our way up to it gradually.

6.1. Motivational discussion. Inspired by Cohen's suggestion, we begin by considering *all possible statements* that might be true of our new structure, and then deciding which ones we want to hold and which one we do not want to hold.

To make the concept of "all possible statements" precise, we must introduce the concept of a *formal language.* Let $\mathfrak{S}$ denote the set of all sentences in the *first-order language of set theory,* i.e., all sentences built out of "atomic" statements such as $x = y$ and xRy (where x and y are *constant symbols* that each represent some fixed element of the domain) using the Boolean connectives OR, AND, and NOT and the quantifiers $\exists$ and $\forall$. The axioms of ZFC can all be expressed in this formal language, as can any theorems (or non-theorems, for that matter) of ZFC. For example, "if A then B" (written $A \to B$) can be expressed as (NOT A) OR B, "A iff B" (written $A \leftrightarrow B$) can be expressed as $(A \to B)$ AND $(B \to A)$, "x is a subset of y" (written $x \subseteq y$) can be expressed as $\forall z\,((zRx) \to (zRy))$, and the powerset axiom can be expressed as

$$\forall x\,\exists y\,\forall z\,((z \subseteq x) \leftrightarrow (zRy)).$$

An important observation is that when choosing which sentences in $\mathfrak{S}$ we want to hold in our new structure, we are subject to certain constraints. For example, if the sentences ϕ and ψ hold, then the sentence ϕ AND ψ must also hold. A natural way to track these constraints is by means of a *Boolean algebra.* The most familiar example of a Boolean algebra is the family 2^S of all subsets of a given set S, partially ordered by inclusion. More generally, a Boolean algebra is any partially ordered set with a minimum element $\mathbf{0}$ and a maximum element $\mathbf{1}$, in which any two elements x and y have a least upper bound $x \vee y$ and a greatest lower bound $x \wedge y$ (in the example of 2^S, $\vee$ is set union and $\wedge$ is set intersection), where $\vee$ and $\wedge$ distribute over each other (i.e., $x \vee (y \wedge z) = (x \vee y) \wedge (x \vee z)$ and $x \wedge (y \vee z) = (x \wedge y) \vee (x \wedge z)$) and every element x has a *complement,* i.e., an element x^* such that $x \vee x^* = \mathbf{1}$ and $x \wedge x^* = \mathbf{0}$.

There is a natural correspondence between the concepts $\mathbf{0}$, $\mathbf{1}$, $\vee$, $\wedge$, and $*$ in a Boolean algebra and the concepts of falsehood, truth, OR, AND, and NOT in logic. This observation suggests that if we know that we want certain statements of $\mathfrak{S}$ to hold in our new structure but are unsure of others, then we can try to record our state of partial knowledge by picking a suitable Boolean algebra $\mathbb{B}$, and mapping every sentence $\phi \in \mathfrak{S}$ to some element of $\mathbb{B}$ that we denote by $[\![\phi]\!]^{\mathbb{B}}$. If ϕ is "definitely true" then we set $[\![\phi]\!]^{\mathbb{B}} = \mathbf{1}$ and if ϕ is "definitely false" then we set $[\![\phi]\!]^{\mathbb{B}} = \mathbf{0}$; otherwise, $[\![\phi]\!]^{\mathbb{B}}$ takes on some intermediate value between $\mathbf{0}$ and $\mathbf{1}$. In a sense, we are developing a kind of "multi-valued logic" or "fuzzy logic"[8] in which some statements are neither true nor false but lie somewhere in between.

It is clear that the mapping $\phi \mapsto [\![\phi]\!]^{\mathbb{B}}$ should satisfy the conditions

$$[\![\phi \text{ OR } \psi]\!]^{\mathbb{B}} = [\![\phi]\!]^{\mathbb{B}} \vee [\![\psi]\!]^{\mathbb{B}} \tag{6.1}$$

$$[\![\phi \text{ AND } \psi]\!]^{\mathbb{B}} = [\![\phi]\!]^{\mathbb{B}} \wedge [\![\psi]\!]^{\mathbb{B}} \tag{6.2}$$

$$[\![\text{NOT } \phi]\!]^{\mathbb{B}} = ([\![\phi]\!]^{\mathbb{B}})^* \tag{6.3}$$

What about atomic expressions such as $[\![x = y]\!]^{\mathbb{B}}$ and $[\![xRy]\!]^{\mathbb{B}}$? Again, if we definitely want certain equalities or membership statements to hold but want to postpone judgment on others, then we are led to the idea of tracking these statements using a structure consisting of "fuzzy sets." To make this precise, let us first observe that an ordinary set may be identified with a function whose range is the trivial Boolean algebra with just two elements $\mathbf{0}$ and $\mathbf{1}$, and that sends the members of the set to $\mathbf{1}$ and the non-members to $\mathbf{0}$. Generalizing, if $\mathbb{B}$ is an arbitrary Boolean algebra, then a "fuzzy set" should take a set of "potential members," which should themselves be fuzzy sets, and assign each potential member y a value in $\mathbb{B}$ corresponding to the "degree" to which y is a member of x. More precisely, we define a a $\mathbb{B}$-*valued set* to be a function from a set of $\mathbb{B}$-valued sets to $\mathbb{B}$. (Defining $\mathbb{B}$-valued sets in terms of $\mathbb{B}$-valued sets might appear circular, but the solution is to note that the empty set is a $\mathbb{B}$-valued set; we can then build up other $\mathbb{B}$-valued sets inductively.)

---

[8]We use scare quotes as these terms, and the term "fuzzy set" that we use later, have meanings in the literature that are rather different from the ideas that we are trying to convey here.

6.2. Construction of $M^{\mathbb{B}}$. We are now in a position to describe more precisely our plan for constructing a new model of ZFC from a given model M. We pick a suitable Boolean algebra $\mathbb{B}$, and we let $M^{\mathbb{B}}$ be the set of all $\mathbb{B}$-valued sets in M. The set $\mathfrak{S}$ should have one constant symbol for each element of $M^{\mathbb{B}}$. We define a a map $\phi \mapsto [\![\phi]\!]^{\mathbb{B}}$ from $\mathfrak{S}$ to $\mathbb{B}$, which should obey equations such as (6.1)–(6.3) and should send the axioms of ZFC to $\mathbf{1}$. The structure $M^{\mathbb{B}}$ will be a so-called *Boolean-valued model* of ZFC; it will not actually be a model of ZFC, because it will consist of "fuzzy sets" and not sets, and if you pick an arbitrary $\phi \in \mathfrak{S}$ and ask whether it holds in $M^{\mathbb{B}}$, then the answer will often be neither "yes" nor "no" but some element of $\mathbb{B}$ (whereas if N is an actual model of ZFC then either N satisfies ϕ or it doesn't). On the other hand, $M^{\mathbb{B}}$ will satisfy ZFC, in the sense that $[\![\phi]\!]^{\mathbb{B}} = \mathbf{1}$ for every ϕ in ZFC. To turn $M^{\mathbb{B}}$ into an actual model of ZFC with desired properties, we will take a suitable quotient of $M^{\mathbb{B}}$ that eliminates the fuzziness.

We have already started to describe $M^{\mathbb{B}}$ and the map $[\![\cdot]\!]^{\mathbb{B}}$, but we are not done. For example, we need to deal with expressions involving the quantifiers $\exists$ and $\forall$. These may not appear to have a direct counterpart in the formalism of Boolean algebras, but notice that another way to say that there exists x with a certain property is to say that either a has the property or b has the property or c has the property or..., where we enumerate all the entities in the universe one by one. This observation leads us to the definition

$$[\![\exists x\, \phi(x)]\!]^{\mathbb{B}} = \bigvee_{a \in M^{\mathbb{B}}} [\![\phi(a)]\!]^{\mathbb{B}} \tag{6.4}$$

Now there is a potential problem with (6.4): In an arbitrary Boolean algebra, an *infinite* subset of elements may not have a least upper bound, so the right-hand side of (6.4) may not be defined. We solve this problem by fiat: First we define a *complete Boolean algebra* to be a Boolean algebra in which arbitrary subsets of elements have a least upper bound and a greatest lower bound. We then require that $\mathbb{B}$ be a complete Boolean algebra; then (6.4) makes perfect sense, as does the equation

$$[\![\forall x\, \phi(x)]\!]^{\mathbb{B}} = \bigwedge_{a \in M^{\mathbb{B}}} [\![\phi(a)]\!]^{\mathbb{B}} \tag{6.5}$$

Equations (6.4) and (6.5) take care of $\exists$ and $\forall$, but we have still not defined $[\![xRy]\!]^{\mathbb{B}}$ or $[\![x = y]\!]^{\mathbb{B}}$, or ensured that $M^{\mathbb{B}}$ satisfies ZFC. The definitions of $[\![x = y]\!]^{\mathbb{B}}$ and $[\![xRy]\!]^{\mathbb{B}}$ are surprisingly delicate; there are many plausible attempts that fail for subtle reasons. The impatient reader can safely skim the details in the next paragraph and just accept the final equations (6.6)–(6.7).

We follow the treatment on pages 22–23 in Bell's book [**3**], which motivates the definitions of $[\![x = y]\!]^{\mathbb{B}}$ and $[\![xRy]\!]^{\mathbb{B}}$ by listing several equations that one would like to hold and inferring what the definitions "must" be. First, we want the axiom of extensionality to hold in $M^{\mathbb{B}}$; this suggests the equation

$$[\![x = y]\!]^{\mathbb{B}} = [\![(\forall w\, (wRx \to wRy)) \text{ AND } (\forall w\, (wRy \to wRx))]\!]^{\mathbb{B}}.$$

Another plausible equation is

$$[\![xRy]\!]^{\mathbb{B}} = [\![\exists w\, ((wRy) \text{ AND } (w = x))]\!]^{\mathbb{B}}.$$

It is also plausible that the expression $[\![\exists w\, ((wRy) \text{ AND } \phi(w))]\!]^{\mathbb{B}}$ should depend only on the values of $[\![\phi(w)]\!]^{\mathbb{B}}$ for those w that are actually in the domain of y

(recall that y, being a $\mathbb{B}$-valued set, is a function from a set of $\mathbb{B}$-valued sets to $\mathbb{B}$, and thus has a domain $\mathrm{dom}(y)$). Also, the value of $[\![wRy]\!]^{\mathbb{B}}$ should be closely related to the value of $y(w)$. We are thus led to the equations

$$[\![\exists w\,(wRy \text{ AND } \phi(w))]\!]^{\mathbb{B}} = \bigvee_{w\in\mathrm{dom}(y)} \left(y(w) \wedge [\![\phi(w)]\!]^{\mathbb{B}}\right)$$

$$[\![\forall w\,(wRy \to \phi(w))]\!]^{\mathbb{B}} = \bigwedge_{w\in\mathrm{dom}(y)} \left(y(w) \Rightarrow [\![\phi(w)]\!]^{\mathbb{B}}\right)$$

where $x \Rightarrow y$ is another way of writing $x^* \vee y$. All these equations drive us to the definitions

$$[\![xRy]\!]^{\mathbb{B}} = \bigvee_{w\in\mathrm{dom}(y)} \left(y(w) \wedge [\![x = w]\!]^{\mathbb{B}}\right) \tag{6.6}$$

$$[\![x = y]\!]^{\mathbb{B}} = \bigwedge_{w\in\mathrm{dom}(x)} \left(x(w) \Rightarrow [\![wRy]\!]^{\mathbb{B}}\right) \wedge \bigwedge_{w\in\mathrm{dom}(y)} \left(y(w) \Rightarrow [\![wRx]\!]^{\mathbb{B}}\right) \tag{6.7}$$

The definitions (6.6) and (6.7) again appear circular, because they define $[\![x = y]\!]^{\mathbb{B}}$ and $[\![xRy]\!]^{\mathbb{B}}$ in terms of each other, but again (6.6) and (6.7) should be read as a joint inductive definition.

One final remark is needed regarding the definition of $M^{\mathbb{B}}$. So far we have not imposed any constraints on $\mathbb{B}$ other than that it be a complete Boolean algebra. But without some such constraints, there is no guarantee that $M^{\mathbb{B}}$ will satisfy ZFC. For example, let us see what happens with the powerset axiom. Given x in $M^{\mathbb{B}}$, it is natural to construct the powerset y of x in $M^{\mathbb{B}}$ by letting

$$\mathrm{dom}(y) = \mathbb{B}^{\mathrm{dom}(x)},$$

i.e., the "potential members" of y should be precisely the maps from $\mathrm{dom}(x)$ to $\mathbb{B}$. Moreover, for each $w \in \mathrm{dom}(y)$, the value of $y(w)$ should be $[\![w \subseteq x]\!]^{\mathbb{B}}$. The catch is that if $\mathbb{B}$ is not in M, then maps from $\mathrm{dom}(x)$ to $\mathbb{B}$ may not be $\mathbb{B}$-valued sets in M. The simplest way out of this difficulty is to require that $\mathbb{B}$ be in M, and we shall indeed require this.[9] Once we impose this condition, we can weaken the requirement that $\mathbb{B}$ be a complete Boolean algebra to the requirement that $\mathbb{B}$ be a complete Boolean algebra *in M*, meaning that infinite least upper bounds and greatest lower bounds *over subsets of B that are in M* are guaranteed to exist, but not necessarily in general. ("Complete," being related to taking powersets, is not absolute.) Examination of the definitions of $M^{\mathbb{B}}$ and $[\![\cdot]\!]^{\mathbb{B}}$ reveals that $\mathbb{B}$ only needs to be a complete Boolean algebra in M, and it turns out that this increased flexibility in the choice of $\mathbb{B}$ is very important.

We are now done with the definition of the Boolean-valued model $M^{\mathbb{B}}$. To summarize, we pick a Boolean algebra $\mathbb{B}$ in M that is complete in M, let $M^{\mathbb{B}}$ be the set of all $\mathbb{B}$-valued sets in M, and define $[\![\cdot]\!]^{\mathbb{B}}$ using equations (6.1)–(6.7).

At this point, one needs to perform a long verification that $M^{\mathbb{B}}$ satisfies ZFC, and that the rules of logical inference behave as expected in $M^{\mathbb{B}}$ (so that, for example, if $[\![\phi]\!]^{\mathbb{B}} = \mathbf{1}$ and ψ is a logical consequence of ϕ then $[\![\psi]\!]^{\mathbb{B}} = \mathbf{1}$). We omit these details because they are covered well in Bell's book **[3]**. Usually, as in the case

---

[9] While choosing $\mathbb{B}$ to be in M suffices to make everything work, it is not strictly necessary. *Class forcing* involves certain carefully constructed Boolean algebras $\mathbb{B}$ that are not in M. However, this is an advanced topic that is not needed for proving the independence of CH.

of the powerset axiom above, it is not too hard to guess how to construct the object whose existence is asserted by the ZFC axiom, using the fact that M satisfies ZFC, although in some cases, completing the argument in detail can be tricky.

6.3. Modding out by an ultrafilter. As we stated above, the way to convert our Boolean-valued model $M^{\mathbb{B}}$ to an actual model of ZFC is to take a suitable quotient. That is, we need to pick out precisely the statements that are true in our new model. To do this, we choose a subset U of $\mathbb{B}$ that contains $[\![\phi]\!]^{\mathbb{B}}$ for every statement ϕ that holds in the new model of ZFC. The set U, being a "truth definition" for our new model, has to have certain properties; for example, since for every ϕ, either ϕ or NOT ϕ must hold in the new model, it follows that for all x in $\mathbb{B}$, U must contain either x or x^*. Similarly, thinking of membership in U as representing "truth," we see that U should have the following properties:

1. $1 \in U$;
2. $0 \notin U$;
3. if $x \in U$ and $y \in U$ then $x \wedge y \in U$;
4. if $x \in U$ and $x \leq y$ (i.e., $x \wedge y = x$) then $y \in U$;
5. for all x in $\mathbb{B}$, either $x \in U$ or $x^* \in U$.

A subset U of a Boolean algebra having the above properties is called an *ultrafilter.*

Given any ultrafilter U in $\mathbb{B}$ (U does not have to be in M), we define the quotient $M^{\mathbb{B}}/U$ as follows. The elements of $M^{\mathbb{B}}/U$ are equivalence classes of elements of $M^{\mathbb{B}}$ under the equivalence relation

$$x \sim_U y \quad \text{iff} \quad [\![x = y]\!]^{\mathbb{B}} \in U.$$

If we write x^U for the equivalence class of x, then the binary relation of $M^{\mathbb{B}}/U$—which we shall denote by the symbol $\in_U$—is defined by

$$x^U \in_U y^U \quad \text{iff} \quad [\![xRy]\!]^{\mathbb{B}} \in U.$$

It is now fairly straightforward to verify that $M^{\mathbb{B}}/U$ is a model of ZFC; the hard work has already been done in verifying that $M^{\mathbb{B}}$ satisfies ZFC.

7. Generic ultrafilters and the conclusion of the proof sketch

At this point we have a powerful theorem in hand. We can take any model M, any complete Boolean algebra $\mathbb{B}$ in M, and any ultrafilter U of M, and form a new model $M^{\mathbb{B}}/U$ of ZFC. We can now experiment with various choices of M, $\mathbb{B}$, and U to construct all kinds of models of ZFC with various properties.

So let us revisit our plan (in Section 5) of starting with a standard transitive model and inserting some missing subsets to obtain a larger standard transitive model. If we try to use our newly constructed machinery to carry out this plan, then we soon find that $M^{\mathbb{B}}/U$ need not, in general, be (isomorphic to) a standard transitive model of ZFC, even if M is. Some extra conditions need to be imposed.

Cohen's insight—perhaps his most important and ingenious one—is that in many cases, including the case of CH, the right thing to do is to require that U be *generic.* The term "generic" can be defined more generally in the context of an arbitrary partially ordered set P. First define a subset D of P to be *dense* if for all p in P, there exists q in D such that $q \leq p$. Then a subset of P is *generic* if it intersects every dense subset. In our current setting, the partially ordered set P is $\mathbb{B}\backslash\{\mathbf{0}\}$, and the crucial condition on U is that it be M-*generic* (or *generic over* M), meaning that U intersects every dense subset $D \subseteq (\mathbb{B}\backslash\{\mathbf{0}\})$ that is a member of M.

If U is M-generic, then $M^{\mathbb{B}}/U$ has many nice properties; it is (isomorphic to) a standard transitive model of ZFC, and equally importantly, it contains U. In fact, if U is M-generic, then $M^{\mathbb{B}}/U$ is the smallest standard transitive model of ZFC that contains both M and U. For this reason, when U is M-generic, one typically writes $M[U]$ instead of $M^{\mathbb{B}}/U$. We have realized the dream of adjoining a new subset of M to obtain a larger model (remember that U is a subset of $\mathbb{B}$ and we have required $\mathbb{B}$ to be in M).

It is, of course, not clear that M-generic ultrafilters exist in general. However, if M is countable, then it turns out to be easy to prove the existence of M-generic ultrafilters; essentially, one just lists the dense sets and hits them one by one. If M is uncountable then the Boolean-valued model machinery still works fine, but M-generic ultrafilters may not exist.[10] Fortunately for us, the idea sketched at the beginning of Section 5 relies on M being countable anyway.

Let us now return to that idea and complete the proof sketch. Start with a countable standard transitive model M of ZFC. If M does not already satisfy $\neg$CH, then let P be the partially ordered set of all *finite partial functions* from $\aleph_2^M \times \aleph_0$ into 2, partially ordered by *reverse* inclusion. (A finite partial function is a finite set of ordered pairs whose first coordinate is in the domain and whose second coordinate is in the range, with the property that no two ordered pairs have the same first element.) There is a standard method, which we shall not go into here, of *completing* an arbitrary partially ordered set to a complete Boolean algebra; we take the completion of P in M to be our Boolean algebra $\mathbb{B}$. Now take an M-generic ultrafilter U, which exists because M is countable. If we blur the distinction between P and its completion $\mathbb{B}$ for a moment, then we claim that $F := \bigcup U$ is a partial function from $\aleph_2^M \times \aleph_0$ to 2. To check this, we just need to check that any two elements x and y of U are consistent with each other where they are both defined, but this is easy: Since U is an ultrafilter, x and y have a common lower bound z in U, and both x and y are consistent with z. Moreover, F is a *total* function; this is because U is generic, and the finite partial functions that are defined at a specified point in the domain form a dense set (we can extend any partial function by defining it at that point if it is not defined already). Also, the sequence of functions from $\aleph_0$ to 2 encoded by F are pairwise distinct; again this is because U is generic, and the condition of being pairwise distinct is a dense condition. The axioms of ZFC ensure that $F \in M[U]$, so $M[U]$ gives us the desired model of $\neg$CH.

There is one important point that we have swept under the rug in the above proof sketch. The set $\aleph_2^M$ is still hanging around in $M[U]$, but it is conceivable that $\aleph_2^M$ may no longer play the role of $\aleph_2$ in $M[U]$; i.e., it may be that $\aleph_2^M \neq \aleph_2^{M[U]}$. Cardinalities are not absolute, and so *cardinal collapse* can occur, i.e., the object that plays the role of a particular cardinal number in M may not play that same role in an extension of M. In fact, cardinal collapse does not occur in this particular case but this fact must be checked.[11] We omit the details, since they are covered thoroughly in textbooks.

[10]This limitation of uncountable models is not a big issue in practice, because typically the Löwenheim-Skolem theorem allows us to replace an uncountable model with a countable surrogate.

[11]The fact that cardinals do not collapse here can be traced to the fact that the Boolean algebra in question satisfies a combinatorial condition called the *countable chain condition* in M.

8. But wait—what about forcing?

The reader may be surprised—justifiably so—that we have come to the end of our proof sketch without ever precisely defining *forcing.* Does "forcing" not have a precise technical meaning?

Indeed, it does. In Cohen's original approach, he asked the following fundamental question. Suppose that we want to adjoin a "generic" set U to M. What properties of the new model will be "forced" to hold if we have only partial information about U, namely we know that some element p of M is in U?

If we are armed with the machinery of Boolean-valued models, then we can answer Cohen's question. Let us informally say that p *forces* ϕ (written $p \Vdash \phi$) if for every M-generic ultrafilter U, ϕ must hold in $M[U]$ whenever $p \in U$. Note that U plays two roles simultaneously; it *is* the generic set that we are adjoining to M, and it also picks out the true statements in $M[U]$. By the definition of an ultrafilter, we see that if $p \le [\![\phi]\!]^{\mathbb{B}}$, then ϕ must be true in $M[U]$ if $p \in U$. Therefore we can give the following formal definition of "$p \Vdash \phi$":

$$p \Vdash \phi \quad \text{iff} \quad p \le [\![\phi]\!]^{\mathbb{B}}. \tag{8.1}$$

The simplicity of equation (8.1) explains why our proof sketch did not need to refer to forcing explicitly. Forcing is actually implicit in the proof, but since $\Vdash$ has such a simple definition in terms of $[\![\cdot]\!]^{\mathbb{B}}$, it is possible in principle to produce a proof of Cohen's result without explicitly using the symbol $\Vdash$ at all, referring only to $[\![\cdot]\!]^{\mathbb{B}}$ and Boolean algebra operations.

Of course, Cohen did not have the machinery of Boolean-valued models available. What he did was to figure out what properties the expression $p \Vdash \phi$ ought to have, given that one is trying to capture the notion of the logical implications of knowing that p is a member of our new "generic" set. For example, one should have $p \Vdash (\phi \text{ AND } \psi)$ iff $p \Vdash \phi$ and $p \Vdash \psi$, by the following reasoning: If we know that membership of p in U forces ϕ to hold and it also forces ψ to hold, then membership of p in U must also force ϕ AND ψ to hold.

By similar but more complicated reasoning, Cohen devised a list of rules analogous to (6.1)–(6.7) that he used to define $p \Vdash \phi$ for any statement ϕ in $\mathfrak{S}$. In this way, he built all the necessary machinery on the basis of the forcing relation, without ever having to introduce Boolean algebras.

Thus there are (at least) two different ways to approach this subject, depending on whether $\Vdash$ or $[\![\cdot]\!]^{\mathbb{B}}$ is taken to be the fundamental concept. For many applications, these two approaches ultimately amount to almost the same thing, since we can use equation (8.1) to pass between them. In this paper I have chosen the approach using Boolean-valued models because I feel that the introduction of "fuzzy sets" and the Boolean algebra $\mathbb{B}$ are relatively easy to motivate. In Cohen's approach one still needs to introduce at some point some "fuzzy sets" (called *names* or *labels*) and a partial order, and these seem (to me at least) to be pulled out of a hat. Also, the definitions (6.1)–(6.7) are somewhat simpler than the corresponding definitions for $\Vdash$.

On the other hand, even when one works with Boolean-valued models, Cohen's intuition about generic sets U, and what is forced to be true if we know that $p \in U$, is often extremely helpful. For example, recall from our proof sketch that the constructed functions from $\aleph_0$ to 2 were pairwise distinct. In geometry, "generically" chosen functions will not be equal; distinctness is a dense condition. Cohen's intuition thus leads us to the (correct) expectation that our generically chosen functions will also be distinct, because no finite $p \in U$ can force two of them to be equal. This kind of reasoning is invaluable in more complicated applications.

9. Final remarks

We should mention that the Boolean-valued-model approach has some disadvantages. For example, set theorists sometimes find the need to work with models of axioms that do not include the powerset axiom, and then the Boolean-valued model approach does not work, because "complete" does not really make sense in such contexts. Also, Cohen's original approach allows one to work directly with an arbitrary partially ordered set P that is not necessarily a Boolean algebra, and a *generic filter* rather than a generic ultrafilter. (A subset F of P is a *filter* if $p \in F$ and $p \leq q$ implies $q \in F$, and every p and q have a common lower bound in F.) In our proof sketch we have already caught a whiff of the fact that in many cases, there is some partially ordered set P lying around that captures the combinatorics of what is really going on, and having to complete P to a Boolean algebra is a technical nuisance; it is much more convenient to work with P directly. If the reader prefers this approach, then Kunen's book [**10**] would be my recommended reference. Note that Kunen helpfully supplements his treatment with an abbreviated discussion of Boolean-valued models and the relationship between the two different approaches.

In my opinion, the weakest part of the exposition in this paper is the treatment of *genericity,* whose definition appears to come out of nowhere. A posteriori one can see that the definition works beautifully, but how would one guess a priori that the geometric concepts of dense sets and generic sets would be so apropos in this context, and come up with the right precise definitions? Perhaps the answer is just that Cohen was a genius, but perhaps there is a better approach yet to be discovered that will make it all clear.

Let us conclude with some suggestions for further reading. Easwaran [**8**] and Wolf [**16**] give very nice overviews of forcing written in the same spirit as the present paper, giving details that are critical for understanding but omitting messy technicalities. Scott's paper [**14**] is a classic exposition written for non-specialists, and Cohen [**7**] gave a lecture late in his life about how he discovered forcing. The reader may also find it helpful to study the connections that forcing has with topology [**5**], topos theory [**12**], modal logic [**15**], arithmetic [**4**], proof theory [**1**] and computational complexity [**9**]. It may be that insights from these differing perspectives can be synthesized to solve the open exposition problem of forcing.

10. Acknowledgments

This paper grew out of an article entitled "Forcing for dummies" that I posted to the USENET newsgroup `sci.math.research` in 2001. However, in that article I did not employ the Boolean-valued model approach, and hence the line of exposition was quite different. Interested readers can easily find the earlier version on the web.

I thank Costas Drossos, Ali Enayat, Martin Goldstern, and Matt Yurkewych for pointing me to references [**14**], [**5**], [**7**], and [**15**] respectively.

I am deeply indebted to Matthew Wiener and especially Andreas Blass for their many prompt and patient answers to my dumb questions about forcing.

Thanks also to Fred Kochman, Miller Maley, Christoph Weiss, Steven Gubkin, James Hirschhorn, and the referee for comments on an earlier draft of this paper that have improved the exposition and fixed some bugs.

References

[1] Jeremy Avigad, Forcing in proof theory, *Bull. Symbolic Logic* **10** (2004), 305–333.
[2] James E. Baumgartner, Independence proofs and combinatorics, in *Relations Between Combinatorics and Other Parts of Mathematics, Proc. Symp. Pure Math.* **XXXIV**, 35–46, American Mathematical Society, 1979.
[3] John L. Bell, *Set Theory: Boolean-Valued Models and Independence Proofs,* 3rd ed., Oxford University Press, 2005.
[4] George S. Boolos and Richard C. Jeffrey, *Computability and Logic,* 3rd ed., Cambridge University Press, 1989.
[5] Kenneth A. Bowen, Forcing in a general setting, *Fund. Math.* **81** (1974), 315–329.
[6] Paul J. Cohen, *Set Theory and the Continuum Hypothesis,* Addison-Wesley, 1966.
[7] Paul J. Cohen, The discovery of forcing, *Rocky Mountain J. Math.* **32** (2002), 1071–1100.
[8] Kenny Easwaran, A cheerful introduction to forcing and the continuum hypothesis, Los Alamos ArXiv, `math.LO/0712.2279`.
[9] Jan Krajíček, *Bounded Arithmetic, Propositional Logic, and Complexity Theory,* Cambridge University Press, 1995.
[10] Kenneth Kunen, *Set Theory: An Introduction to Independence Proofs,* North Holland, 1983.
[11] Michael Monastyrsky, *Modern Mathematics in the Light of the Fields Medals,* 2nd ed., A. K. Peters, 1998.
[12] Saunders Mac Lane and Ieke Moerdijk, *Sheaves in Geometry and Logic,* Springer-Verlag, 1992.
[13] Donald J. Newman, *Analytic Number Theory,* Springer-Verlag, 1998.
[14] Dana Scott, A proof of the independence of the continuum hypothesis, *Math. Sys. Theory* **1** (1967), 89–111.
[15] Raymond M. Smullyan and Melving Fitting, *Set Theory and the Continuum Problem,* Oxford University Press, 1996.
[16] Robert S. Wolf, *A Tour Through Mathematical Logic,* Mathematical Association of America, 2005.

Center for Communications Research, 805 Bunn Drive, Princeton, NJ 08540
E-mail address: `tchow@alum.mit.edu`
URL: `http://alum.mit.edu/www/tchow`

Contemporary Mathematics
Volume **479**, 2009

Higher Order Necessary Conditions in Smooth Constrained Optimization

Elena Constantin

ABSTRACT. Our goal is to give some higher order necessary conditions for a constrained minimization problem with smooth data using the theory of tangent sets. We illustrate the applicability of our results by analyzing some examples for which the second derivative test fails.

We are dealing with the following constrained minimization problem

$$F(\bar{x}) = \text{Local Minimum } F(x), \quad x \in D, \tag{P}$$

where X is a real linear normed space of norm $||\cdot||$ and $F : U \subseteq X \to \mathbb{R}$ is a function of class C^p on the open set U, $\bar{x} \in D \subseteq U$, p positive integer.

Our goal is to present some higher order necessary conditions for $\bar{x}$ to be a local minimizer to problem (P). We apply our results to some examples for which the second derivative test fails. Such examples appear in multivariable calculus textbooks in the sections on extrema of functions of two variables and on Lagrange multipliers. In some easier cases of problem (P), our results are accessible to calculus students.

Recall that $F : X \to Y$ is said to be Fréchet differentiable at $\bar{x}$ if it is possible to represent it near $\bar{x}$ in the form

$$F(\bar{x} + h) = F(\bar{x}) + \Lambda h + \alpha(h)||h||,$$

where Λ is a linear continuous operator from the linear normed space X into the linear normed space Y and

$$\lim_{||h|| \to 0} ||\alpha(h)|| = ||\alpha(0)|| = 0.$$

The operator Λ is called the Fréchet derivative of F at $\bar{x}$, and is denoted by $F'(\bar{x})$.

In terms of the ϵ, δ formalism, the above relations can be stated thus: given an arbitrary $\epsilon > 0$, there is $\delta > 0$ for which the inequality

$$||F(\bar{x} + h) - F(\bar{x}) - \Lambda h|| \leq \epsilon ||h||$$

holds for all h such that $||h|| < \delta$.

2000 *Mathematics Subject Classification.* Primary 90C30, 90C48 ; Secondary 49K27, 49K30.

Key words and phrases. constrained optimization, Fréchet differentiability, higher order tangent vectors.

The higher order derivatives are defined by induction:

$$F^{(n)}(x) = (F^{(n-1)})'(x) \in \underbrace{L(X, ..., L(X, Y)...)}_{n \text{ times}}.$$

We say that $F^{(p)}$ exists at $\bar{x} \in U$ if $F'(x)$ $F''(x), ..., F^{(p-1)}(x)$ exist in a neighborhood of $\bar{x}$ and $F^{(p)}(\bar{x})$ exists. If $F^{(p)}(x)$ exists at each point $x \in U$ and $x \to F^{(p)}(x)$ is continuous in the uniform topology of the space $L(X, ..., L(X, Y)...)$ (generated by the norm), then F is said to be a mapping of class C^p on U.

Throughout the paper, if a function F is of class C^p on an open set U, then $F^{'}(x)$, $F^{''}(x)$, $F^{'''}(x)$, $F^{(p)}(x)$, $p \geq 4$, denote its first, second, third, and p-th order Fréchet derivatives at $x \in U$, and

$$F^{(p)}(x)[y]^p = F^{(p)}(x)\underbrace{(y) \cdots (y)}_{p \text{ times}}.$$

Recall that an arbitrary real-valued function $F : U \subseteq X \to \mathbb{R}$ is said to have a local minimum on $D \subseteq U$ at $\bar{x} \in D$ if there exists $\delta > 0$ such that $F(x) \geq F(\bar{x})$, for all $x \in D$ satisfying $0 < ||x - \bar{x}|| < \delta$. If the inequality is strict, $\bar{x}$ is a strict local minimum point of F on D.

We will formulate our higher order necessary conditions for a general constraint set D and then we will analyze in more detail the case when D is the null-set of a sufficiently often Fréchet differentiable mapping $G : X \to Y$, i.e., $D = D_G = \{x \in X; \, G(x) = 0\}$.

It is known the method Lagrange developed for approaching such functional constrained optimization problems.

THEOREM 1. *[Lagrange Multipliers Method]*
Let $F : U \subseteq \mathbb{R}^n \to \mathbb{R}$ and $G = (G_1, \ldots, G_k) : U \to \mathbb{R}^k$ be of class C^1 on the open set $U \subseteq \mathbb{R}^n$. If F has an extremum on $D = \{x \in U; \, G_1(x) = 0, G_2(x) = 0, \ldots, G_k(x) = 0\}$, at $\bar{x} \in D$, and $G'_1(\bar{x}), G'_2(\bar{x}), \ldots, G'_k(\bar{x})$ are linearly independent, then there exists a unique vector $\lambda = (\lambda_1, \lambda_2, \ldots, \lambda_k)$, called a Lagrange multiplier vector, such that $F'(\bar{x}) = \lambda G'(\bar{x}) = \lambda_1 G'_1(\bar{x}) + \lambda_2 G'_2(\bar{x}) + \ldots + \lambda_k G'_k(\bar{x})$.
Suppose in addition that F and G are on class C^2 on U.
If $\bar{x}$ is a local minimum of F on D, then $[F''(\bar{x}) - \lambda G''(\bar{x})][y]^2 \geq 0$, for all y such that $G'(\bar{x})(y) = 0$.
If $\bar{x}$ is a local maximum of F on D, then $[F''(\bar{x}) - \lambda G''(\bar{x})][y]^2 \leq 0$, for all y such that $G'(\bar{x})(y) = 0$.

When $\bar{x}$ is a constrained critical point and the above second order expression is equal to zero in any direction y such that $G'(\bar{x})(y) = 0$, we can not distinguish between a candidate $\bar{x}$ for a local minimizer or a local maximizer based on the above second order necessary conditions but we will be able to do so using our higher order necessary conditions (see Example 2).

Necessary conditions are useful in narrowing down the field of candidates for local extrema. To guarantee that a given point is a local extremum, we need sufficient conditions for optimality. We recall the following classical second order sufficient conditions.

THEOREM 2. *Assume that* $F : U \subseteq I\!R^n \to I\!R$, *and* $G = (G_1, \ldots, G_k) : U \to I\!R^k$ *are of class* C^2 *on the open set* U, *and let* $\bar{x} \in D = \{x \in U;\ G_1(x) = 0,\ G_2(x) = 0, \ldots, G_k(x) = 0\}$ *and* $\lambda \in I\!R^k$ *satisfy* $F'(\bar{x}) = \lambda G'(\bar{x})$. *Then*

i) If $[F''(\bar{x}) - \lambda G''(\bar{x})][y]^2 > 0$, *for all* $y \neq 0$ *such that* $G'(\bar{x})(y) = 0$, *then* $\bar{x}$ *is a strict local minimum of* F *on* D.

ii) If $[F''(\bar{x}) - \lambda G''(\bar{x})][y]^2 < 0$, *for all* $y \neq 0$ *such that* $G'(\bar{x})(y) = 0$, *then* $\bar{x}$ *is a strict local maximum of* F *on* D.

Characterizing optimal solutions by means of second order conditions is a problem of continuing interest in the theory of mathematical programming constrained problems with twice continuously differentiable or twice Fréchet differentiable data, and many good results have been established for such problems with equality, inequality and set constraints (see, for example, the books [**3**], [**4**], [**10**], [**13**], [**19**], [**7**], [**1**]).

While there are a lot of publications on second order conditions in nonsmooth optimization, the number of the publications concerning conditions of order higher than two is rather limited (see, for example, [**20**], [**9**], [**12**], [**11**]).

The following notion of tangent vector to an arbitrary set D at $x \in D$ will play a basic role in the sequel. An element $v \in X$ is called a tangent vector to D at x if

$$\lim_{t \downarrow 0} \frac{1}{t} d(x + tv; D) = 0, \tag{1}$$

(see [**21**], [**8**]).

The set of all tangent vectors to D at $x \in D$ is denoted by $T_x D$ and is called the tangent cone to D at x. It is also known as the intermediate or adjacent cone (see [**2**]). $T_x D$ is a closed cone in X and it is always nonempty containing at least $0 \in X$.

The definition of the tangent cone has been extended by some authors who obtained some different definitions for higher order tangent sets (see [**2**], [**12**], [**18**], [**16**]).

Our higher order necessary conditions are given in terms of the higher order tangent vectors introduced by N.H. Pavel and C. Ursescu (see [**18**], [**16**]).

An element $v_n \in X$ is called a n-th order tangent vector to D at $x \in D$ if there are some $v_i \in X$, $i = 1, 2, \cdots, n-1$, $n \geq 2$, such that

$$\lim_{t \downarrow 0} \frac{1}{t^n} d(x + tv_1 + \frac{t^2}{2!} v_2 + \frac{t^3}{3!} v_3 + \ldots + \frac{t^n}{n!} v_n; D) = 0, \tag{2}$$

where $d(z; D) = \inf\{||z - y||; y \in D\}$.

The vectors v_i, $i = 1, 2, \ldots n-1$, are said to be the correspondent vectors of v_n (or associated with v_n).

The set of all n-th order tangent vectors to D at $x \in D$ is denoted by $T_x^n D$.

It is easy to check that $T_x^n D$, $n \geq 2$, is a cone in X.

It is obvious that if x belongs to the interior of D, then $T_x^n D = X$, $n \geq 1$, where $T_x^1 D = T_x D$.

It is known the following characterization of a n-th order tangent vector.

PROPOSITION 1. *i) The fact that* v *belongs to* $T_x D$ *is equivalent to the existence of a function* $\gamma : (0, \infty) \to X$ *with* $\gamma(t) \to 0$ *as* $t \downarrow 0$ *and*

$$x + t(v + \gamma(t)) \in D,\ \forall t > 0.$$

ii) The fact that v_n belongs to $T_x^n D$ with the corespondent vectors $v_i \in X$, $i = 1, 2, ..., n-1$, $n \geq 2$, as in (2), is equivalent to the existence of a function $\gamma_n : (0, \infty) \to X$ with $\gamma_n(t) \to 0$ as $t \downarrow 0$ and

$$x + tv_1 + \frac{t^2}{2} v_2 + ... + \frac{t^n}{n!}(v_n + \gamma_n(t)) \in D, \ \forall t > 0.$$

PROPOSITION 2. *If $v_n \in T_x^n D$ then its associated vectors v_i, $1 \leq i \leq n-1$, belong to $T_x^i D$, respectively.*

The n-th order tangent set $T_x^n D$ is nonempty, containing at least $v_n = 0$ (where one can take some arbitrary $v_i \in T_x^i D$, $i = 1, 2, ..., n-1$, $n \geq 2$).

We introduce a notation that will be used to formulate our main results Theorems 3 and 6. Given a function F of class C^m near $\bar{x}$ and m vectors $v_1, v_2, \ldots, v_m \in X$, m positive integer, we denote by $S_m^F(\bar{x}; v_1, \ldots, v_m)$ the expression given below.

$$S_m^F(\bar{x}; v_1, \ldots, v_m) = \sum_{k=1}^{m} \frac{m!}{k!} \left[\sum_{\substack{i_1, \ldots, i_k \in \{1, \ldots, m\} \\ i_1 + \ldots + i_k = m}} \frac{1}{i_1! i_2! \ldots i_k!} F^{(k)}(\bar{x})(v_{i_1}) \cdots (v_{i_k}) \right].$$

In particular,
for $m = 1$, $S_1^F(\bar{x}; v_1) = F'(\bar{x})(v_1)$
for $m = 2$, $S_2^F(\bar{x}; v_1, v_2) = F'(\bar{x})(v_2) + F''(\bar{x})[v_1]^2$,
for $m = 3$, $S_3^F(\bar{x}; v_1, v_2, v_3) = F'''(\bar{x})[v_1]^3 + 3F''(\bar{x})(v_1)(v_2) + F'(\bar{x})(v_3)$.

Next we give our main result concerning higher order necessary conditions of extremum for the smooth constrained optimization problem (P) (see Theorem 2.2, [**6**]).

THEOREM 3. *Let $\bar{x}$ be a local minimum point of $F : X \to \mathbb{R}$ on D, i.e.,*

$$F(\bar{x}) = \text{Local Minimum } F(x), \ \text{subject to } x \in D, \tag{P}$$

where D is a nonempty subset of the linear normed space X.

Then

i) If F is of class C^1 near $\bar{x}$, then

$$F'(\bar{x})(v_1) \geq 0, \ \forall v_1 \in T_{\bar{x}} D. \tag{3}$$

ii) If F is of class C^2 near $\bar{x}$, then

$$F'(\bar{x})(v_2) + F''(\bar{x})[v_1]^2 \geq 0, \tag{4}$$

$\forall v_2 \in T_{\bar{x}}^2 D$ with the associated vector $v_1 \in T_{\bar{x}} D$ such that $F'(\bar{x})(v_1) = 0$.

iii) If F is of class C^3 near $\bar{x}$, then

$$F'''(\bar{x})[v_1]^3 + 3F''(\bar{x})(v_1)(v_2) + F'(\bar{x})(v_3) \geq 0, \tag{5}$$

$\forall v_3 \in T_{\bar{x}}^3 D$ with associated vectors $v_1 \in T_{\bar{x}} D$, and $v_2 \in T_{\bar{x}}^2 D$ such that

$$F'(\bar{x})(v_1) = 0 \text{ and } F'(\bar{x})(v_2) + F''(\bar{x})[v_1]^2 = 0. \tag{6}$$

In general, if F is sufficiently smooth, there must exist a positive integer p with the property that

$$S_p^F(\bar{x}; v_1, \ldots, v_p) \geq 0, \ \forall v_1, v_2, ..., v_p \in X \text{ such that}$$
$$\bar{x} + tv_1 + \frac{t^2}{2} v_2 + ... + \frac{t^p}{p!}(v_p + \gamma(t)) \in D, \ \forall t > 0, \ \gamma(t) \to 0 \text{ as } t \downarrow 0, \text{ and}$$
$$S_n^F(\bar{x}; v_1, \ldots, v_n) = 0, \text{ for every } n, \ 0 < n \leq p-1.$$

PROOF. We use the fact that $v_p \in T_{\bar{x}}^p D$ and $v_i \in T_{\bar{x}}^i D, 1 \leq i \leq p-1$ are its associated vectors, is equivalent to the existence of a mapping $\gamma_p(t) \to 0$ as $t \downarrow 0$, and

$$\bar{x} + tv_1 + \frac{t^2}{2}v_2 + \frac{t^3}{3!}v_3 + \ldots + \frac{t^p}{p!}(v_p + \gamma_p(t)) \in D, \forall t > 0.$$

Since $\bar{x}$ is a local minimum point of F on D, there is $\varepsilon > 0$ such that

$$F(\bar{x} + tv_1 + \frac{t^2}{2}v_2 + \frac{t^3}{3!}v_3 + \ldots + \frac{t^p}{p!}(v_p + \gamma_p(t))) - F(\bar{x}) \geq 0,\ 0 < t \leq \varepsilon.$$

In view of Taylor's formula, the right-hand side can be written as

$$F'(\bar{x})(tv_1 + \frac{t^2}{2}v_2 + \frac{t^3}{3!}v_3 + \ldots + \frac{t^p}{p!}(v_p + \gamma_p(t))) + \frac{1}{2}F''(\bar{x})[tv_1 + \frac{t^2}{2}v_2 + \frac{t^3}{3!}v_3 + \ldots$$
$$+\frac{t^p}{p!}(v_p + \gamma_p(t))]^2 + \frac{1}{3!}F'''(\bar{x})[tv_1 + \frac{t^2}{2}v_2 + \frac{t^3}{3!}v_3 + \ldots + \frac{t^p}{p!}(v_p + \gamma_p(t))]^3 + \ldots$$
$$+\frac{1}{p!}F^{(p)}(\bar{x})[tv_1 + \frac{t^2}{2}v_2 + \frac{t^3}{3!}v_3 + \ldots + \frac{t^p}{p!}(v_p + \gamma_p(t))]^p + t^p\alpha(t)$$
$$= tF'(\bar{x})(v_1) + \frac{t^2}{2}[F'(\bar{x})(v_2) + F''(\bar{x})[v_1]^2] + \frac{t^3}{3!}[F'''(\bar{x})[v_1]^3 + 3F''(\bar{x})(v_1)(v_2) +$$
$$F'(\bar{x})(v_3)] + \ldots + \frac{t^p}{p!}S_p^F(\bar{x}; v_1, \ldots, v_p) + t^p\alpha(t) \geq 0,\ 0 < t \leq \varepsilon,$$

where $\alpha(t) \to 0$ as $t \downarrow 0$.

Dividing by t and letting t go to zero through positive values, one obtains $F'(\bar{x})(v_1) \geq 0$, for $v_1 \in T_{\bar{x}}D$. If $F'(\bar{x})(v_1) = 0$, one divides by t^2 and gets (4) when $t \downarrow 0$. Taking into account that (6) holds, after dividing by t^3 and letting t go to 0, it can be seen that (5) occurs too. Now, the general case follows in the same way since if the coefficients of $t, t^2, \ldots, t^{p-1}$ are equal to zero, then the coefficient of t^p must be nonnegative. This ends the proof. □

REMARK 4. i) If $\bar{x}$ is a local maximizer of $F : X \to \mathbb{R}$ on D, $D \subseteq X$, $D \neq \emptyset$, and F is a function of class C^p in a neighborhood of $\bar{x}$, then it can be shown analogously to the proof of the previous theorem that $S_p^F(\bar{x}; v_1, \ldots, v_p) \leq 0$, for all $v_p \in T_{\bar{x}}^p D$ with associated vectors $v_1, v_2, \ldots, v_{p-1}$ such that $S_j^F(\bar{x}; v_1, \ldots, v_j) = 0$, for all $1 \leq j \leq p-1$.

ii) If $\bar{x}$ is a critical point of F on D, i.e., $F'(\bar{x})(v_1) = 0$, for any $v_1 \in T_{\bar{x}}D$, and if there exists a positive integer p such that the expression $S_p^F(\bar{x}; v_1, \ldots, v_p)$ does not have a constant sign for all $v_1, \ldots, v_p$ as above, then $\bar{x}$ can not be a local extremum point of F on D. This allows us to exclude some of the critical points which can not be candidates for either maximum or minimum points.

EXAMPLE 1. We consider the function
$F(x_1, x_2) = x_2^4 - x_1^4 - x_2^8 + x_1^{10}$, $F : \mathbb{R}^2 \to \mathbb{R}$.

The determinant of the Hessian matrix of this function at $(0, 0)$, $\det HF(0, 0) = 0$ as all the second order partial derivatives of F are equal to zero. Thus the second derivative test can not determine the behaviour of F at $(0, 0)$.

We notice that $(0, 0)$ is a local maximizer along the line $x_1 = x_2$ as $F(x_1, x_1) = x_1^8(x_1^2 - 1) \leq 0$, $|x_1| \leq 1$, but $(0, 0)$ is a local minimizer of F along the parabola $x_1 = x_2^2$ since $F(x_2^2, x_2) = x_2^4(1 - 2x_2^4 + x_2^{16}) \geq 0$, $x_2 \leq .85$.

Thus $(0, 0)$ is neither a local minimizer nor a local maximizer of F on $\mathbb{R}^2$.

We can arrive at the same conclusion by means of our higher order necessary conditions which represent a general criterion for determining the behavior of such a function rather than an ad-hoc method as above.

We will use our fourth order necessary conditions given in Theorem 3. The set D is $\mathbb{R}^2$ and since $(0,0)$ is an interior point of D, the tangent cones of any order to D at $(0,0)$ coincide with $\mathbb{R}^2$.

We can see that the first, second, and third order derivatives of F at $(0,0)$ in any direction are equal to zero, and the fourth order expression reduces to $F^{(4)}(0,0)[y]^4 = 4!(y_2^4 - y_1^4)$, for all $y = (y_1, y_2) \in \mathbb{R}^2$, which does not have a constant sign on $\mathbb{R}^2$. We conclude that the origin is a saddle point of F on $\mathbb{R}^2$. This means that the behavior of F can be determined completely by means of our fourth order necessary conditions.

Next we present several situations when we can determine the tangent cones to the null-set of a mapping G, i.e., $D = D_G = \{x \in X;\, G(x) = 0,\ G : X \to Y\}$.

DEFINITION 1. A mapping $G : X \to Y$ is said to be strictly differentiable at a point $\bar{x}$ if there exists a linear continuous operator Λ from the linear normed space X into the linear normed space Y with the property that for any $\epsilon > 0$ there is $\delta > 0$ such that for all x_1 and x_2 satisfying the inequalities $||x_1 - \bar{x}|| < \delta$ and $||x_2 - \bar{x}|| < \delta$, the following inequality holds

$$||G(x_1) - G(x_2) - \Lambda(x_1 - x_2)|| \leq \epsilon ||x_1 - x_2||.$$

A strictly differentiable function G at a point $\bar{x}$ is Fréchet differentiable at $\bar{x}$ and $\Lambda = G'(\bar{x})$.

In 1934 Lyusternik described the tangent space to the null-set of a strictly differentiable functional (see, for example, [**1**]).

THEOREM 5. *[Lyusternik Theorem] Let X, Y be Banach spaces, let U be a neighborhood of a point $\bar{x} \in X$, and let $G : U \to Y$, $G(\bar{x}) = 0$.*

If G is strictly differentiable at $\bar{x}$ and $G'(\bar{x})$ is onto, then the tangent space to the set $D_G = \{z \in X;\, G(z) = 0\}$ at the point $\bar{x}$ is given by

$$T_{\bar{x}} D_G = \{y \in X; G'(\bar{x})(y) = 0\}.$$

N.H. Pavel and C. Ursescu showed that when Y is finite dimensional for the conclusion of Lyusternik's Theorem to remain valid it is enough G to be only Fréchet differentiable at $\bar{x}$ with $G'(\bar{x})$ onto, and continuous near $\bar{x}$ (see Theorem 3.1, [**18**]). Also they characterized the second order tangent cone under similar hypothesis (see Theorem 3.2, [**18**]) and we extended their result as follows (see Corollary 1.1, [**6**]).

THEOREM 6. *Assume that $G : X \to \mathbb{R}^m$ is a function of class C^p in a neighborhood of a point $\bar{x} \in X$ with $G(\bar{x}) = 0$, and $G'(\bar{x}) : X \to \mathbb{R}^m$ is onto, $p \geq 1$.*

Then, $v_p \in T^p_{\bar{x}} D_G$ with the associated vectors $v_n \in T^n_{\bar{x}} D_G$, $n = 1, ..., p-1$, if and only if

$$S_n^G(\bar{x}; v_1, \ldots, v_n) = 0,\ \forall\, 1 \leq n \leq p, \tag{7}$$

where $D_G = \{z \in X, G(z) = 0\}$ is nonempty.
In particular,
for $p = 1$, Lyusternik's result is recovered under weaker hyphothesis,

for $p = 2$, $v_2 \in T^2_{\bar{x}} D_G$ *with associated vector* $v_1 \in T_{\bar{x}} D_G$ *iff*

$G^{'}(\bar{x})(v_2) + G^{''}(\bar{x})[v_1]^2 = 0,$

$G^{'}(\bar{x})(v_1) = 0,$

for $p = 3$, $v_3 \in T^3_{\bar{x}} D_G$ *with associated vectors* $v_1 \in T_{\bar{x}} D_G$, $v_2 \in T^2_{\bar{x}} D_G$ *iff*

$G^{'''}(\bar{x})[v_1]^3 + 3G^{''}(\bar{x})(v_1)(v_2) + G^{'}(\bar{x})(v_3) = 0,$

$G^{'}(\bar{x})(v_2) + G^{''}(\bar{x})[v_1]^2 = 0,$

$G^{'}(\bar{x})(v_1) = 0.$

REMARK 7. If $X = IR^n$, n positive integer, the condition $G'(\bar{x})$ is onto is equivalent to the fact that the components of $G'(\bar{x})$ are linear independent.

REMARK 8. If $X = IR^n$, n positive integer, and $\bar{x}$ is an extremum point of F on D_G, then $F^{'}(\bar{x})(v_1) = 0$, for all $v_1 \in T_{\bar{x}} D_G = \{y \in X; G'(\bar{x})(y) = 0\}$ because $F^{'}(\bar{x}) = \lambda G^{'}(\bar{x})$. This is an analogous statement to the "zero gradient condition" of unconstrained optimization.

REMARK 9. In view of the characterization of the second order tangent cone given in Theorem 6 under the hypothesis $G : \mathbb{R}^n \to \mathbb{R}^m$ is of class C^2, and $G'(\bar{x})$ is onto, the second order necessary conditions given in Theorem 3 become equivalent to the classical second order necessary conditions of Theorem 1.

Under the same conditions upon G, the classical second order sufficient conditions for $\bar{x}$ to be a local minimum of F subject to $G(x) = 0$

$$[F''(\bar{x}) - \lambda G''(\bar{x})][y]^2 > 0, \forall\, y \text{ such that } G'(\bar{x})(y) = 0,$$

become equivalent to the following second order sufficient (Theorem 2.2, [**17**])

$$F^{'}(\bar{x})(v_2) + F^{''}(\bar{x})[v_1]^2 > 0, \tag{8}$$

for all (v_1, v_2) different from zero satisfying

$$G'(\bar{x})(v_1) = 0, G^{'}(\bar{x})(v_2) + G^{''}(\bar{x})[v_1]^2 = 0. \tag{9}$$

Indeed, in view of the hypothesis that $x \to G'(\bar{x})(x)$ is onto, for each $v_1 \in IR^n$ with $G'(\bar{x})(v_1) = 0$ there is $v_2 \in IR^n$ such that

$$G'(x)(v_2) = -G''(x)(v_1)(v_1).$$

In the case of a local maximum point, similar second order conditions are valid (with the inequalities reversed).

EXAMPLE 2. Consider the function $F(x_1, x_2) = x_2^6 + x_1^3 + 4x_1 + 4x_2$, subject to $G(x_1, x_2) = x_1^5 + x_2^4 + x_1 + x_2 = 0$, $F, G : \mathbb{R}^2 \to \mathbb{R}$.

We notice that $\bar{x} = (0, 0)$ is a constrained critical point because $G(0, 0) = 0$, and $(\lambda, x_1; x_2) = (4; 0, 0)$ is a solution of the equation $F^{'}(x_1, x_2) = \lambda G^{'}(x_1, x_2)$.

Also the gradient of G at $(0, 0)$ is different from zero, $G'(0, 0) = (1, 1)$.

The second derivative test for constrained extrema is not applicable to this example because $[F''(0, 0) - 4G''(0, 0)][y]^2 = 0$, for all $y \in \mathbb{R}^2$, since all the second order partial derivatives of F and G at $(0, 0)$ are equal to zero.

Thus the origin can not be classified as a local minimizer or a local maximizer of F on the null-set of G using the classical second order sufficient conditions (Theorem 2) but it can not be rejected as a local extremum point either using the classical second order necessary conditions (Theorem 1). Therefore the classical criteria give us no information and we must use another method to determine whether or not the constrained critical point $(0, 0)$ is an extremum point.

Next we will apply our higher order necessary conditions (Theorem 3) to the critical point $(0,0)$.

Clearly, the only nonzero partial derivatives of F at $(0,0)$ are $F_{x_1}(0,0) = F_{x_2}(0,0) = 4$, and $F_{x_1x_1x_1}(0,0) = 6$, and all the other second and third order partial derivatives of F evaluated at any $(x_1,x_2) \in \mathbb{R}^2$ are equal to zero.

Also, all the second and the third order partial derivatives of G at $(0,0)$ are equal to zero. Obviously, $G_{x_1}(0,0) = 0$ and $G_{x_2}(0,0) = 1$, and thus $G'(0,0)$ is onto and the tangent cones can be characterized by means of Theorem 6.

The first, second and third order tangent cones to D_G at $(0,0)$ are given below.

$T_{(0,0)}D_G = \{v_1 = (v_{11}, v_{12}) \in \mathbb{R}^2; v_{11} + v_{12} = 0\}$,

$T^2_{(0,0)}D_G = \{v_2 = (v_{21}, v_{22}) \in \mathbb{R}^2; v_{21} + v_{22} = 0\}$,

$T^3_{(0,0)}D_G = \{v_3 = (v_{31}, v_{32}) \in \mathbb{R}^2; v_{31} + v_{32} = 0\}$.

In this situation,

$F^{'}(0,0)(v_1) = 4(v_{11} + v_{12}) = 0, \forall\, v_1 \in T_{(0,0)}D_G$.

The second order expression becomes

$F^{'}(0,0)(v_2) + F^{''}(0,0)(v_1)(v_1) = 4(v_{21} + v_{22}) = 0$, for any $v_2 \in T^2_{(0,0)}D_G$ with associated vector $v_1 \in T_{(0,0)}D_G$.

The third order expression can be simplified to

$F^{'''}(0,0)[v_1]^3 + 3F^{''}(0,0)(v_1)(v_2) + F^{'}(0,0)(v_3) = 6v_{11}^3 + 4(v_{31} + v_{32}) = 6v_{11}^3$, which does not have a constant sign whenever $v_3 \in T^3_{(0,0)}D_G$ with correspondent vectors $v_2 \in T^2_{(0,0)}D_G$ and $v_1 \in T_{(0,0)}D_G$.

According to our third order necessary conditions, the origin can not be a local minimizer because this expression is not nonnegative and it can not be a local maximum either because the expression is not less than or equal to zero.

This ensures us that $(0,0)$ is not a local extremum point of F on D_G.

REMARK 10. Making use of Theorems 3 and 6, it can be shown that the constrained critical point $(0,0)$ of the function $F(x_1,x_2) = x_2^{2n+2} + x_1^{2n-1} + 4x_1 + 4x_2$, subject to the constraint $G(x_1,x_2) = x_1^{2n+1} + x_2^{2n} + x_1 + x_2 = 0$, $(x_1,x_2) \in \mathbb{R}^2$, $p \geq 3$, is a saddle point (take $p = 2n-1$ in Theorem 3).

EXAMPLE 3. Consider the function $F(x_1,x_2) = x_2^6 + x_1^3 + 2x_1^2 - x_2^2 + 4x_1 + 4x_2$, subject to the same constraint as in Example 2.

It can be checked that $(0,0)$ is a constrained critical point with the Lagrange multiplier $\lambda = 4$.

The origin is a local minimizer because the sufficient conditions of Theorem 2 are satisfied as $[F''(0,0) - 4G''(0,0)][y]^2 = 2y_1^2 - y_2^2 = y_2^2 > 0$, for all $y = (y_1,y_2) \neq 0$ such that $G'(0,0)(y) = 0$. Here $G''(0,0) = 0$.

We can draw the same conclusion by verifying the sufficient conditions given by (8) and (9). Indeed, $F^{'}(\bar{x})(v_2) + F^{''}(\bar{x})[v_1]^2 = 2v_{11}^2 - v_{12}^2 + 4(v_{21} + v_{21}) = v_{11}^2 > 0$, for all $v_1 \neq 0$ as $v_{11} + v_{12} = 0$ and $v_{21} + v_{22} = 0$.

In conclusion, using the theory of tangent sets, we formulated some higher order necessary conditions for a constrained optimization problem with C^p data, that are useful for treating situations where classical criteria such as the second derivative test can not be applied.

Acknowledgements. The author would like to thank the referee for some helpful comments and suggestions.

References

[1] V.M. Alekseev & V.M. Tikhomirov & S.V. Fomin, *Optimal Control*, New York, Consultants Bureau, 1987.
[2] J.P. Aubin & H. Frankowska, *Set-valued Analysis*, Birkhäuser, Boston, 1990.
[3] M. Avriel, *Nonlinear Programming: Analysis and Methods*, Prentice - Hall, Englewood Cliffs, New Jersey, 1976.
[4] M.S. Bazaraa & O.M. Shetty, *Foundations of Optimization*, Springer - Verlag, Berlin, Heidelberg, New York, 1976.
[5] D. P. Bertsekas, *Nonlinear Programming*, Athena Scientific, Belmont, Massachusetts, 1995.
[6] E. Constantin, *Higher Order Necessary and Sufficient Conditions for Optimality*, PanAmerican Math. J., 14 (3) (2004), 1-25.
[7] G. Giorgi, A. Guerraggio & J. Thierfelder, *Mathematics of Optimization: Smooth and Nonsmooth Case*, Elsevier, Amsterdam, 2004.
[8] I.V. Girsanov, *Lectures on Mathematical Theory of Extremum Problems*, Spring-Verlag, Berlin·Heidelberg·New York, 1972.
[9] K.H. Hoffmann & H.J. Kornstaedt, *Higher Order Necessary Conditions in Abstract Mathematical Programming*, J.Optim. Theory Appl., 26, 1978, 533-569.
[10] A.D. Ioffe & V.M. Tihomirov, *Theory of Extremal Problems*, North - Holland, Amsterdam, 1979.
[11] P.T. Kien & D.V. Luu, *Higher Order Optimality Conditions for Isolated Local Minima*, Nonlinear Funct. Anal. & Appl., 8, 1, 2003, 35-48.
[12] U. Ledzewicz & H. Schaettler, *High-Order Tangent Cones and Their Applications in Optimization*, Nonlinear Analysis, Theory, Methods & Applications, 30 (4) (1997), 2449-2460.
[13] O.L. Mangasarian, *Nonlinear Programming*, McGraw - Hill Book Company, New York, 1969.
[14] D. Motreanu & N.H. Pavel, *Tangency, Flow Invariance for Differential Equations, and Optimization Problems*, Marcel Dekker, Inc., New York, 1999.
[15] N.H. Pavel, *Second Order Differential Equations on Closed Subsets of a Banach Space*, Boll. Un. Mat. Ital., 12 (1975), 348-353.
[16] N.H. Pavel, J.K. Huang & J.K. Kim, *Higher Order Necessary Conditions for Optimization*, Libertas Math. 14 (1994), 41-50.
[17] N.H. Pavel, F. Potra, *Flow-invariance and Nonlinear Optimization Problems via First and Second Order Generalized Tangential Cones*, Analele Ştiinţifice ale Universitaţii 'Al.I. Cuza', Iaşi, 51 (2) (2005), 281-292.
[18] N.H. Pavel & C. Ursescu, *Flow-invariant Sets for Autonomous Second Order Differential Equations and Applications in Mechanics*, Nonlinear Anal., 6 (1982), 35-77.
[19] E. Polak, *Optimization: Alghorithms and Consistent Approximations*, Springer - Verlag, New York, 1997.
[20] M. Studniarski, *Necessary and Sufficient Conditions for Isolated Local Minima of Nonsmooth Functions*, SIAM J. Control and Optimization, 24, 5, 1986, 1044-1049.
[21] C. Ursescu, *Tangent Sets' Calculus and Necessary Conditions for Extramality*, SIAM Journal on Control and Optimization, 20 (1982), 563-574.

Department of Mathematics, University of Pittsburgh at Johnstown, Johnstown, PA 15904
E-mail address: constane@pitt.edu

Contemporary Mathematics
Volume **479**, 2009

Hamiltonian Paths and Hyperbolic Patterns

Douglas Dunham

Abstract. In 1978 I thought it would be possible to design a computer algorithm to draw repeating hyperbolic patterns in a Poincaré disk based on Hamiltonian paths in their symmetry groups. The resulting successful program was capable of reproducing each of M.C. Escher's four "Circle Limit" patterns. The program could draw a few other patterns too, but was somewhat limited. Over the years I have collaborated with students to develop algorithms that are more general and more sophisticated. I describe these algorithms and show some of the patterns they produced.

1. Introduction

For more than a century mathematicians have been drawing patterns to explain concepts in hyperbolic geometry. In the hyperbolic plane, tessellations of congruent triangles are often used to explain the notion of symmetry and kaleidoscopes for that surface. Figure 1 shows one such pattern chosen by the Canadian mathematician H.S.M. Coxeter for this purpose in his paper [**Cox57**]. In 1958 Coxeter sent the Dutch artist M.C. Escher a reprint of that paper. When Escher saw the hyperbolic pattern, he said that it "gave me quite a shock", since it showed him how to make an infinite pattern within the confines of a finite disk. (Some of this Coxeter-Escher correspondence is recounted in [**Cox79**].) Later that year, with the pattern of Figure 1 as inspiration, Escher created his first artistic hyperbolic pattern, *Circle Limit I*, a rendition of which is shown in Figure 2. In my 2003 Mathematics Awareness Month essay [**Dun03**], I offer one explanation of how Escher might have constructed *Circle Limit I* from the pattern of Figure 1. Over the next two years Escher went on to create three more *Circle Limit* patterns. Since he was working by hand, for each pattern carving a woodblock with a sector of the pattern, then repeatedly rotating and printing the woodblock to fill out the whole pattern, this was time consuming, and probably the reason he stopped at four patterns.

In the early 1970's I became intrigued by Escher's mathematical art. In 1977, soon after I joined the faculty of University of Minnesota Duluth (UMD), I discussed the group-theoretical aspects of Escher's patterns with colleague Joe Gallian. At

2000 *Mathematics Subject Classification.* Primary 05C45, 51M20; Secondary 05C25, 51M10.
Key words and phrases. Hamiltonian paths, hyperbolic geometry, repeating patterns.
The author was partly supported by a Summer 2007 VDIL (Visualization and Digital Imaging Lab) Research grant at the University of Minnesota Duluth.

FIGURE 1. A hyperbolic triangle pattern.

that time he had undergraduate research students who were working on finding Hamiltonian cycles and paths in finite groups. Several months later it occurred to me that following a Hamiltonian path in the symmetry group of a hyperbolic pattern could lead to an algorithm to draw such patterns. Indeed, this turned out to be the case, as will be elaborated below.

In what follows, I start with a brief discussion of hyperbolic geometry, repeating patterns, and regular tessellations. Then I explain how Hamiltonian paths in hyperbolic symmetry groups were utilized in a computer algorithm to draw repeating hyperbolic patterns. That method was the inspiration for other algorithms for drawing such patterns. Finally, I discuss possible directions of future work. Along the way I show several hyperbolic patterns.

2. Hyperbolic Geometry, Repeating Patterns, and Regular Tessellations

Axioms for the hyperbolic plane can be taken to be those of the Euclidean plane except that the Euclidean parallel axiom is replaced by one of its negations: given a line and a point not on it, there is more than one line through the point not meeting the original line [**Gre07**]. In 1901 David Hilbert proved that the entire hyperbolic plane has no smooth, isometric embedding in Euclidean 3-space [**Hil01**]. Thus, we must rely on Euclidean *models* of hyperbolic geometry in which distance is measured differently and concepts such as hyperbolic lines have interpretations as Euclidean constructs. Following Coxeter's illustration in Figure 1, Escher used the *Poincaré disk model* of hyperbolic geometry. In this model, hyperbolic points are just the (Euclidean) points within a Euclidean bounding circle. Hyperbolic lines are represented by circular arcs (including diameters) orthogonal to the bounding circle. For example, the backbone lines of the fish in Figure 2 lie along hyperbolic lines. Figure 3 shows that this model satisfies the hyperbolic parallel axiom.

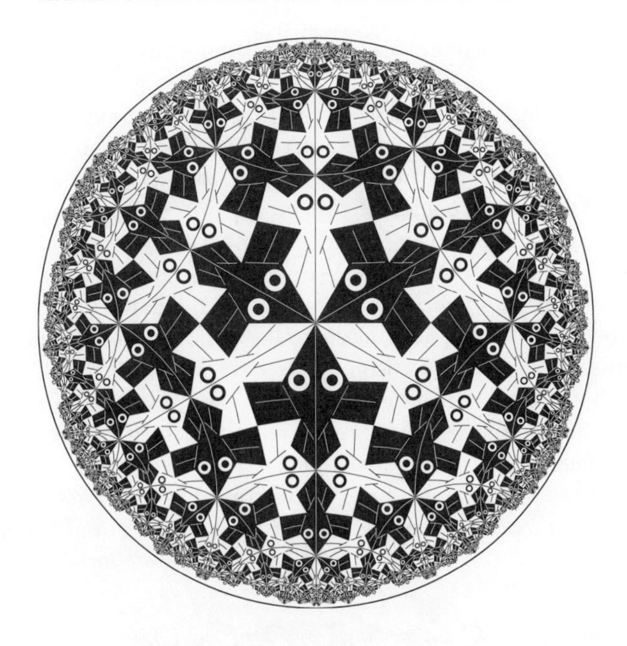

FIGURE 2. A rendition of Escher's *Circle Limit I* pattern.

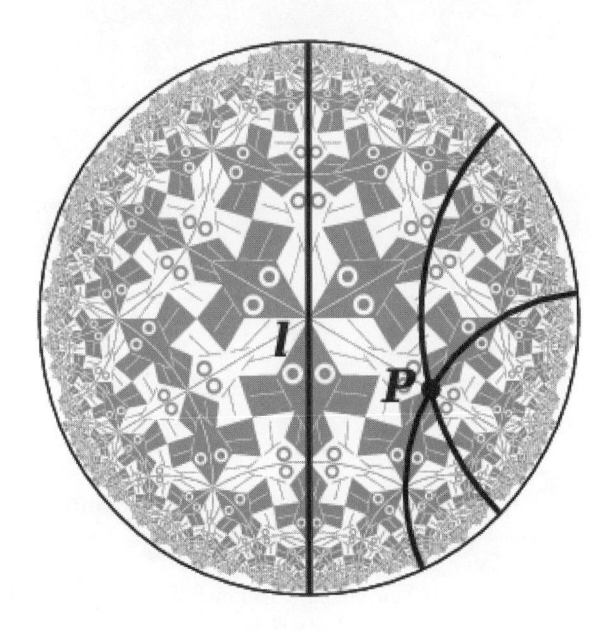

FIGURE 3. An example of the hyperbolic parallel property: a line l, a point P not on l, and two lines through P not meeting l.

The Poincaré disk model is also *conformal*: the hyperbolic measure of an angle is the same as its Euclidean measure. As a consequence, all the black fish in *Circle Limit I* have roughly the same Euclidean shape, and the same is true of the white fish. However, equal hyperbolic distances correspond to ever smaller Euclidean distances as one travels toward the edge of the disk. So all the black fish in *Circle*

Limit I are the same (hyperbolic) size, as are all the white fish. Note that the white fish are not isometric to the black fish since the nose angle of the white fish is $90°$ and the tail angle is $60°$, the reverse of the nose and tail angles of the black fish. The Poincaré disk model is appealing to artists (and appealed to Escher) since an infinitely repeating pattern can be enclosed in a bounded area and shapes remain recognizable even for small copies of the motif, due to conformality. (Of course Escher was more interested in the Euclidean properties of the disk model than the fact that it could be interpreted as hyperbolic geometry.)

A *repeating pattern* in hyperbolic geometry is a regular arrangement of copies of a basic subpattern or *motif.* The copies should not overlap, and a characteristic of Escher's patterns is that there are also no gaps between motifs. Half of one white fish plus half of an adjacent black fish form a motif for the pattern in Figure 2. Similarly, a white triangle and an adjacent black triangle form a motif for the pattern in Figure 1. One special repeating pattern is the *regular tessellation*, $\{p, q\}$, by regular p-sided polygons or *p-gons*, q of which meet at each vertex. Figures 4 and 5 show the $\{6, 4\}$ tessellation superimposed on renditions of *Circle Limit I* and *Circle Limit IV.*

FIGURE 4. The $\{6, 4\}$ tessellation superimposed on the *Circle Limit I* pattern.

Similarly, Figures 6 and 7 show the $\{8, 3\}$ tessellation superimposed on renditions of *Circle Limit II* and *Circle Limit III.* Note that for $\{p, q\}$ to be a tessellation of the hyperbolic plane, it is necessary that $(p-2)(q-2) > 4$, otherwise one obtains one of the finitely many Euclidean or spherical tessellations. Doris Schattschneider's book *Visions of Symmetry* [**Sch04**] is the definitive reference for Escher's repeating patterns.

3. Symmetry Groups for Hyperbolic Patterns

A *symmetry* of a repeating pattern is an isometry (distance-preserving transformation) that transforms the pattern onto itself. In the Poincaré disk model,

FIGURE 5. The $\{6,4\}$ tessellation superimposed on the *Circle Limit IV* pattern.

FIGURE 6. The $\{8,3\}$ tessellation superimposed on the *Circle Limit II* pattern.

reflections across hyperbolic lines are inversions in the circular arcs representing those lines; reflections across diameters are ordinary Euclidean reflections. And in hyperbolic geometry, just as in Euclidean geometry, a translation is the composition of successive reflections across two lines having a common perpendicular; the

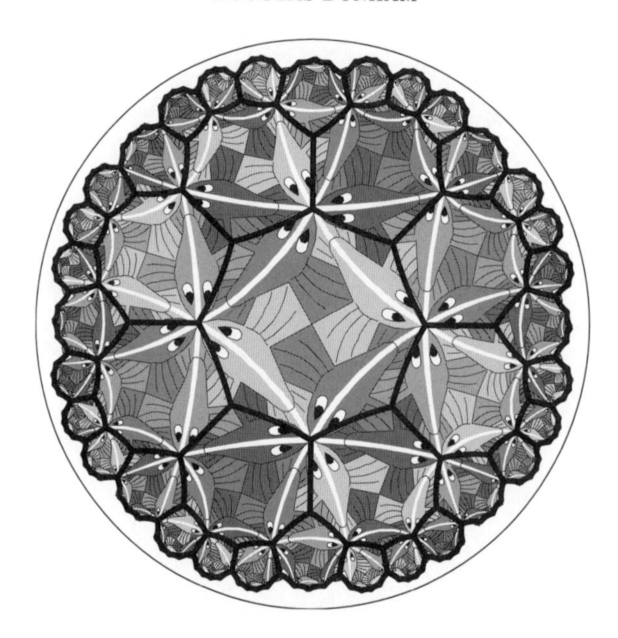

FIGURE 7. The $\{8, 3\}$ tessellation superimposed on the *Circle Limit III* pattern.

composition of reflections across two intersecting lines produces a rotation about the intersection point by twice the angle of intersection. In *Circle Limit I* (Figure 2), hyperbolic reflections across the backbone lines are symmetries of the repeating pattern. Other symmetries of *Circle Limit I* include rotations by 120° about points where three black fish noses meet, rotations of 180° about white fish noses and also about the points where the trailing edges of fin-tips meet, and translations by four fish-lengths along backbone lines.

The *symmetry group* of a pattern is the set of all symmetries of the pattern. The symmetry group of the tessellation $\{p, q\}$ is denoted $[p, q]$ and can be generated by reflections across the sides of a right triangle with acute angles of $180/p$ degrees and $180/q$ degrees. For example, $[6, 4]$ can be generated by reflections across the sides of any one of the triangles in Figure 1. The orientation-preserving subgroup of $[p, q]$ (of index 2), consists of symmetries composed of an even number of reflections, and is denoted $[p, q]^+$. Figure 8 shows a hyperbolic pattern with symmetry group $[5, 5]^+$ (ignoring color). This pattern uses a fish motif like that of Escher's Notebook Drawing Number 20 (p. 131 of [**Sch04**]) and his carved sphere with fish (p. 244 of [**Sch04**]); those fish patterns have symmetry groups $[4, 4]^+$ and $[3, 3]^+$ respectively.

There is another index 2 subgroup of $[p, q]$ that is denoted $[p^+, q]$ and is generated by a rotation of $360/p$ degrees about the center of a p-gon and a reflection in one of its sides, where q must be even so that the reflections across the sides of the p-gon match up. The symmetry group of *Circle Limit IV* (Figure 5) is an instance of $[4^+, 6]$ since rotating a joined half-angel and half-devil by 90° three times about their common wing tips will fill out a square; reflecting the rotated design across an edge of that square and continuing to repeat the process will produce the whole pattern (see [**Dun86**]). The symmetry group of *Circle Limit II* (Figure 6) is an

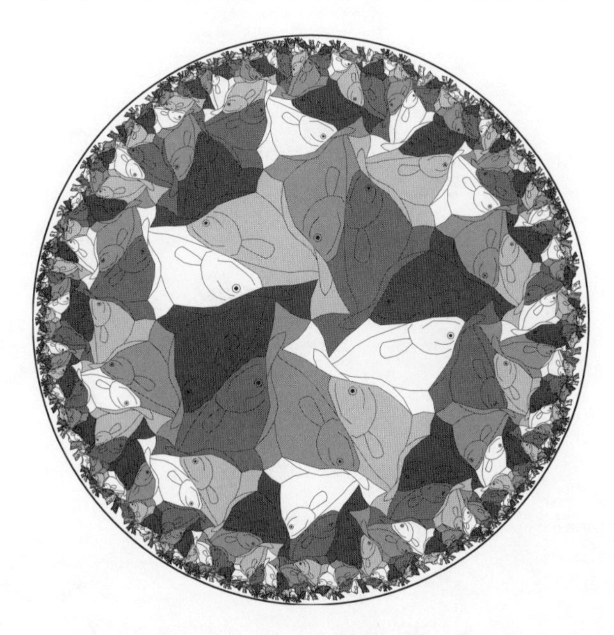

FIGURE 8. A hyperbolic pattern with symmetry group $[5,5]^+$ using a fish motif like that of Escher's Notebook Drawing Number 20.

instance of $[3^+,8]$. Figure 9 shows a pattern of 5-armed crosses with symmetry group $[3^+,10]$ that is similar to Escher's *Circle Limit II*. In both patterns, 120° rotation centers are to the left and right of the ends of each cross arm, and $q/2$ reflection lines pass through the center of the crosses (and the center of the bounding circle). In addition to *Circle Limit IV*, Escher used the group $[p^+,q]$ for two other "angel and devil" patterns: Notebook Drawing Number 45 and *Heaven and Hell* on a carved maple sphere, with symmetry groups $[4^+,4]$ and $[3^+,4]$ respectively (see pages 150 and 244 of [**Sch04**]). The angel and devil pattern is the only one Escher implemented on all three surfaces: the sphere, the Euclidean plane, and the hyperbolic plane.

If both p and q are even, there is yet another index 2 subgroup of $[p,q]$, denoted $cmm_{p/2,q/2}$, that is generated by reflections in two adjacent sides of a rhombus with angles of $360/p$ degrees and $360/q$ degrees, and a 180° rotation about its center. This notation generalizes the Euclidean case in which $cmm_{4/2,4/2} = cmm$. Thus *Circle Limit I* (Figure 2) has symmetry group $cmm_{3,2}$. Figure 10 shows a pattern with symmetry group $cmm_{3,3}$.

The symmetry group of the Escher's *Circle Limit III* pattern (Figure 7) is generated by three rotations: a 90° rotation about the right fin tip, a 120° rotation about the left fin tip, and a 120° rotation about the nose of a fish. The two different kinds of 3-fold points alternate around the vertices of 8-gons of the $\{8,3\}$ tessellation. This symmetry group is often denoted $(3,3,4)$. Figure 11 shows a pattern with symmetry group $(3,3,5)$ that is based on the $\{10,3\}$ tessellation. For more on these "Circle Limit III" patterns, see [**Dun07b**].

FIGURE 9. A hyperbolic pattern of 5-armed crosses with symmetry group $[3^+, 10]$.

FIGURE 10. A *Circle Limit I* fish pattern with symmetry group $cmm_{3,3}$.

4. Hamiltonian Paths and An Algorithm for Creating Repeating Hyperbolic Patterns

Given a generating set for a group, the *Cayley digraph* for that group is the graph whose vertices are group elements and whose edges are labeled by the generators that take one vertex to another (i.e. there is a directed edge from u to v

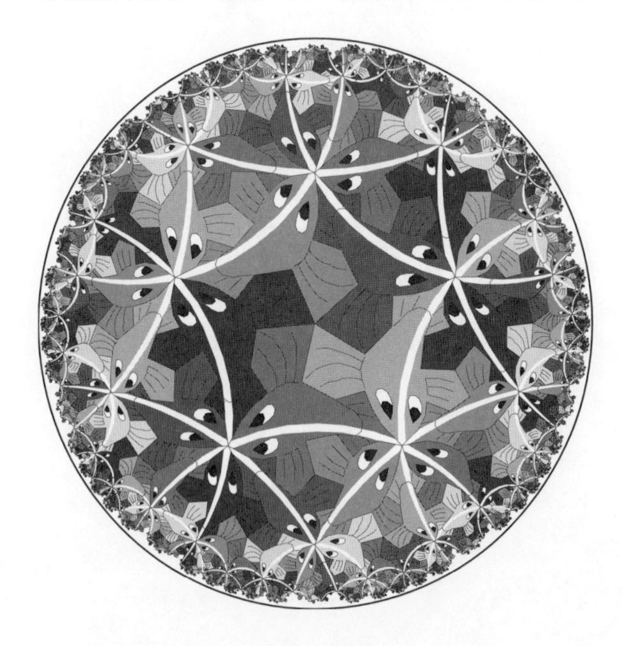

FIGURE 11. A "Circle Limit III" pattern with symmetry group $(3,3,5)$.

if there is a generator g such that $v = gu$). A *Hamiltonian path* in (the Cayley digraph of) a group is a path that meets each group element/vertex exactly once, and does not traverse any edge twice. In the infinite symmetry groups of hyperbolic patterns, we are interested in "one-way" (infinite) Hamiltonian paths that start at a vertex (usually the identity) and traverse all the group elements/vertices. The motivation is that by traversing such a path, we identify a sequence of generator transformations to apply to successive copies of a motif that can theoretically generate the entire hyperbolic pattern. For this to succeed, the identified motif must be such that the pattern is motif-transitive (isohedral), that is, the symmetry group of the pattern acting on a single motif produces the whole pattern (said another way, the orbit of a single motif is the whole pattern).

Figure 12 shows in fine lines the edges of the triangle pattern of Figure 1, and in heavier gray and black lines the Cayley graph of its symmetry group $[6,4]$. The Cayley graph is undirected since the three reflections that generate the group $[6,4]$ are their own inverses. A one-way Hamiltonian path for $[6,4]$ is traced out in the heavier black lines. Note that the Cayley graph forms the dual of the triangle pattern: to produce the Cayley graph, one triangle is designated as the generating motif, and a dot/vertex representing the identity element is placed within it. Each of the three generators (which are reflections in the edges of that triangle) acts on that generating motif (and the dot/vertex within it) to produce neighboring copies; these new vertices are joined to the first, and the process repeats until the whole graph is produced. The Cayley graph is itself a hyperbolic pattern of 12-gons, 8-gons, and 4-gons.

Where did the path of Figure 12 come from? In early 1980 I posed the problem of finding such a path to David Witte (Morris) via Joe Gallian. Witte was one of the students in Joe Gallian's 1979 REU program at the University of Minnesota Duluth (UMD). Witte obtained a solution to that problem, and returned to UMD

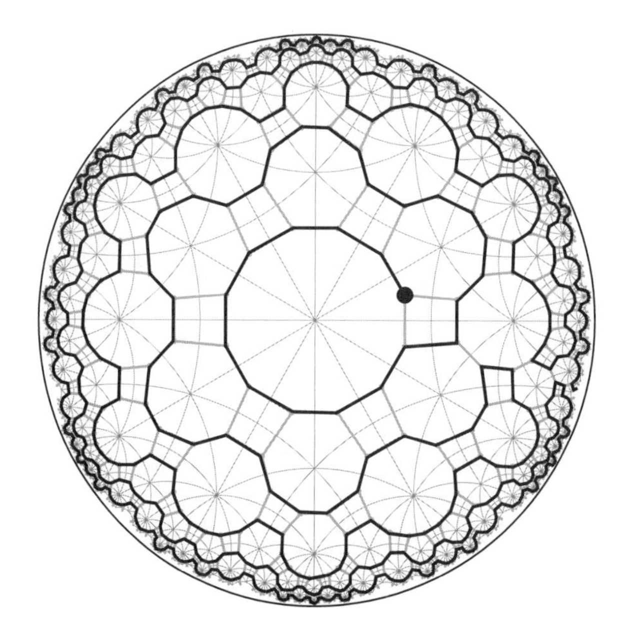

FIGURE 12. A Hamiltonian path in the group $[6, 4]$.

in the summer of 1980 as an advisor to the REU program. He had also solved the related problems of finding Hamiltonian paths in the symmetry groups of all four of Escher's Circle Limit patterns. Initially Witte did not think such paths existed and set about trying to prove that for the simpler case of the Cayley graph of the full symmetry group of the icosahedron. But, by continuing to explore the possibilities, instead he found that such a path existed for that case. Thus encouraged, Witte tackled the original problem of finding a Hamiltonian path in the group $[6, 4]$, and he succeeded.

The key idea was to notice that the light triangles of Figure 12 could be arranged in what we called *layers* (others call them coronas). The first layer consists of all the triangles with a vertex at the origin. Inductively, the triangles of layer $n + 1$ are those not in any previous layer and share an edge or vertex with some triangle in the nth layer. It is easy enough to find a partial Hamiltonian path in any layer. The trick is to connect the partial paths in successive layers, the jump from the first layer to the second layer being a special case. Also, the even numbered layers are different than the odd numbered layers. Witte found the connections between layers as can be seen in Figure 12.

Witte's method was actually more general in that it produced Hamiltonian paths in the groups $[p, q]$ (where, as mentioned above, $(p - 2)(q - 2) > 4$ for the group to be hyperbolic). The cases $p = 3$ or $q = 3$ needed special treatment. Using Witte's method, it was not too hard to find one-way Hamiltonian paths in the groups $[p, q]^+$, $[p^+, q]$, $cmm_{3,2}$, and $(3, 3, p/2)$, using appropriate generators. These groups included the symmetry groups of Escher's four "Circle Limit" patterns. In fact Witte, Douglas Jungreis, one of Gallian's 1985 REU students, and I, with help from other undergraduate research students over a few years, published existence and non-existence results for one-way and two-way paths in both directed and undirected Cayley graphs [**Dun95**].

As suggested above, the motivation for finding Hamiltonian paths was to incorporate them into an algorithm to draw repeating hyperbolic patterns. Our first goal was to be able to draw each of Escher's four Circle Limit patterns using a computer, thus avoiding the tedious hand work that Escher had to go through. In addition, his method of producing a sector of the circular pattern and then printing that several times did not seem to be an efficient method for a computer algorithm. In examining the Hamiltonian path in Figure 12, one notices that it circles around the center until it is one edge short of completing the 12-gon, then turns right, following along edges of the 4-gons and 8-gons in the next layer outward, continues around the next layer until it is one edge short of running into itself, then circles in the opposite direction, and continues this process of following edges in a spiraling meander, in theory forever, visiting every vertex in the Cayley graph. This was exactly the kind of pattern we wanted for the path, since if we used this sequence of transformations to draw copies of the motif as we followed the path, we would fill up the Poincaré disk from the center outward without gaps or overlaps. We wanted to avoid gaps to obtain a complete pattern, and we wanted to avoid overlaps not only for efficiency, but because at that time our main printing device was a pen plotter that would tear through the paper if it drew over the same spot too many times. Of course we could only draw a finite number of the outward layers of encircling motifs, but this was sufficient to give the idea of the entire infinite pattern. David Witte formulated what were essentially substitution rules for creating the path on a subsequent layer, given the choices on the current layer; as mentioned above, transitions from one layer to the next also had to be figured out.

At this point John Lindgren, a University of Minnesota Duluth undergraduate, entered the picture to do the programming. So the setup was: Witte would come up with the rules for the algorithm, I would translate them to pseudocode, and Lindgren would implement the algorithm in FORTRAN. The program that we created could generate patterns for several kinds of symmetry groups: $[p,q]$, $[p,q]^+$, $[p^+,q]$, $cmm_{3,2}$, and $(3,3,p/2)$ (I think it may have worked for general $cmm_{p/2,q/2}$ too). Thus we could generate all four of Escher's Circle Limit patterns, which was our goal. We published the results in SIGGRAPH '81 [**Dun81**], generating the figures for that paper using our FORTRAN program, and presenting a recursive algorithm. Unfortunately there was an error in that algorithm, but it was fixed in a later paper [**Dun86**].

5. Other Hyperbolic Pattern-Generating Algorithms

The first simplification of our algorithm was motivated by the observation that many of Escher's repeating patterns, including all the spherical and hyperbolic ones, have an underlying regular $\{p,q\}$ tessellations. The idea was to unify all the motifs within a p-gon into a *supermotif* (which we called a *p-gon pattern* in previous articles), then transform the supermotif to fill the hyperbolic plane. Figure 13 shows the supermotif for *Circle Limit I.* To produce an algorithm to transform the supermotif, we found Hamiltonian paths in what we call "Cayley coset graphs" of the desired symmetry groups. The stabilizer H of the supermotif is used to define cosets of the symmetry group; the cosets of H in the symmetry group can be represented visually by p-gons in the $\{p,q\}$ tessellation. In the Cayley coset graph, vertices are the cosets, and there is an edge from coset xH to coset yH if there is a generator g such that $gxH = yH$. Finding Hamiltonian paths in these graphs

FIGURE 13. The supermotif for *Circle Limit I.*

and programming an algorithm based on them seemed to be easier than those for the full group. Figure 14 shows as light circular arcs the $\{6, 4\}$ tessellation and in heavier gray and black straight lines the coset graph edges for the group $[6, 4]$; the Hamiltonian path consists of the black line segments of the coset graph.

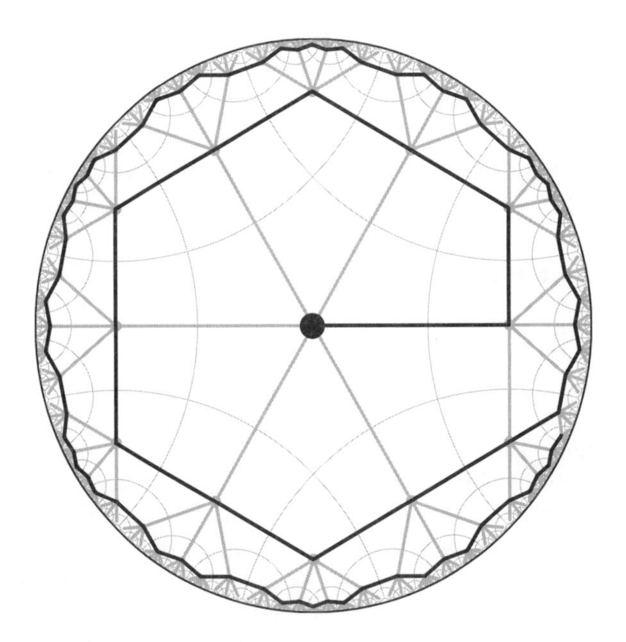

FIGURE 14. A Hamiltonian path in the coset graph of $[6, 4]$.

The Hamiltonian path methods led to roundoff errors due to successively multiplying many matrices together to get the current transformation matrix. We actually didn't notice this until we replaced an old but high-precision computer with a newer one with half the precision. The roundoff error problem was solved by devising a recursive algorithm that essentially traversed a spanning tree in the coset graph. Figure 15 shows a spanning tree in the coset graph of the group $[6, 4]$. Again, it was fairly easy to devise spanning trees and to program them. One such algorithm was presented in [**Dun86**].

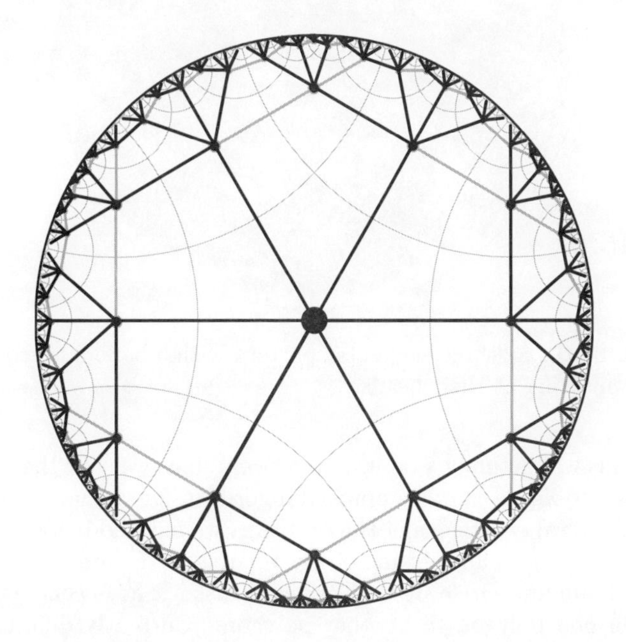

FIGURE 15. A spanning tree in the coset graph of $[6, 4]$.

Recently a more general recursive algorithm has been devised that removes the restriction that the pattern be based on a $\{p, q\}$ tessellation [**Dun07a**]. This method assumes that the motif is contained in a finite convex polygon that tiles the hyperbolic plane, and thus has rational angles at its corners. The algorithm proceeds by transforming the motif across sides of copies of that polygon. These transformations are reflections across the sides or 180° rotations about the midpoints of the sides of the polygon or conjugates of those reflections or rotations by elements of the symmetry group of the polygon. This algorithm seems more combinatorial than group-theoretic in nature. Figure 16 shows a pattern produced by this algorithm — it is based on Escher's "three element" pattern, Notebook Drawing Number 85 (page 184 of [**Sch04**]). The three elements are lightly shaded bats for air, gray lizards for earth, and dark fish for water (Escher's colors are yellow, red, and blue for the bats, lizards, and fish, respectively).

6. Conclusion and Future Work

Hamiltonian paths in Cayley graphs of symmetry groups were the key in creating our first programs to draw hyperbolic patterns; these in turn led to other

FIGURE 16. A "three elements" pattern with 3 bats, 5 lizards, and 4 fish meeting at their heads.

algorithms. This was actually a bootstrap process since we used the patterns drawn by one program to figure out an improved algorithm for the next program.

There are several extensions of the problems we have addressed that are yet to be solved. The first is to allow some of the vertices of the motif-enclosing polygon to be on the bounding circle of the Poincaré disk. The second is to transform a motif within one polygon to another polygon. Currently different versions of our programs can produce patterns with (perfect) color symmetry, but the color permutations must be determined ahead of time and hand-coded into the motif data files. This leads to the third, and seemingly most difficult problem: automating the process of generating patterns with color symmetry, one of the hallmarks of Escher's patterns.

Acknowledgment

I would like to thank the reviewer for many helpful suggestions, which significantly improved the exposition.

References

[Cox57] H.S.M. Coxeter, *Crystal symmetry and its generalizations*, Transactions of the Royal Society of Canada, (3), **51** (1957), 1–13.

[Cox79] H.S.M. Coxeter, *The Non-Euclidean Symmetry of Escher's Picture 'Circle Limit III'*, Leonardo, **12** (1979), 19–25.

[Dun81] D. Dunham, J. Lindgren, D. Witte, *Creating repeating hyperbolic patterns*, Computer Graphics, vol. 15, no. 3, (1981), (Proceedings of SIGGRAPH '81), 215–223.

[Dun86] D. Dunham, *Hyperbolic symmetry*, Computers & Mathematics with Applications vol. 12**B**, nos.(1,2) (1986) 139–153.

[Dun95] D. Dunham, D. Jungreis, David Witte (Morris), *Infinite Hamiltonian paths in Cayley digraphs of hyperbolic symmetry groups*, Discrete Mathematics, **143** (1995), 1–30.

[Dun03] D. Dunham, *Hyperbolic Art and the Poster Pattern*: `http://www.mathaware.org/mam/03/essay1.html`, on the Mathematics Awareness Month 2003 web site: `http://www.mathaware.org/mam/03/`.

[Dun07a] D. Dunham, *An Algorithm to Generate Repeating Hyperbolic Patterns*, in Proceedings of ISAMA 2007 (eds. Ergun Akleman and Nat Friedman), College Station, Texas, (2007), 111–118.

[Dun07b] D. Dunham, *A "Circle Limit III" Calculation*, in Bridges Donostia: Mathematical Connections in Art, Music, and Science, (ed. Reza Sarhangi), San Sebastian, Spain, (2007), 451–458.

[Gre07] M. Greenberg, *Euclidean & Non-Euclidean Geometries: Development and History,* 4th Ed., W. H. Freeman, Inc., New York, 2007. ISBN 0716799480

[Hil01] D. Hilbert, *Über Flächen von konstanter gausscher Krümmung*, Trans. of the Amer. Math. Soc., pp. 87–99, 1901.

[Sch04] D. Schattschneider, *M.C. Escher: Visions of Symmetry*, 2nd Ed., Harry N. Abrams, Inc., New York, 2004. ISBN 0-8109-4308-5

Department of Computer Science, University of Minnesota Duluth, Duluth, MN 55812-3036, USA

E-mail address: `ddunham@d.umn.edu`

Contemporary Mathematics
Volume **479**, 2009

When graph theory meets knot theory

Joel S. Foisy and Lewis D. Ludwig

ABSTRACT. Since the early 1980s, graph theory has been a favorite topic for undergraduate research due to its accessibility and breadth of applications. By the early 1990s, knot theory was recognized as another such area of mathematics, in large part due to C. Adams' text, The Knot Book. In this paper, we discuss the intersection of these two fields and provide a survey of current work in this area, much of which involved undergraduates. We will present several new directions one could consider for undergraduate work or one's own work.

1. Introduction

This survey considers three current areas of study that combine the fields of graph theory and knot theory. Recall that a *graph* consists of a set of vertices and a set of edges that connect them. A *spatial embedding* of a graph is, informally, a way to place the graph in space. Historically, mathematicians have studied various graph embedding problems, such as classifying what graphs can be embedded in the plane (which is nicely stated in Kuratowski's Theorem [**26**]), and for non-planar graphs, what is the fewest number of crossings in a planar drawing (which is a difficult question for general graphs and still the subject of ongoing research, see [**24**] for example). A fairly recent development has been the investigation of graphs that have non-trivial links and knots in every spatial embedding. We say that a graph is *intrinsically linked* if it contains a pair of cycles that form a non-splittable link in every spatial embedding. Similarly, we say that a graph is *intrinsically knotted* if it contains a cycle that forms a non-trivial knot in every spatial embedding. Conway, Gordon [**9**], and Sachs [**32**] showed the complete graph on six vertices, K_6, is intrinsically linked. We refer the reader to a very accessible proof of this result in Section 8.1 of The Knot Book [**1**]. Conway and Gordon further showed that K_7 is intrinsically knotted. These results have spawned a significant amount of work, including the complete classification of minor-minimal examples for intrinsically linked graphs by Robertson, Seymour, and Thomas [**31**]. After the completion of this classification, work has turned to finding graphs in which every embedding has a more complex structure such as finding other minor-minimal intrinsically knotted graphs [**18**],[**17**], graphs with cycles with high linking number in every

Key words and phrases. spatial graph, straight-edge embedding, intrinsically linked, complete graph, convex polyhedra, linking.

spatial embedding [**14**], as well as graphs with complex linking patterns [**11**] (see Section 2 for a bit more on what this means).

Recall that a natural generalization of an intrinsically linked graph is an intrinsically n-linked graph, for an integer $n \geq 2$. A graph is *intrinsically n-linked* if there exists a non-split n-component link in every spatial embedding. In Section 2, we will try to survey known results about intrinsically 3-linked graphs, and we present a few less-technical proofs. In particular, we discuss Flapan, Naimi and Pommersheim's [**13**] result that K_{10} is the smallest complete graph that is intrinsically 3-linked. We also talk about other examples of intrinsically 3-linked graphs that are minor-minimal or possibly minor-minimal.

In Section 3, we restrict our attention to embeddings of graphs with straight edges. Conway and Gordon's work guarantees a 2-component link in any embedding of K_6 and a knot in any embedding of K_7, but says nothing of the number of such embeddings. Due to the restrictive nature of straight-edge embeddings, we can determine the possible *number* of links and knots in such embeddings. Theorem 3.1 and 3.2 characterize the number of 2-component links in K_6 and K_7. Table 1 characterizes the number of stick knots occurring in a large class of straight-edge embeddings of K_7. This work is of interest to molecular chemists who are trying to synthesize topologically complex molecules. One could imagine that the vertices of these graphs represent atoms and the edges are the bonds of a molecule.

In Section 4, we expand the work of Conway and Gordon by showing in Theorem 4.1 [**4**] that every K_n ($n \geq 7$) contains a knotted Hamiltonian cycle in every spatial embedding.

While we make every effort to explain the machinery necessary for the following results in each section, we refer the reader to The Knot Book [**1**] and Introduction to Graph Theory [**7**]. A number of open questions will be posed throughout the sections. For easy reference, the questions will be listed again in Section 5.

2. Intrinsically 3-linked graphs

We start this section with a quick introduction to the linking number. Recall that given of a link of two components, L_1 and L_2 (two disjoint circles embedded in space), one computes the linking number of the link by examining a projection (with over and under-crossing information) of the link. Choose an orientation for each component of the link. At each crossing between two components, one of the pictures in Figure 1 will hold. We count $+1$ for each crossing of the first type (where you can rotate the over-strand counter-clockwise to line up with the under-strand) and -1 for each crossing of the second type. To get the linking number, $lk(L_1, L_2)$, take the sum of $+1$s and -1s and divide by 2. One can show that the absolute value of the linking number is independent of projection, and of chosen orientations (see [**1**] for further explanation). Note that if $lk(L_1, L_2) \neq 0$, then the associated link is non-split. The converse does not hold. That is, there are non-split links with linking number 0 (the Whitehead link is a famous example, see again [**1**]). Any linking numbers we use will be the ordinary linking number, taken mod 2.

In this section, we survey the known results about intrinsically 3-linked graphs, and we present a few results. Before doing so, we introduce some more terminology. Recall that a graph H is said to be a *minor* of the graph G if H can be obtained from G by a sequence of edge deletions, edge contractions and/or vertex deletions.

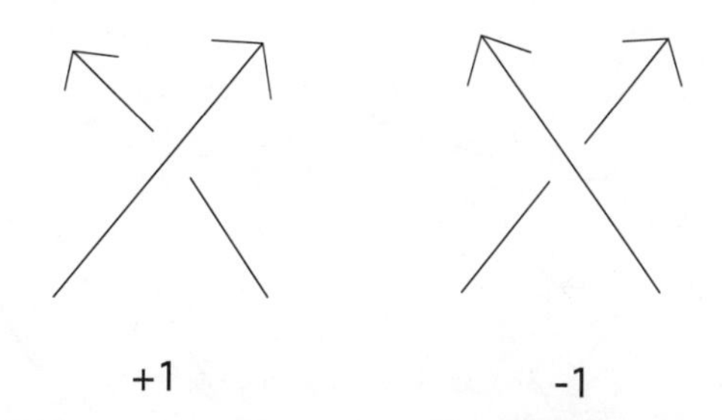

FIGURE 1. Computing the linking number.

A graph G is said to be *minor-minimal* with respect to a property, if G has the property, but no minor of G has the property. It follows from the result of Robertson and Seymour [**30**] that there are only finitely many minor-minimal intrinsically n-linked graphs, and since having a n-linkless embedding is preserved by minors (see [**27**], [**13**]), a graph is intrinsically n-linked if and only if it contains a member of a finite list (not yet determined) of graphs as a minor.

The study of intrinsically 3-linked graphs first appeared briefly in a student paper [**20**], where the authors showed a 3-component linkless embedding of $K_{3,3,3}$. Soon after that paper was written, the first author spoke with Erica Flapan about the problem of finding intrinsically 3-linked graphs. She became interested in determining the lowest value of n such that K_n is intrinsically n-linked. The first author had the more modest goal of finding an intrinsically 3-linked graph. As a result of this conversation, we formulated the following pasting type lemma, which first appeared in [**12**], and is easily proven. Recall that if the cycles C_2 and C_3 intersect along an arc, then we may form a new cycle, $C_2 + C_3$ by using the edges that are only in C_2 or only in C_3.

LEMMA 2.1. *If C_1, C_2, and C_3 are cycles in an embedded graph, C_1 disjoint from C_2 and C_3, and $C_2 \cap C_3$ is an arc, then $lk(C_1, C_2) + lk(C_1, C_3) = lk(C_1, C_2 + C_3)$.*

This leads to:

LEMMA 2.2. [**12**] *Let G be a spatially embedded graph that contains simple closed curves C_1, C_2, C_3 and C_4. Suppose that C_1 and C_4 are disjoint from each other and both are disjoint from C_2 and C_3, and $C_2 \cap C_3$ is an arc. If $lk(C_1, C_2) = 1$ and $lk(C_3, C_4) = 1$, then G contains a non-split 3-component link.*

PROOF. If $lk(C_1, C_3) = 1$ or if $lk(C_2, C_3) = 1$, then C_1, C_2 and C_3 form a non-split 3-component link. Similarly, if $lk(C_1, C_4) = 1$, then C_1, C_3 and C_4 form a non-split 3-component link. Finally, if $lk(C_1, C_4) = lk(C_2, C_4) = lk(C_1, C_3) = lk(C_1, C_4) = 0$, then by Lemma 2.1, $lk(C_1, C_2 + C_3) = lk(C_1, C_2) + lk(C_1, C_3) = 1$ and $lk(C_4, C_2 + C_3) = lk(C_4, C_2) + lk(C_4, C_3) = 1$. Thus $C_1, C_2 + C_3, C_4$ forms a non-split 3-component link. □

One can use this Lemma to show that various graphs are intrinsically 3-linked. For example (see [**13**]), let J be the graph obtained by pasting two copies of $K_{4,4}$ along an edge (see Figure 2). Sachs [**33**] showed that for every spatial embedding of $K_{4,4}$, every edge of the graph is contained in a cycle that is non-split linked to

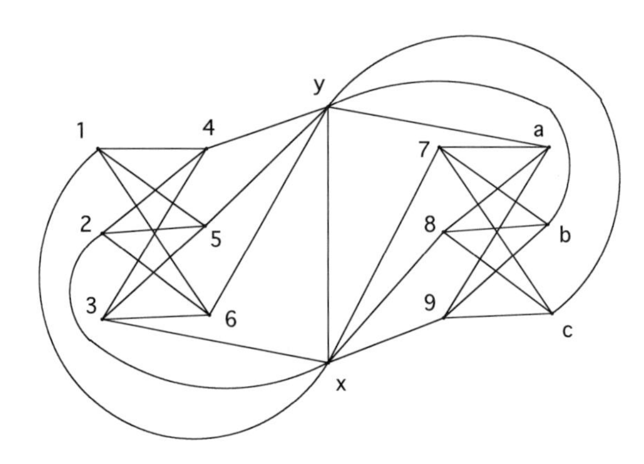

FIGURE 2. The graph J.

another cycle. Consider an arbitrary embedding of J. In one copy of $K_{4,4}$ there are a pair of cycles with non-zero linking number, call them C_1 and C_2, and by Sachs' result, we may assume one of the cycles, say C_2, uses the edge shared by the two copies of $K_{4,4}$. In the other copy of $K_{4,4}$, there are another pair of cycles with non-zero linking number, call them C_3 and C_4, and again, we may assume that one of the cycles, say C_3, uses the shared edge. Thus by Lemma 2.2, there is a 3-component link in this embedding of J. It follows that J is intrinsically 3-linked. It is not known if J is minor-minimal with respect to this property. At the time the paper was being written, the authors of [**13**] believed that J is either minor-minimal, or the graph obtained from removing the shared edge from J is minor-minimal with respect to being intrinsically 3-linked, though a proof of this was never written down.

The fact that J is intrinsically 3-linked was later generalized in [**5**] to include the graph obtained from two copies of K_7 pasted along an edge, as well as the graph obtained from $K_{4,4}$ and K_7 pasted along an edge. We quickly sketch a proof here. We first need the following lemma:

LEMMA 2.3. [**5**] *Let G be a spatial embedding of K_7, then every edge of G is in a non-split linked cycle.*

PROOF. First embed K_7, then consider an edge $e_1 = (v_1, v_2)$ in K_7. The vertices of $G - v_2$ induce a K_6. Then vertex v_1 is in a linked cycle in this embedded K_6, say (v_1, v_3, v_4) is linked to cycle C. By Lemma 2.2, $lk((v_1, v_3, v_4), C) = lk((v_1, v_3, v_2), C) + lk((v_1, v_2, v_3, v_4), C)$, and thus e_1 is in a linked cycle. □

The proof of the following result is similar to the proof that J is intrinsically 3-linked.

THEOREM 2.1. [**5**] *Let G be a graph formed by identifying an edge of a graph G_1 with an edge from another graph G_2, where G_1 and G_2 are either K_7 or $K_{4,4}$. Then every such G is intrinsically 3-linked.*

At this time, we do not know whether the graphs described by this theorem are minor-minimal or not. Before we go further, we review one important definition.

Let $a, b,$ and c be vertices of a graph G such that edges $(a, b), (a, c)$ and (b, c) exist. Then a $\triangle - Y$ *exchange* on a triangle (a, b, c) of graph G is as follows. Vertex v is added to G, edges $(a, b), (a, c)$ and (b, c) are deleted, and edges $(a, v), (b, v)$ and (c, v) are added. Given the graph G in Figure 3, the illustration in the Figure depicts the result of $\triangle - Y$ expansion on triangle abc. A $Y - \triangle$ exchange is the reverse operation.

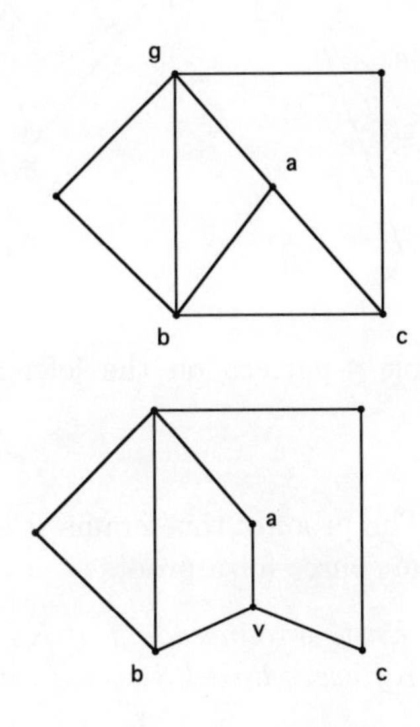

FIGURE 3. A graph G and the results of a Δ-Y exchange.

In [**13**], the authors were able to find a minor-minimal intrinsically $(n + 1)$-linked graph $G(n)$, for every integer $n > 2$. By showing that J is not obtainable from $G(2)$ by a sequence of $\Delta - Y$ and $Y - \Delta$ moves, they also showed that the set of all minor-minimal intrinsically 3-linked graphs cannot be obtained from one of the graphs in the set by a sequence of $\Delta - Y$ and $Y - \Delta$ moves–unlike the set of minor-minimal intrinsically linked graphs which can all be obtained from K_6 by $\Delta - Y$ and $Y - \Delta$ moves.

In [**12**], Flapan, Naimi and Pommersheim were able to determine that K_{10} was intrinsically 3−linked. By exhibiting a 3-linkless embedding of K_9, they also established that $n = 10$ is the smallest n for which K_n is intrinsically 3-linked. In order to prove their result for K_{10}, the authors used a careful examination of linking patterns of triangles in spatial embeddings of K_9, as well as Lemma 2.2. We will briefly discuss those patterns here.

A *4-pattern* within an embedded graph, G, consists of a 3-cycle, B, that is linked with four other 3-cycles that can be described as follows. For vertices q,r in G, each 3-cycle linked to B is of the form (q, r, x) where x is one of any four vertices of G other than B, q, and r (see Figure 4).

A *6-pattern* within an embedded graph, G, consists of a 3-cycle, B, that is linked with six other 3-cycles that can be described as follows. For vertices p,q,r in G, each 3-cycle linked to B is either of the form (p, q, x) or (p, r, x) where x is one of any three vertices of G other than B, p, q, and r (see Figure 4). We may now

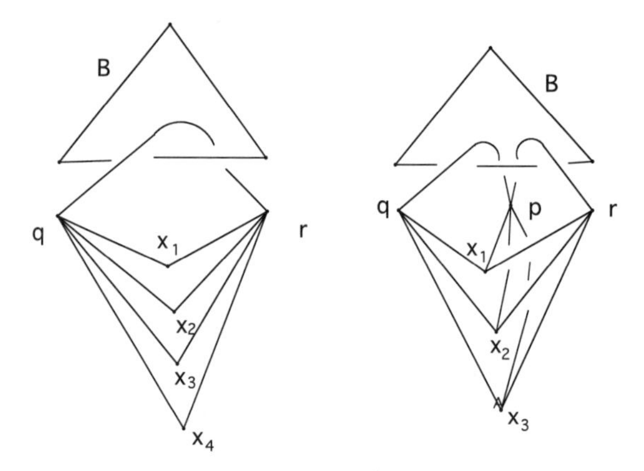

FIGURE 4. A possible 4-pattern on the left, and a possible 6-pattern on the right

state the following Lemma. The proof of this lemma is somewhat technical, so we refer the reader to the original source for a proof.

LEMMA 2.4. [**12**] *There exists an embedding of K_9 without any 3-component links. For any embedding of K_9 every linked 3-cycle is in a 4-pattern, a 6-pattern, or a 3-component link.*

More recently, O'Donnol [**28**] has used a clever examination of linking patterns in complete bipartite graphs to show that every embedding of $K_{2n+1,2n+1}$ contains a non-split link of n-components. O'Donnol further showed that for $n \geq 5$, K_{4n+1} is intrinsically n-linked. Even more recently, Drummund-Cole and O'Donnol [**10**] improved this result by showing that for every $n > 1$, every embedding of $K\lfloor\frac{7}{2}n\rfloor$ contains a non-split link of n-components. It would be a good project to determine if this is the best one can do for low values of n. In particular, is 17 the fewest vertices of an intrinsically 5-linked graph (this number could be as low as 15)? For $n = 4$, 14 is currently the fewest number of vertices need to guarantee K_n is intrinsically 4 linked, but this number could be as low as 12. Drummund-Cole and O'Donnol further showed that there exists a function $f(n)$ such that $\lim_{n\to\infty} \frac{f(n)}{n} = 3$ and, for every n, $K_{f(n)}$ is intrinsically linked. As 3 vertices are the fewest possible for a link component, this asymptotic result is the best possible.

The quest for finding a complete set of minor-minimal intrinsically 3-linked graphs is still very much alive–there remains much work to be done. In [**13**], there are two families of intrinsically 3- linked graphs presented. As we mentioned earlier in the paper, one is the single member family consisting of the triangle-free graph J (or possibly some minor of J. If this minor had a 3-cycle, then the family would be more than one member). The other family consists of the graph $G(2)$ described in [**13**], as well as the other two graphs that can be obtained from $G(2)$ by $Y - \Delta$ exchanges (one can readily argue that they are intrinsically 3-linked, using the same arguments given in [**13**]). Moreover, since $G(2)$ is minor-minimal intrinsically 3-linked, so are these graphs. This follows from the following lemma, which makes for a good exercise in graph theory. The curious and/or frustrated reader can look up the proof online if they are interested.

LEMMA 2.5. [**29**], [**6**] *Let P be a graph property that is preserved by $\Delta - Y$ exchanges, and let G' be a graph obtained from G by a sequence of $\Delta - Y$ moves. If G has property P, and if G' is minor-minimal with property P, then G is also minor-minimal with property P.*

The graph J, the graph obtained by pasting two copies of K_7 along an edge, and the graph obtained by pasting an edge of K_7 to an edge of $K_{4,4}$ may also lead to new families of minor-minimal intrinsically 3-linked graphs–we just do not know yet if these graphs are themselves minor-minimal, or if they can be pared down.

As we mentioned earlier, the authors in [**12**] showed that K_{10} is intrinsically 3-linked. Bowlin and the first author [**5**] later showed, using techniques similar to those used in [**12**], that the subgraph obtained from K_{10} by removing 4 edges incident to a common vertex is also intrinsically 3-linked; they also showed that the subgraph obtained from K_{10} by removing two non-adjacent edges is also intrinsically 3-linked. They were not able to prove that these graphs are minor-minimal (the first author strongly suspects at least the former is). If they were, then by $\Delta - Y$ exchanges, they would yield two new families of graphs for our set. Finally, Bowlin and Foisy showed that any graph obtained by joining two graphs from the Petersen family by a 6−cycle that has vertices that alternate between copies of the two graphs is intrinsically 3-linked:

THEOREM 2.2. [**5**] *Let G be a graph containing two disjoint graphs from the Petersen family, G_1 and G_2, as subgraphs. If there are edges between the two subgraphs G_1 and G_2 such that the edges form a 6-cycle with vertices that alternate between G_1 and G_2, then G is intrinsically 3-linked.*

The proof of this theorem requires the use of the following lemma, whose proof is similar to the proof of Lemma 2.2:

LEMMA 2.6. [**5**] *In an embedded graph with mutually disjoint simple closed curves, C_1,C_2,C_3, and C_4, and two disjoint paths x_1 and x_2 such that x_1 and x_2 begin in C_2 and end in C_3, if $lk(C_1, C_2) = lk(C_3, C_4) = 1$ then the embedding contains a non-split 3-component link.*

PROOF (of Theorem 2.2). Let $\{a_1, a_2, a_3, b_1, b_2, b_3\}$ be the set of vertices that make up the 6-cycle described in the statement of the theorem, where $\{a_1, a_2, a_3\}$ are in G_1 and $\{b_1, b_2, b_3\}$ are in G_2. Embed G. By the pigeonhole principle, at least two vertices in the set $\{a_1, a_2, a_3\}$ are in a linked cycle within the embedded G_1 (without loss of generality, a_1 and a_2), and likewise we may assume that the vertices b_1 and b_2 are in a linked cycle in G_2. Because of the edges between $\{a_1, a_2, a_3\}$ and $\{b_1, b_2, b_3\}$, we know that there are two disjoint edges between the sets $\{a_1, a_2\}$ and $\{b_1, b_2\}$. By Lemma 2.6, a 3-component link is present in the embedding. □

We shall henceforth call a 6−cycle as in the statement of Theorem 2.2 an *alternating* 6−*cycle*. We suspect that many of the graphs obtained by joining Petersen graphs by an alternating 6−cycle are minor-minimal with respect to being intrinsically 3-linked. For example, consider two copies of K_6 joined by an alternating 6−cycle, which we will denote by S. We examine the embedding pictured in Figure 5. In the embedded shown, in the K_6 on the left, the only

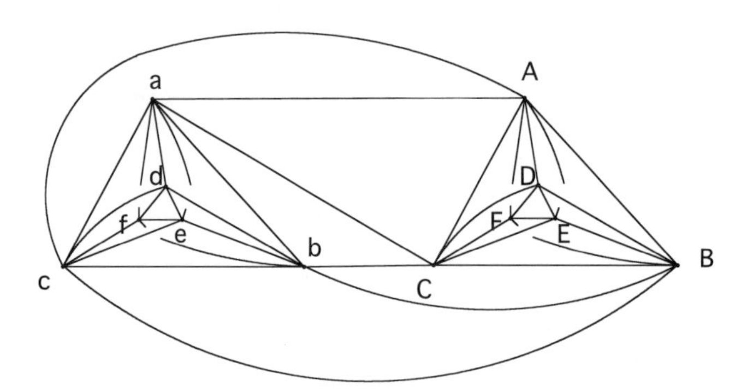

FIGURE 5. The graph S is minor-minimal.

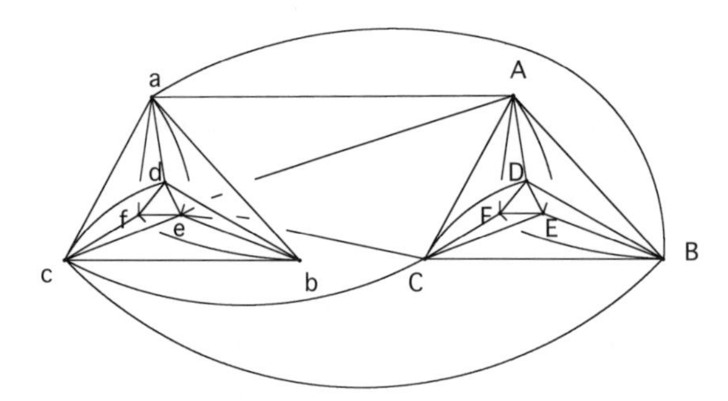

FIGURE 6. Removing edge (a, c) results in a 3-linkless embedding.

linked cycles are (a, c, e) and (b, d, f). Similarly, for the K_6 on the right, only (A, C, E) is linked with (B, D, F). The only 3-component link in this embedding is $(b, d, f), (a, e, c, A, E, C), (B, D, F)$. If we remove any one of the edges (c, A), (d, f) or (a, e), then the resulting graph has a 3-linkless embedding. If we contract any one of the edges (b, B), (a, b), (d, e), (a, e), then the resulting embedding is 3-linkless.

It remains to show that removing an edge in the class of (a, c) results in a graph with a 3-linkless embedding. This can be seen by examining the embedding depicted in Figure 6 (note that the vertices have been re-labelled slightly).

It will take some time and effort to enumerate exactly what graphs are in the family of all Petersen graphs joined by an alternating 6−cycle. There is, up to isomorphism, only one way to connect copies of K_6, but for all of the other graphs in the Petersen family, there are multiple ways to connect them. Perhaps Lemma 2.5 might be helpful in efficiently demonstrating that some of these graphs are minor-minimal.

Up to this point in time, all of the minor-minimal intrinsically 3-linked graphs have been shown to be intrinsically 3-linked by using some sort of analogy to Lemma 2.1. For such graphs, the guaranteed 3-link contains at least one cycle that was pasted together from two smaller cycles. (Though it is interesting that Drummund-Cole and O'Donnol [**10**] have recently shown that every embedding of K_{14} contains a 3-link of triangles.) Recently some students worked on a related problem, and their work might suggest that there will be some minor-minimal intrinsically 3-linked graphs that cannot be proven to be intrinsically 3-linked using an analogy to Lemma 2.1. We briefly describe this work now.

An S^1 *embedding* of a graph G is an injective map of the vertices of G into S^1. A *0-sphere* in an S^1 embedding of a graph G is composed of any two vertices that are the endpoints of a simple path in G. We denote a 0-sphere by writing the endpoints of the associated path as an ordered pair. Just as a pair of disjoint cycles forms a link in a spatial embedding, a pair of disjoint 0-spheres (with disjoint underlying paths) forms a link in an S^1 embedding. A link (a, b) and (c, d) is said to be *split* if a and b lie on the same component of $S^1 - \{c, d\}$. Thus the link is *non-split* if a and b lie on different components of $S^1 - \{c, d\}$. For S^1 embeddings, the *mod 2 linking number* of two 0-spheres (a, b) and (c, d), denoted $lk((a, b), (c, d))$, is 0 if and only if (a, b) and (c, d) are split linked and is 1 if and only if (a, b) and (c, d) are non-split linked. An S^1 *n-link* in an S^1 embedding of a graph G is a set of n disjoint 0-spheres in the embedding of G. An n-link in an S^1 embedding is said to be *split* if there are two points, x and y, on the circle such that both components of $S^1 - \{x, y\}$ contains at least one vertex involved in the n-link and every 0-sphere in the link lies entirely on one component of $S^1 - \{x, y\}$. Just as some graphs are intrinsically linked in space, some graphs are intrinsically S^1 linked. A graph is intrinsically S^1 linked if every S^1 embedding contains a non-split link. It was shown by Cicotta et al. that the complete minor-minimal set of intrinsically S^1 linked graphs is K_4 and $K_{3,2}$ [**8**]. A graph is said to be intrinsically S^1 n-linked if every S^1 embedding of the graph contains a non-split n-link.

The students easily proved the following analog of Lemma 2.1:

LEMMA 2.7. [**6**] *Given 0-spheres $(a, b), (c, d), (c, e)$, and (d, e) in an S^1 embedding of graph G, $lk_2((a, b), (c, e)) = lk_2((a, b), (c, d)) + lk_2((a, b), (d, e))$.*

They also proved the following analog of Theorem 2.1:

THEOREM 2.3. [**6**] *Let G be a graph formed by pasting together graphs A and B, where A and B are each either a K_4 or $K_{3,2}$, at a vertex. The graph G is intrinsically S^1 3-linked.*

They went on to find 28 minor-minimal intrinsically S^1 3-linked graphs, 6 of which were shown to be intrinsically S^1 3-linked using Lemma 2.7. The other 22 graphs were shown to be intrinsically S^1 3-linked by using other ad hoc methods (it is possible to analyze such graphs by using combinatorics and case checking since there are only finitely many non-equivalent S^1embedding classes of a given graph). By comparison, all of the intrinsically 3-linked graphs in space have been shown to be intrinsically 3-linked by using some sort of analogy to Lemma 2.7. The work in [**6**] is thus interesting because it suggests that the intrinsically 3-linked graphs thus far discovered may only be the tip of the iceburg. It is also interesting because it provides a more tractable analogous problem. Hopefully, someone will soon prove that the 28 graphs (or possibly a superset) forms the complete set of minor-minimal intrinsically S^1 3-linked graphs.

In summary, the quest for a complete minor-minimal set of intrinsically 3-linked graphs is going to require some time-consuming methodical work, as well as some breakthroughs. We thus feel it is well-suited to eager and persistent students who have fresh ideas.

We briefly mention some other related results that might be of interest. One could also look for graphs that contain, in every spatial embedding, multi-component links with various patterns. The authors in [**13**] were also able to show the existence of an "n-necklace" (a link $L_1 \cup L_2 \cup ... \cup L_n$, such that for each $i = 1, ..., n-1$, $L_i \cup L_{i+1}$ is non-split and $L_n \cup L_1$ is non-split) in every embedding of the graph they call $F(n)$. Flapan, Mellor and Naimi [**11**] came up with a powerful generalization of this result to show that, given any n and α, every embedding of any sufficiently large complete graph in $\mathbb{R}^3$ contains an oriented link with components $Q_1, ..., Q_n$ such that, for every $i \neq j$, $|lk(Q_i, Q_j| \geq \alpha$. The first author [**16**] has also shown the existence of a graph that, for every spatial embedding, contains either a 3-component link or a knotted cycle, but it has a knotless embedding and a 3-component linkless embedding.

Finally, we mention one more related open question.

QUESTION 2.1. What is the smallest n, such that, for every straight edge embedding of K_n, there is a non-split link of 3 components?

We know n is at most 10, but could be 9.

3. Links and knots in straight-edge embeddings of graphs.

In this section, we consider complete graphs composed of straight edges or sticks. A *stick knot* is a knot formed out of rigid straight sticks. Molecular chemists are interested in this type of knot because at the molecular level, molecules are more like rigid sticks than flexible rope, Figure 7. With this application in mind, the following two questions were posed at a knot theory workshop in 2004:

(1) Does there exist a straight-edge embedding of K_6 with 9 (3-3) links?
(2) Given a straight-edge embedding of K_7, how many and what types of knots occur?

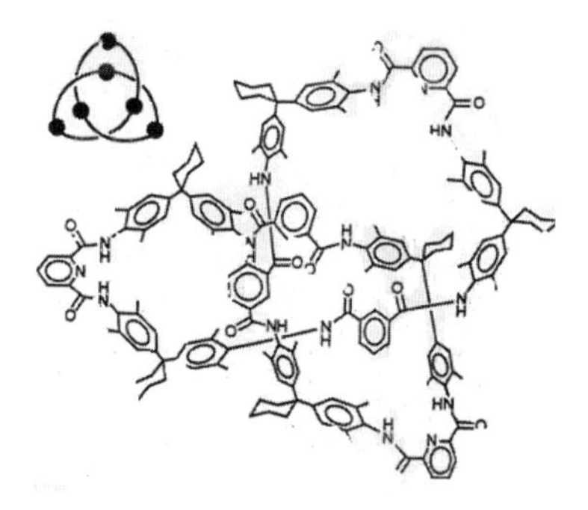

FIGURE 7. The trefoil knot and a knotted molecule

The first question was motivated by Conway-Gordon and Sachs' proof that K_6 is intrinsically linked. Any three vertices and adjoining edges form a 3-cycle. In K_6 there are 10 disjoint pairs of 3-cycles. If the edges were allowed to bend and stretch, one could place the vertices and edges of K_6 in space such that all 10 pairs of triangles were linked. But what would happen to the number of links if the edges had to remain straight as in a molecular bond? Due to the techniques used in their proof, it was known that the number of linked pairs had to be odd. Hence, the question asked if the maximum 9 pairs could be attained.

In regards to the second question, K_7 has 360 Hamiltonian cycles consisting of 7 edges. It is well known that only two non-trivial knots, the trefoil and the figure-8, can be made with 7 sticks. The minimum number of sticks needed to make a knot is 6 and this only occurs for the trefoil. So, to answer the second question, one must not only consider the Hamiltonian cycles on K_7, but all cycles of length 6 as well.

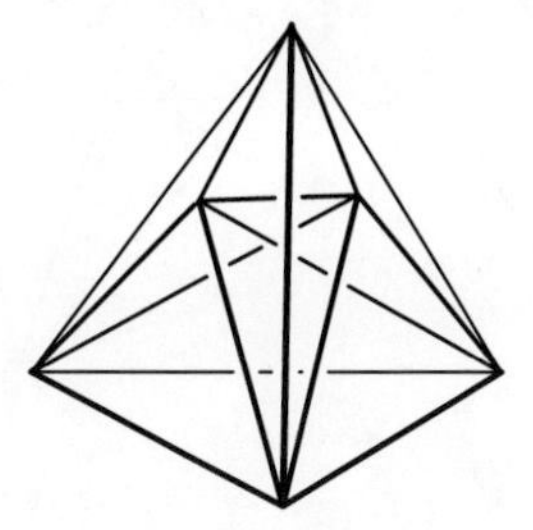

FIGURE 8. K_6 with two internal vertices

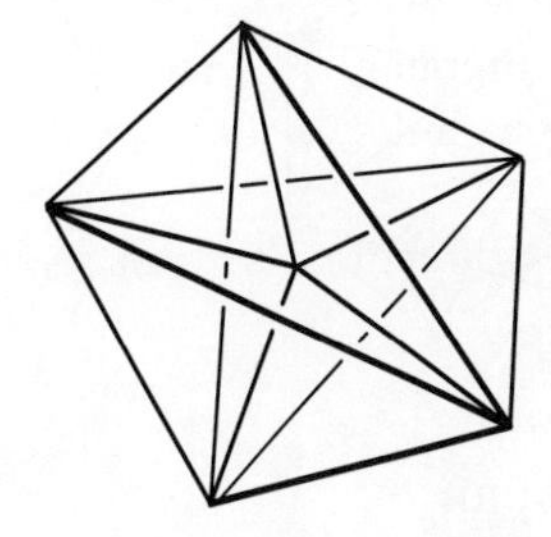

FIGURE 9. K_6 with one internal vertex

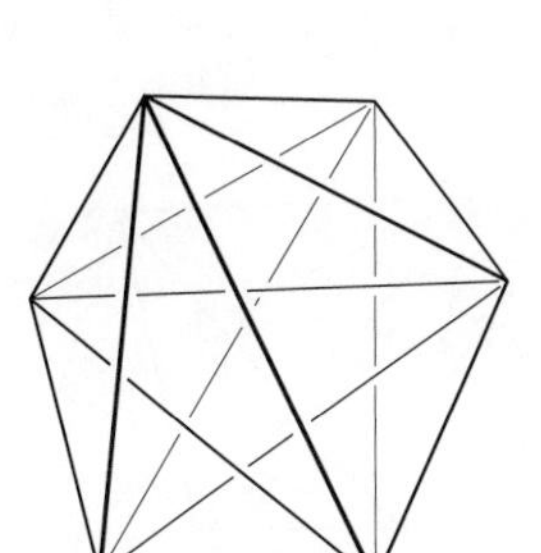

FIGURE 10. K_6 with no internal vertices, version 1

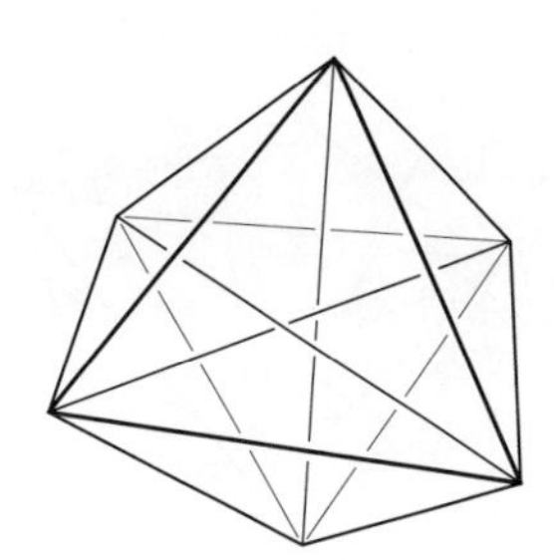

FIGURE 11. K_6 with no internal vertices, version 2

In 2004 a student of the second author, C. Hughes, showed that any straight-edge embedding of K_6 contains either 1 or 3 disjoint 2-component links, thus answering the first question [**21**]. To do this, she considered the four distinct convex polyhedra that form straight-edge embeddings of K_6 [**35**] (see Figure 8–11). It was shown that Figures 8 and 9 are ambient isotopic to Figure 10 (note the isotopies preserve the linearity of the edges). Through a series of geometric arguments, Hughes then showed Figure 10 has one 2-linked component and Figure 11 has three distinct 2-linked components, again up to ambient isotopy that preserves the linearity of the edges. Interestingly, in 2007 Huh and Jeon independently showed these same results as well as proving Figure 11 is the only straight-edge embedding of K_6 that contains a knot, a single trefoil [**22**].

THEOREM 3.1. *A straight-edge embedding of K_6 has either one or three 2-component links.*

In 2006, the second author and P. Arbisi extended the work of Hughes, Huh, and Jeon by classifying all the 2-component links in certain straight-edge embeddings of K_7. This was a challenging task as there are five distinct embeddings of K_7 that form convex polyhedra. In addition, unlike K_6 that has 10 pairs of disjoint 3-cycles, K_7 has 70 pairs of disjoint 3-cycles. With one extra vertex, there is now the possibility of 3-cyles linking with 4-cycles. There are 35 such pairs.

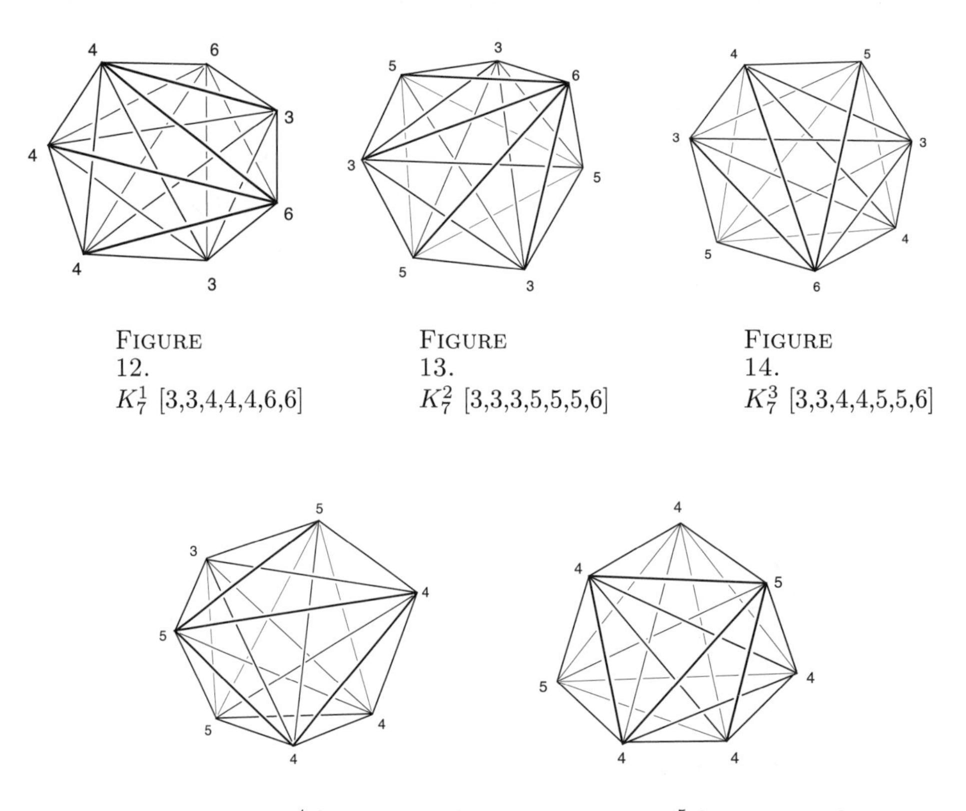

FIGURE 12. K_7^1 [3,3,4,4,4,6,6]

FIGURE 13. K_7^2 [3,3,3,5,5,5,6]

FIGURE 14. K_7^3 [3,3,4,4,5,5,6]

FIGURE 15. K_7^4 [3,4,4,4,5,5,5]

FIGURE 16. K_7^5 [4,4,4,4,4,5,5]

While Hughes was able to argue that three of the embeddings of K_6 were equivalent under ambient isotopies that preserve linearity of edges, this was not readily apparent with K_7. Instead, the second author and Arbisi focused on the 5 distinct straight-edge embeddings of K_7 that form convex polyhedra [**35**](see Figures 12–16). In order to distinguish the five embeddings, the figures are labeled with the *external* degree of each vertex. That is, the number associated with each vertex represents the number of edges on the hull that are incident to it. For the remainder of the section, when we refer to the degree of a vertex, we actually mean the number of edges from the hull incident to that vertex, unless otherwise stated.

THEOREM 3.2. *The minimum number of linked components in any straight-edge embedding of K_7 which forms a convex polyhedron of seven vertices is twenty-one, and the maximum number of linked components in K_7 is forty-eight. Specifically,* *we have the following:*

	K_7^1	K_7^2		K_7^3		K_7^4				K_7^5		
(3-3)	7	7	9	7	9	9	11	13	15	13	15	17
(3-4)	14	14	18	14	18	18	22	23,26	27,30	23,26	27,30	31

The main method employed for the results in Theorem 3.2 was to consider a specific embedding of K_7, systematically remove a vertex and its adjoining edges, then determine which version of K_6 remained: K_6^1 (Figure 10) with one (3-3) link or K_6^2 (Figure 11) with three (3-3) links. K_7^1–K_7^3 were relatively straight-forward. If a vertex was removed that was not of external degree of 6, then the resulting embedding of K_6 had only one (3,3) link (K_6^1). Removing the degree 6 vertex was a bit more challenging and would result in either K_6^1 or K_6^2, depending on the arrangement of the internal edges in the given embedding of K_7. This resulted in a varying number of links for K_7^2 and K_7^3.

Without any degree 6 vertices, K_7^4 and K_7^5 were considerably more challenging. Thankfully, we were able to determine the arrangement of the vertices in various embeddings of K_7^4 and K_7^5 via Steinitz's Theorem which states that a graph G is isomorphic to the vertex-edge graph of a 3-D polyhedron if and only if G is planar and 3-connected. However, we were still finding some inconsistencies in these two embeddings compared to the previous three. Specifically, the number of (3,4) links was not always twice the number of (3,3) links as with the prior cases. This lead us to the following results.

PROPOSITION 3.1. *In a straight-edge embedding of K_7, every 3-cycle is contained in 0, 2, or 4 (3,3) links.*

To see why this is true, consider a 3-cycle, A, in a straight-edge embedding of K_7. Either A is contained in a link or not. Suppose that A is contained in a (3,3) link. The four vertices not contained in A form a straight-edge embedding of K_4. There are only two possible ways that A will link with one of the faces of the tetrahedron, please see Figure 17.

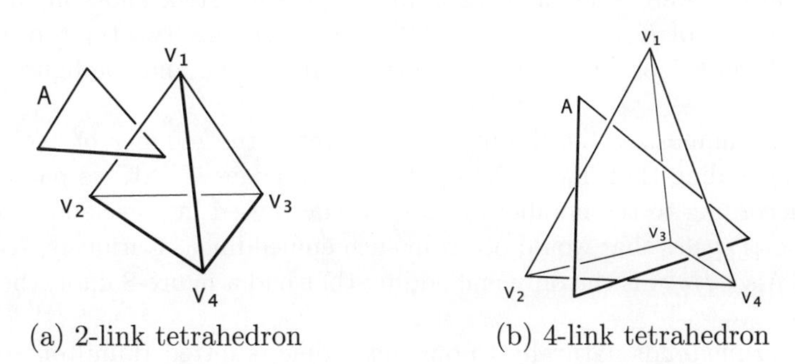

(a) 2-link tetrahedron (b) 4-link tetrahedron

FIGURE 17. The two ways a 3-cycle can appear in a link.

Figure 17(b) is particularly interesting as such a linking will create 4 (3,3) links, but only 1 (3,4) linking. In contrast, Figure 17(a) creates 2 (3,3) links and 2 (3,4) links. The reason that K_7^4 and K_7^5 were more challenging cases is due to the following.

PROPOSITION 3.2. *In a straight-edge embedding of K_7, only K_7^4 or K_7^5 can have a 4-link tetrahedron.*

There are obvious directions in which this work could continue. For the problem under consideration, each embedding of K_7 formed a convex polyhedron with seven vertices. What about an embedding of K_7 which forms a convex polyhedron with 4 vertices? That is, four of the vertices form the hull of the polyhedron and the other three vertices are internal. It seems reasonable that such embeddings are isomorphic to one of the five cases with seven external vertices, but this is not obvious. Moreover, as the number of vertices increases, it seems that one could construct an example of an embedding of a complete graph on n vertices where at least one of the vertices is internal and can not be passed to the hull of the embedding via ambient isotopies that preserve the linearity of the edges. Recently the second author and his student, R. Grotheer, constructed a subgraph of K_9 with one internal vertex that can not be passed to the surface of the hull via ambient isotopies that preserve the linearity of the edges [**19**]. So one may ask, given a straight-edge embedding G of K_n, $7 \leq n \leq 8$, with $4 \leq k \leq 8$ external vertices and $m = n - k$ internal vertices, is G always isomorphic to an embedding of K_n with n external vertices?

Another direction of study is to consider K_n, $n \geq 7$. While K_6 has only 10 disjoint triangle pairs to consider, K_7 has 70, and K_8 has 280. Moreover, with K_6 there were only (3-3) links. K_7 introduced (3-4) links and for K_8, one would have to consider (3-3), (3-4), (3-5), and (4-4) links. Whereas there were only 5 distinct convex polyhedral embeddings of K_7, it is well known there are 14 for K_8 (see, for example [**35**]). So, given a straight-edge embedding of K_n, how many (k, m) links does it contain, where $3 \leq k \leq n-3$ and $3 \leq m \leq n-k$? Clearly this is an ambitious question. Possibly a more attainable question is the following: Given a straight-edge embedding of K_n, what is an upper or lower bound for the number of (k,m) links it contains, $3 \leq k \leq n-3$ and $3 \leq m \leq n-k$?

We now turn our attention to the second question posed at the knot theory workshop: Given a straight-edge embedding of K_7, how many and what types of knots occur? Using the insight gained from the work with Arbisi, the second author and R. Grotheer were able enumerate all the possible stick knots in the straight-edge embeddings of K_7, Figure 12–16[**19**]. There are only two types of knots that can be made with 7 or fewer sticks: the trefoil requires 6 and the figure-8 requires 7.

Table 1 summarizes the findings. We counted the number of cycles possible for each embedding that had 0 though 6 internal edges. Next, we partitioned the problem according to the number of internal edges used in a cycle and then found the number of cycles that would occur in such embeddings. Curiously, K_7^1 has only *one* knot. Also, K_7^5 was the only embedding that had a figure-8 knot, the rest were all trefoils.

This work extends naturally to our next topic, knotted Hamiltonian cycles in spatial embeddings of graphs.

Internal Edges	Cycles	Knots	Cycles	Knots	Cycles	Knots	Cycles	Knots	Cycles	Knots
0	14	0	18	0	17	0	24	0	30	0
1	80	0	72	0	92	0	96	0	90	0
2	164	0	174	0	143	0	123	0	120	0
3	88	1	78	1,3	91	0,1	90	2,3	90	1,2,3,4,5
4	14	0	18	0,2	16	0,1,2	24	0,1	20	2, 4
5	0	0	0	0	1	0, 1	3	0	10	1, 5
6	0	0	0	0	0	0	0	0	0	0
	K_7^1		K_7^2		K_7^3		K_7^4		K_7^5	

TABLE 1. The number of knots appearing in a straight-edge embedding of K_7.

4. Knotted Hamiltonian Cycles in Spatial Embeddings of Graphs.

Conway and Gordon's beautiful proof that K_7 is intrinsically knotted also shows that K_7 has a knotted Hamiltonian cycle in every spatial embedding. What other graphs have this quality? As Kohara and Suzuki [**25**] point out, of the graphs obtained from K_7 by $\Delta - Y$ exchanges, all except the graph they call C_{14} are known to have embeddings without Hamiltonian knots.

In [**4**], the authors show that every embedding of K_n, for $n \geq 7$ contains a knotted Hamiltonian cycle. Here we will present the proof of this result (for background on Arf invariant, see [**1**] and [**23**]).

LEMMA 4.1. [**4**] *In every spatial embedding of K_7, there exists an edge of K_7 that is contained in an odd number of Hamiltonian cycles with non-zero Arf invariant.*

PROOF. Consider an arbitrary embedding of K_7. By Conway-Gordon's result [**9**], the sum of the Arf invariants of all Hamiltonian cycles in an arbitrary embedding of K_7 must be odd. Thus, in the given embedding there must be an odd number of Hamiltonian cycles with non-zero Arf invariant. Let's say the number of such cycles is $2n+1$. Now, if we count up the edges of such cycles, we get that a grand total of $7(2n+1)$ edges (counting multiplicities) are in a cycle with non-zero Arf invariant. On the other hand, if we number the edges of K_7 as $e_1, \ldots e_{21}$, and let n_i, $i = 1, 2, \ldots, 21$ stand for the number of Hamiltonian cycles with nonzero Arf invariant that contain e_i, then we must have that $\sum_{i=1}^{21} n_i = 7(2n+1)$, thus $\sum_{i=1}^{21} n_i$ must be odd. It follows that at least one of the n_i must be odd, and our lemma is proven.

□

THEOREM 4.1. [**4**] *Every K_n, for $n \geq 7$ contains a knotted Hamiltonian cycle in every spatial embedding.*

PROOF. We will prove the theorem for K_8. The proof for general n is similar. Embed K_8. Consider the embedding of the subgraph G_7 induced by seven vertices of K_8, and let v denote the eighth vertex, and let G_7 denote the subgraph on 7 vertices. By the previous lemma, the embedded G_7 contains an edge that is contained in an odd number of Hamiltonian cycles with non-zero Arf invariant; we denote this edge e, and let w_1 and w_2 denote the vertices of e. Now, we ignore the edge e, and consider the subdivided K_7 that results from replacing e with the edges (v, w_1) and (v, w_2). We denote this subdivided K_7 by G'_7. Ignoring the degree 2 vertex v, the embedded G'_7 must have an odd number of Hamiltonian cycles with non-zero Arf invariant. Since there was an odd number of Hamiltonian cycles of G_7 through the edge e with non-zero Arf invariant, there is an even number of Hamiltonian cycles in G_7 that do not contain e and with non-zero Arf invariant. The Hamiltonian cycles of G_7 not containing e are exactly the same as the Hamiltonian cycles in G'_7 not containing the edges (v, w_1) and (v, w_2). Thus, in the embedding of G'_7, there must be an odd number of Hamiltonian cycles through the edges (v, w_1) and (v, w_2) with non-zero Arf invariant. Such a cycle is a Hamiltonian cycle in K_8. Thus, in the original embedded K_8, there must be a knotted Hamiltonian cycle. □

We note here that Susan Beckhardt [**3**], a student at Union College, has been able to adapt Conway and Gordon's proof for K_7 to prove that K_8 has a knotted Hamiltonian cycle in every spatial embedding. She was not able to extend her result to K_9. We also note here that the proof of Theorem 4.1 can be used to show that every edge of K_9 is contained in at least two knotted Hamiltonian cycles in every spatial embedding of K_9. This can be seen by removing an edge, call it e, from K_9. The vertices disjoint from e induce a K_7 subgraph. In an arbitrary embedding of K_9, consider the embedded sub-K_7. One of its edges must lie in an odd number of Hamiltonian cycles with non-zero Arf invariant. We denote this edge f. The edges e and f are connected by 4 different edges, which we shall denote e_1, e_2, e_3, e_4. Without loss of generality, e_1 and e_2 share no vertex, and neither do e_3 and e_4. If we replace the edge f with the 4$-$ (vertex) path (e_1, e, e_2), then there is a knotted Hamiltonian cycle through the 4$-$path. Similarly, there is a knotted Hamiltonian cycle through the 4$-$path (e_3, e, e_4). Thus, there are at least two different knotted Hamiltonian cycles through the edge e. One can use an analogous argument to show that every 3-path in K_{10} is contained in at least two knotted Hamiltonian cycles in every spatial embedding, and in general, every $(n-7)-$path in K_n is contained in at least two knotted Hamiltonian cycles in every spatial embedding, for $n \geq 9$.

This reasoning allows us to estimate a minimum number of knotted Hamiltonian cycles in every spatial embedding of K_n for $n > 8$. One need only compute the number of paths of length $(n-7)$, then multiply by 2 and divide by n (because every Hamiltonian cycle in K_n contains exactly n paths of length $(n-7)$). To get double the number of paths of length $(n-7)$ in K_n, one merely computes $n(n-1)(n-2)....(8)$. Dividing by n gives our lower bound:

THEOREM 4.2. [**4**] *For $n > 8$, the minimum number of knotted Hamiltonian cycles in every embedding of K_n is at least $(n-1)(n-2)...(9)(8)$.*

QUESTION 4.1. Can the lower bound on the number of knotted Hamiltonian cycles in every spatial embedding of K_n given in Theorem 4.2 be improved?

The lower bound of at least 1 Hamiltonian knotted cycle in every spatial embedding of K_8 was reported to be improved to 3 in [4], using techniques of Shimabara [**34**]. There are, unfortunately, errors in [4] and [**34**], (thanks to Masakazu Tergaito and Kouki Taniyama for pointing them out, see [**15**]), and the best known lower bound is currently 1.

Finally, we bring up the question of whether or not every spatial embedding of $K_{3,3,1,1}$ contains a knotted Hamiltonian cycle? Kohara and Suzuki [**25**] show an embedding of $K_{3,3,1,1}$ with exactly one knotted Hamiltonian cycle in the form of a trefoil knot, and they show another embedding of $K_{3,3,1,1}$ with exactly two knotted Hamiltonian cycles, each in the form of a trefoil. Foisy's proof [**18**] that $K_{3,3,1,1}$ is intrinsically knotted does not prove that there exists a knotted Hamiltonian cycle in every spatial embedding. It is also unknown at this time if $K_{3,3,1,1}$ contains a knotted Hamiltonian cycle in every straight-edge embedding.

5. Questions and Acknowledgments

We conclude with a listing of the open questions presented in the article.

QUESTION 5.1. Determine the complete set of minor-minimal intrinsically 3-linked graphs. Are the subgraphs of K_{10} described in [**5**] minor-minimal intrinsically 3-linked?

QUESTION 5.2. Is K_{14} the smallest complete graph that contains a 3-link of triangles in every spatial embedding [**10**]? (At this point, K_{10} has not been ruled out.)

QUESTION 5.3. What is the smallest n, such that, for every straight edge embedding of K_n, there is a non-split link of 3 components? (n is at most 10, but could be 9.)

QUESTION 5.4. Given a straight-edge embedding G of K_n, $7 \leq n \leq 8$, with $4 \leq k \leq 8$ external vertices and $m = n - k$ internal vertices, is G always isomorphic to an embedding of K_n with n external vertices?

QUESTION 5.5. Given a straight-edge embedding of K_n, how many (k, m) links does it contain, where $3 \leq k \leq n - 3$ and $3 \leq m \leq n - k$?

Clearly this is an ambitious question. Possibly a more attainable question is the following:

QUESTION 5.6. Given a straight-edge embedding of K_n, what is an upper or lower bound for the number of (k,m) links it contains, $3 \leq k \leq n - 3$ and $3 \leq m \leq n - k$?

QUESTION 5.7. What is the minimum number of knotted Hamiltonian cycles in every spatial embedding of K_8? Every straight edge embedding?

QUESTION 5.8. Does every spatial embedding of $K_{3,3,1,1}$ contain a knotted Hamiltonian cycle? Every straight-edge embedding?

Finally, we would like to thank the organizers for their hard work in making the conference and this publication possible. We would also like to thank Joe for the inspiration he has given us and for making undergraduate research in mathematics a common practice.

References

[1] C. Adams, *The Knot Book*, American Mathematical Society, Providence, RI, 2004.

[2] P. Arbisi, L. Ludwig, *Linking in straight-edge embeddings of* K_7, preprint.

[3] S. Beckhardt, presentation at the Hudson River Undergraduate Mathematics Conference, Sienna College, April 21, 2007.

[4] P. Blain, G. Bowlin, J. Foisy, Joel, J. Hendricks, J. LaCombe, *Knotted Hamiltonian cycles in spatial embeddings of complete graphs*, New York J. Math. 13 (2007), 11–16 (electronic).

[5] G. Bowlin and J. Foisy, *Some new intrinsically 3-linked graphs*, J. Knot Theory Ramifications, **13**, no. 8 (2004), 1021–1027.

[6] A. Brouwer, R. Davis, A. Larkin, D. Studenmund, C. Tucker, *Intrinsically* S^1 *3-linked graphs and other aspects of* S^1 *embeddings*, Rose-Hulman Undergraduate Mathematics Journal **8**, no. 2 (2007).

[7] G. Chartrand and P. Zhang, *Introduction to Graph Theory*, McGraw-Hill, New York, NY, 2005.

[8] C. Cicotta, J. Foisy, T. Reilly, S. Rezvi, B. Wang, and A. Wilson. *Two analogs of intrinsically linked graphs,* preprint.

[9] J. H. Conway and C. McA. Gordon, *Knots and links in spatial graphs*, J. Graph Theory, **7**, no. 4 (1983), 445–453.

[10] G.C. Drummund-Cole and D. O'Donnol, *Intrinsically* n*-linked complete graphs*, preprint.

[11] E. Flapan, B. Mellor, and R. Naimi, *Intrisically linking and knotting are arbitrarily complex*, preprent, arXiv:math/0610501v5

[12] E, Flapan, R.Naimi and J. Pommersheim, J. *Intrinsically triple linked complete graphs*, Top. App., **115** (2001), 239–246.

[13] E. Flapan, J. Foisy, R. Naimi, J. Pommersheim, *Intrinsically* n*-linked graphs*, J. Knot Theory Ramifications **10**, no. 8 (2001), 1143-1154.

[14] T. Fleming, and A. Diesl, *Intrinsically linked graphs and even linking number*, Algebr. Geom. Topol., **5** (2005), 1419–1432.

[15] J. Foisy *Correction to "Knotted Hamiltonian cycles in spatial embeddings of complete graphs"*, arXiv:0807.1483v1.

[16] J. Foisy *Graphs with a knot or a three-component link in every spatial embedding*, J. Knot Theory Ramifications **15**, no. 9 (2006), 1113-1118.

[17] J. Foisy, *A newly recognized intrinsically knotted graph*, J. Graph Theory, **43**, no. 3 (2003), 199–209.

[18] J. Foisy, *Intrinsically knotted graphs*, J. Graph Theory **39**, no. 3 (2002), 178–187.

[19] R. Grotheer and L. Ludwig, L. *The complete story of stick knots in* K_7, preprint.

[20] J. Hespen, T. Lalonde, K. Sharrow, and N. Thomas, *Graphs with (edge) disjoint links in every spatial embedding*, preprint 1999.

[21] C. Hughes, *Linked triangle pairs in a straight edge embedding of* K_6, Pi Mu Epsilon J., **12**, no. 4 (2006), 213–218.

[22] Y. Huh, and C. B. Jeon, *Knots and links in linear embeddings of* K_6, J. Korean Math. Soc., **44**, no. 3 (2007), 661-671.

[23] L. Kauffman, *Formal Knot Theory*, Mathematical Notes, 30, Princeton University Press, Princeton, New Jersey, (1983), MR0712133 (85b:57006), Zbl 0537.57002. Reprinted by Dover (2006).

[24] E. de Klerk, J. Maharry, D. Pasechnik, R.B. Richter, G. Salazar, *Improved bounds for the crossing numbers of* $K_{m,n}$ *and* K_n, SIAM Journal of Discrete Mathematics **20**, no. 1 (2006), 189-202.

[25] T. Kohara and S. Suzuki, *Some remarks on knots and links in spatial graphs*, Knots 90 (Osaka, 1990), 435–445.

[26] K. Kuratowski, *Sure le problème des courbes gauches en topologie*, Fundamenta Mathematicae **15** (1930), 271-283.

[27] R. Motwani, A. Raghunathan, and H. Saran, *Constructive results from graph minors: Linkless embeddings*, 29th Annual Symposium on Foundations of Computer Science, IEEE, 1988, pp. 398-409.

[28] D. O'Donnol, *Intrinsically* n*-linked complete bipartite graphs*, preprint, arXiv:math/0512205v2.

[29] M. Ozawa, Y. Tsutsumi, *Primitive spatial graphs and graph minors*, Rev. Mat. Complut. **20**. no. 2 (2007), 391-406.
[30] N. Robertson, P. Seymour, *Graph minors XX. Wagner's conjecture*, J. Combin Theory Ser B **92** no. 2 (2004), 325–357.
[31] N. Robertson, P. Seymour,and R. Thomas, *Sachs' linkless embedding conjecture*, J. Combin. Theory Ser. B, **64**, no. 2 (1995), 185–227.
[32] H. Sachs, *On spatial representation of finite graphs*, (Proceedings of a conference held in Lagow, February 10-13, 1981, Poland), Lecture Notes in Math., 1018, Springer-Verlag, Berlin, Heidelberg, New York, and Tokyo (1983).
[33] H. Sachs, *On spatial representations of finite graphs, finite and infinite sets*, (A. Hajnal, L. Lovasz, and V. T. Sós, eds), colloq. Math. Soc. János Bolyai, vol. 37, North-Holland, Budapest, (1984), 649-662.
[34] M. Shimabara, *Knots in certain spatial graphs* Tokyo J. Math. Vol. 11, No 2 (1988), 405-413.
[35] N. J. A. Sloane, (2007), The On-Line Encyclopedia of Integer Sequences, published electronically at www.research.att.com/~njas/sequences/.

(Foisy) Mathematics Department, SUNY Potsdam, Potsdam, NY 13676 foisyjs@potsdam.edu

(Ludwig) Department of Mathematics and Computer Science, Denison University, Granville, OH 43023 ludwig_l@denison.edu

Contemporary Mathematics
Volume **479**, 2009

Can an Asymmetric Power Structure Always be Achieved?

Jane Friedman and Cameron Parker

Dedicated to Joseph Gallian, in appreciation for all he has done for a generation of young mathematicians.

"Power to the People"
– John Lennon

Much of the public attention focused on voting theory relates to Arrow's impossibility theorem [**1**] and similar theorems that have been developed in the fifty years since that landmark paper. Those results involve the paradoxes that occur when a voting system is created to decide between three or more candidates or proposals where all voters have equal weight. This paper looks at a different question; we consider the situation where we are deciding between only two possible choices and an unequal apportionment of power is desired.

Why would an asymmetric apportionment of power result in a fair system? We give two examples below to answer that question. In the first example, the voting system attempts to give more power to the individuals who have more at stake, and in the second the system should favor the individuals who are representing more constituents. After we determine that asymmetric power is needed, we investigate when it is actually possible.

1. Introductory Examples

Three friends–Carl Moneybags, Barbara Average, and Andrew Pauper–form an investing club. Their plan is to pool their money and democratically decide how to invest the funds. Due to their differing financial situations, they contribute different amounts of money. Carl contributes more than Barbara, who contributes more than Andrew. It then seems only fair that Carl should have more say than Barbara about how the funds are invested,

2000 *Mathematics Subject Classification.* Primary 91B12; secondary 91A80.

We would like to thank the editors and referees for their helpful suggestions. This paper is greatly improved thanks to their efforts.

and Barbara should have more say than Andrew. They can then take a proposed investment and vote up or down on it. But how will they decide? If each has one vote and majority rules, any investment with two out of three "yes" votes will pass, and Andrew will have just as much power as Carl. It might instead seem fair to give each investor a number of votes proportional to their investment. For example, assume their contributions are such that Andrew is given one vote, Barbara two votes and Carl three votes. Suppose we say now that an investment will be made if it gets four votes, a majority. In this case, an investment will be made only if it is favored by Carl and either one of the others, so Barbara and Andrew have the same power. Suppose instead we require five votes to approve an investment. To achieve five votes, both Carl and Barbara must vote yes, and what Andrew thinks will never make a difference. So Carl and Barbara will have the same power and poor Andrew will have none! The simplest approaches to achieving the desired power structure for the investors do not seem to work. Perhaps a more clever scheme is needed.

The problem revealed in the investors example also appears when dividing up power among members of an elected body. Imagine a region which is divided into two districts. District A contains 1000 voters, and district B contains 10,000 voters. Suppose the voters of each district elect a representative to the governing council. Each person has one vote, but the votes of voters of district A seem to be worth ten times as much as those of the district B voters. Indeed, this was quite common during much of American history, when large urban districts and small rural districts each elected one member of Congress. This practice was ended by the 1964 decision of *Reynolds v. Sims* which introduced the principle of one man/one vote (or better yet, one person/one vote).

The most common response to the *Reynolds v. Sims* decision was to create districts of equal population. This method has some problems. The most notable is gerrymandering, or the shaping of districts for political purposes. Another problem arises when certain districts are too fundamental to break up, such as when countries send delegates to an international body. In this case, each country may want its views represented by a single delegate.

This problem leads to another possible solution: weighting the votes so that representatives' votes are proportional to their districts' populations. We might give the representative of district B, in the example above, ten times as many votes as the representative of district A. If power is proportional to the number of votes, then assigning a number of votes proportional to the district's population should ensure that we have satisfied the one person/one vote principle.

Unfortunately, it turns out that power is not proportional to the number of votes. This important fact was highlighted for the political and legal communities by John Banzhaf in his seminal paper [**2**]. In this paper Banzhaf discusses what has become a famous example, the Nassau County Board of

Supervisors. This example has become ubiquitous in the literature, particularly in books and articles aimed at general audiences (see for example [**18**] pg. 61 or [**5**] pg. 359).

In the 1960s, Nassau County in Long Island was governed by a Board of Supervisors. As described in [**2**], the county was divided into six districts, and each had a representative with votes proportional to the district's population. A majority of the votes (58 votes) was required to pass a motion.

- Hempstead 1 - 31 votes
- Hempstead 2 - 31 votes
- Oyster Bay - 28 votes
- North Hempstead - 21 votes
- Long Beach - 2 votes
- Glen Cove - 2 votes

The above voting system is designed to give Hempstead 1 and Hempstead 2 equal power, but more power than Oyster Bay, which in turn has more power than North Hempstead, which has more than Long Beach and Glen Cove. Was this goal achieved? No! Notice that to get a proposition to pass, 58 votes are needed. This can only occur if at least two of the three largest districts vote for the proposition. How the three smaller districts vote will never matter.

So, all the power is held by the three most populous districts, and none is held by the three least populous districts. North Hempstead with twenty-one votes has just as much power (no power) as Long Beach with two votes. In addition, Oyster Bay has the same power as the two Hempstead representatives, even though it has fewer votes.

2. Weighted and Simple Voting Games

The Nassau County Board of Supervisors is an example of a real situation that can be modeled by a weighted voting game (WVG). A WVG consists of a finite set of voters (or players), $N = \{P_1, P_2, \ldots, P_n\}$. Each player controls a number of votes called that player's weight. Let $w(P_i) = w_i$ denote the weight of player i. The convention is to label the players so that $w_1 \geq w_2 \geq \ldots \geq w_n$. The players vote yes or no on a bill or measure and may not abstain (see [**8**] and [**10**] for a discussion of some of the issues involved in allowing abstentions). The quota q, where $0 < q \leq w_1 + w_2 + \ldots + w_n$, is the minimal number of votes needed to pass a measure. We denote such a system by $[q : w_1, w_2, \ldots, w_n]$. Thus, the Nassau County Board of Supervisors can be represented by $[58 : 31, 31, 28, 21, 2, 2]$.

In a WVG, a coalition is any subset of N. A winning coalition is one which controls at least q votes and a losing coalition is one which controls less than q votes. We can extend the definition of the weight function w so that if X is a coalition, then $w(X)$ is the sum of the weights of the players in X. Hence X is winning if $w(X) \geq q$ and losing otherwise. Banzhaf's idea of how to measure power (the Banzhaf power index[1]) is that a player is powerful when that player is part of a winning coalition which would become losing should that player defect. Such a player is called a *critical player* in that coalition. Banzhaf's index calculates a player's power by counting the number of times that player is critical in any coalition, and divides this by the total number of times all players are critical. If we look back at the Nassau County Board of supervisors, we see that Glen Cove will never be a critical player and thus would have zero power, which is not proportional to its weight.

As we see from the Nassau County example, before we even consider whether each player has the correct proportion of power, we should first decide if a player who is supposed to have more power than another player actually does. This is addressed in the following definition.

DEFINITION 2.1. Player A has more power than player B, denoted $A > B$, if there exists a losing coalition X, such that $X \cup A$ is winning, but $X \cup B$ is losing. We say that X distinguishes A from B. Additionally, we say that A and B have the same power if for all coalitions X containing neither A or B, $X \cup \{A\}$ is winning if and only if $X \cup \{B\}$ is winning. If A has the same or more power than B, we write $A \geq B$.

In a WVG, if two players have the same weight, then they have the same power. But, as we have seen, it is possible for two players to have different weights but still have the same power. A WVG that satisfies the additional property that $A > B$ if and only if $w(A) > w(B)$ is said to be *transparent.*

Banzhaf's measure and most other commonly used measures, including the important Shapely-Shubik measure,[2] [**16**] agree on the relative power of players as determined by the above definition (see [**7**] or [**24**]). With this definition, we can talk about the hierarchy which is induced by a WVG. Suppose in a given WVG, P_1 and P_2 have the same power, which is more than P_3 and P_4 which have the same power, all of which have more power than P_5. This WVG would induce the hierarchy $=>=>$. So a hierarchy for an n-player game is a sequence of $m = n - 1$ symbols, all of which are either $>$ or $=$.

EXAMPLE 2.2. $[25 : 16, 8, 4]$. Here all three players are needed to meet the quota, so all three have equal power. The hierarchy induced by this WVG is $==$.

EXAMPLE 2.3. $[10 : 8, 6, 2, 2]$. $P_1 > P_2$, since if $X = \{P_4\}$, then $X \cup \{P_1\} \in \mathcal{W}$ but $X \cup \{P_2\} \notin \mathcal{W}$. But all the other players have equal power. So the hierarchy induced here is $>==$.

---

[1]The methodology of this paper was first proposed in [**14**].

[2]For a detailed discussion of Shapely-Shubik measure see [**15**]. See [**17**] for a comparison of the Shapely-Shubik and Banzahf power measures. For other examples of power measures see [**9**], [**21**] or [**22**].

Notice that in the definition of "greater power," no mention of players' weights is needed. The definition only relies on knowing which coalitions are winning and which are losing. Thus, this definition and hence the definition of hierarchy makes sense in a more general context, that of simple voting games.

DEFINITION 2.4. A simple voting game (SVG) is a set of players N, together with a set of winning coalitions (subsets of N) denoted by $\mathcal{W}$, satisfying the following axioms:

(1) $N \in \mathcal{W}$
(2) $\emptyset \notin \mathcal{W}$
(3) if $X \subset Y$ and $X \in \mathcal{W}$ then $Y \in \mathcal{W}$.

It is easy to see that every WVG is an SVG. Since the quota q is such that $0 < q \leq w(N)$, axioms 1 and 2 are satisfied. Also if $X \in \mathcal{W}$ and $X \subseteq Y$ then $w(Y) \geq w(X) \geq q$ so $Y \in \mathcal{W}$.

Greater power as defined in Definition 2.1 is not well-defined for general SVGs, as can be seen in the following example.

EXAMPLE 2.5. Consider a system with four players, A, B, C, D. Suppose the winning coalitions are those which contain either $\{A, C\}$ or $\{B, D\}$. So $A > B$ by the distinguishing set $X = \{C\}$ but also $B > A$ by the distinguishing set $Y = \{D\}$.

We must impose an additional constraint on SVGs to make $>$ well defined. The following is exactly what is required.

DEFINITION 2.6. A swap robust SVG is an SVG such that if one player from each of two (not necessarily disjoint) winning coalitions are exchanged, at least one of the two resulting coalitions remains winning.

This additional condition outlaws SVGs like the one in Example 2.5. In that example, $\{A, C\}$ and $\{B, D\}$ are both winning, but if we exchange A, D we get the two losing coalitions $\{C, D\}$ and $\{A, B\}$. Therefore this SVG is not swap robust.

Note that every WVG is a swap robust SVG, since in a trade between two winning coalitions, either we trade two players with the same weight and thus both resulting coalitions are winning, or the coalition that receives the player with the larger weight will be winning after the trade. However, there are swap robust SVGs that are not WVGs[3](see [**21**] pg. 189 for an example or [**20**] for an infinite class of examples).

If we have a swap robust SVG, we can then, independently of the power index chosen, order the players by relative power. So we can extend the definition of hierarchy to this larger class of games in a natural way. This definition of a hierarchy for an SVG is from [**13**] and appears implicitly in [**4**].

[3]WVGs are characterized by a stronger property than swap robust, called trade robust (see [**19**] for details).

We can characterize $B \geq A$ in terms of power in another way. For all coalitions X which contain A and not B, if we form a coalition X by swapping B for A but making no other changes (i.e. $Y = X \backslash \{A\} \cup \{B\}$), then $B \geq A$ implies that if X is winning then Y must be winning. This is because otherwise $X \backslash \{A\}$ would distinguish A from B and we would have $A > B$.[4] In this situation, where Y is formed from X by swapping players A and B with $B \geq A$, we would also say that coalition $Y \geq X$.

3. Possible and Impossible Hierarchies

Returning to our three investors, they would like to set up a voting system with a $>>$ hierarchy. We will let C stand for Carl, B for Barbara and A for Andrew. Since $B > A$, we must have a losing coalition X which contains neither B nor A, such that $X \cup \{B\}$ is winning but $X \cup \{A\}$ is losing. There are only two possibilities for X.

Case 1: $X = \{\}$. This would mean that $\{B\}$ is winning, and hence all coalitions containing B must be winning (B is said to be a passer) and thus $C > B$ is impossible.

Case 2: $X = \{C\}$. In this case $\{C, A\}$ is losing. Thus any coalition not containing B is losing (B is said to be vetoer), and thus $C > B$ is impossible.

The hierarchy $>>$ therefore cannot be achieved by any SVG.

* * *

Suppose that Doris Supperrich (we will call her D) wishes to join the investment group. She contributes even more money than Carl. The group hopes her infusion of cash will allow more lucrative investments, but her entrance into the group is contingent on having the most influence on decisions. Perhaps they can create a voting system which will reflect the power structure they now desire, namely $>>>$.

First notice that no one-person coalition, except possibly $\{D\}$, can be winning.

Since D and C are old friends, the others are concerned that they will always vote together and therefore skew the decisions. They wonder if it would be possible to construct the hierarchy $>>>$, in such a way that $S = \{D, C\}$ is a losing coalition. This would imply that all two-person coalitions are losing, since all of them have less influence than S. Then the only possible coalition which could distinguish C from B would be $\{D, A\}$ and the only possible coalition which could distinguish B and A is $\{D, C\}$. But

[4]For coalitions X and Y, another way to express that Y has at least as much power as X is to let 1_X be the indicator function on whether or not X is a winning coalition, then X has at least as much power as Y if $1_X \geq 1_Y$. If the hierarchy is known, then it can be shown that X has at least as much power as Y if X can be formed from Y by a sequence of, adding players, swapping players from Y with players with at least as much power and removing players with no power. See either [**3**] or [**22**] for details.

this is contradictory, since the first implies that $\{D, C, A\}$ wins, while the second implies that this coalition loses. So S must be a winning coalition.

We might wonder if there are any other two-person winning coalitions. If $\{D, B\}$ is losing, then all two-person coalitions other than S will be losing, since they all have less influence than $\{D, B\}$.

Let's assume that $\{D, B\}$ is winning. If we wish to distinguish C from B, the only possible distinguishing set would be $\{A\}$, which implies that $\{C, A\}$ and thus $\{D, A\}$ wins. But now there is no set which can distinguish B from A.

So we see that $\{D, B\}$ is a losing coalition, and thus S is the only two-person winning coalition and there are no one-person winning coalitions.

Given this, can we distinguish B from A? The distinguishing set must have at least two players and must be a subset of $S = \{D, C\}$. But the only such set is S, which is a winning coalition and thus cannot be the distinguishing set. We cannot achieve $>>>$ with any SVG.

* * *

Ethel Uberwealth has heard about the group and asks to join and contribute even more money than Doris Superrich. They tell her that they have decided to disband, since it is impossible to come up with a voting structure which will accurately reflect the money they have contributed. "Wait," says Ethel, displaying the insight that led to her enormous fortune. "If I join, it is possible! Suppose we have a quota of 9 and we assign 5,4,3,2,1 votes to each of us, creating the WVG $[9 : 5, 4, 3, 2, 1]$ then we will have the desired hierarchy." And amazingly, she is right.

In fact, the n-player strict hierarchy[5], $>^{n-1}$ (where $>^{n-1}$ denotes a string of $n - 1$ $>$'s) is always possible for all $n \geq 5$. Are $>>$ and $>>>$ the only impossible hierarchies? No, indeed if Ethel had wished to contribute the same as Doris and thus wanted equal power (or if Ethel, Doris and Carl all contributed the same) their desired hierarchy would again turn out to be impossible and the investing club would once again have to fold. As the following theorem shows, there are two infinite classes of impossible hierarchies.

THEOREM 3.1. *For all $m \geq 0$ the hierarchies:*

$$=^m>> \quad \text{and} \quad =^m>>> \tag{1}$$

cannot be achieved by any SVG (where $=^m$ denotes a string of m $=$'s).

PROOF. We will only prove that $=^m>>$ is impossible. (The proof that $=^m>>>$ is impossible is similar and can be found in [**13**].) Suppose that we have a larger group of investors. We have $n = m + 1$ investors who all

[5]In [**23**], Tolle constructs all Banzhaf power distributions for 4-player games. In particular, he shows that the 4-player strict hierarchy is impossible. He also shows that strict hierarchies exist for five and six player games, leaving the existence of strict hierarchies with more players as an open question.

contribute exactly the same amount of money and thus should have equal say in the investment decisions. We will call all these players C's. Suppose there is also a player B who contributes less money than any C player and another player A who contributes even less money than player B.

We must be able to distinguish player B and player A, thus there must be a coalition consisting only of C players, call this coalition X such that $X \cup \{B\}$ is winning but $X \cup \{A\}$ is losing. Suppose X consists of exactly k C players where $0 \leq k \leq n$. Then, if $k < n$, the coalition Y consisting of exactly $k+1$ C players must also be winning. Thus a coalition is winning if and only if it contains at least $k+1$ players which are all C's or B. If $k = n$, then any winning coalition must contain all the C players and player B. In either case, B and C are symmetric in determining if a coalition is winning, so we cannot distinguish the C's from B. □

We have two classes of hierarchies which are impossible in any SVG and therefore in any WVG. Indeed these are the only impossible hierarchies. In fact, [**13**] contains a constructive proof that all other hierarchies are possible in some WVG and hence for SVGs.

THEOREM 3.2. *All hierarchies other than those in (1) can be achieved by a WVG*[6].

PROOF OUTLINE. We will outline the proof from [**13**] here.

Throughout this proof we will consider games with n players and therefore the hierarchy will contain $n-1$ symbols.

Step 1. We show that for $n \geq 5$ and for $n = 2$ the strict hierarchy $>^{n-1}$ is possible. The WVGs $[2 : 2, 1]$, $[9 : 5, 4, 3, 2, 1]$ and $[12 : 6, 5, 4, 3, 2, 1]$ all induce strict hierarchies.

For $n > 6$. Consider the WVG $[q_n : n, n-1, \ldots, 1]$, (so $w(P_i) = n-i+1$) where:

$$q_n = \begin{cases} j(j+1) & \text{if } n = 2j \\ j^2 & \text{if } n = 2j-1. \end{cases}$$

Define $X = \{P_1, P_3, P_5, \ldots, P_{2j-1}\}$ then $w(X) = q_n$. If we replace P_{2k-1} by P_{2k} in X we will have a losing coalition. Therefore $X/\{P_{2k-1}\}$ distinguishes P_{2k-1} from P_{2k}. Consider $Y = X/\{P_{2k+1}, P_{2k+3}\} \cup \{P_{2k}, P_{2k+4}\}$. Since $w(P_{2k}) = w(P_{2k+1}) + 1$ and $w(P_{2k+4}) = w(P_{2k+3}) - 1$, $w(Y) = q_n$. So $Y/\{P_{2k}\}$ distinguishes P_{2k} from P_{2k+1} when $k < j-2$. A similar argument works for $k \in \{j-2, j-1\}$. Thus all strict hierarchies except $>>$ and $>>>$ are possible.

Step 2. We show that if a particular hierarchy, $\mathcal{H}$ can be achieved, then so can any hierarchy that can be formed from $\mathcal{H}$ by adding a single "=" anywhere in the hierarchy. This is done by adding a player of equal weight to one of the other players (in the position where the "=" is to occur) and

[6]In [**12**] it is shown that all hierarchies except four are achievable in the class of weakly linear SVGs which contains the class of swap robust SVGs.

then adding the same amount to the quota. It can be shown that this method works as long as the original WVG satisfies two properties:

(1) The set of weights (with the multiplicities removed) are consecutive integers starting at 1. WVGs satisfying this property are called *weight minimal.*
(2) The WVG is transparent.

For example $[4 : 3, 2, 1]$ induces $>=$, so this WVG is weight minimal but not transparent.

If we add an "=" to a transparent, weight minimal WVG, the resulting WVG is also transparent and weight minimal. All the hierarchies created in step 1 are transparent and weight minimal.

For example, since $[9 : 5, 4, 3, 2, 1]$ is a transparent, weight minimal WVG which induces the hierarchy $>>>>$, it follows that $[13 : 5, 4, 4, 3, 2, 1]$, is a transparent, weight minimal WVG which induces the hierarchy $>=>>>$.

Step 3. If we wish to create any hierarchy with $m \geq 4$ or $m = 1, >$'s, we can do so by starting with the strict hierarchy $>^m$, and adding equals as in step 2.

The same method works for hierarchies with two or three $>$'s with different base cases. If the desired hierarchy ends with (farthest to the right) either $>=$ or $==$, then we use the base case $[5 : 3, 2, 1, 1]$ which induces the hierarchy $>>=$. If instead the desired hierarchy ends with $=>$, we use $[5 : 3, 2, 2, 1]$ which induces the hierarchy $>=>$. The only other way the hierarchy could end is $>>$, but then it is in the family of hierarchies in (1), and so is impossible to achieve.

A similar argument works for three $>$'s (see [**13**] for details). □

EXAMPLE 3.3. To create $=>=>==$ we can start with $[5 : 3, 2, 1, 1]$, which induces the hierarchy $>>=$, and then add equals as in step 2. This gives $[11 : 3, 3, 2, 2, 1, 1, 1]$, which is transparent and weight minimal and induces the hierarchy $=>=>==$.

4. Gridlock and Majority Quotas

The construction used in the proof of Theorem 3.2 leads to WVGs where the quota is an arbitrarily large proportion of the total weight.

EXAMPLE 4.1. Suppose we use the construction in Theorem 3.2 to create a WVG for the hierarchy $=^{100}>>>>$. We will get a WVG with 101 players of weight five and a quota of 509, which is 99% of $w(N)$. In contrast, in the base WVG $[9 : 5, 4, 3, 2, 1]$, the quota is 60% of $w(N)$. If these systems represented legislative bodies, it is clear that it would be much harder to get the support to pass a bill in the constructed WVG than the base WVG.

WVGs with quotas which are a large proportion of $w(N)$ also have an increased probability of gridlock. Gridlock occurs when the players in a WVG can be partitioned into two disjoint coalitions, neither of which has enough votes to pass a motion. Suppose one coalition contains voters who wish to vote yes on some bill B and its complement contains voters who wish to vote yes on the bill not B. We will have a situation where neither B nor not B can be enacted. This is gridlock. The higher the quota is with respect the weights, the harder it is to pass a motion, and the more likely that gridlock will occur. In the WVG constructed in Example 4.1 gridlock is very likely; since unless at least 100 of the 101 players of weight five agree, gridlock is ensured.

In some cases, a high probability of gridlock is desired. For example, many states and municipalities require a 2/3 vote by either the legislative body or the voters to pass a tax increase. Such a rule makes tax increases hard to pass and gives preference to the status quo. In other cases gridlock is undesirable; this is especially the case if a decision must be made and there is no status quo which should be given special preference.

A majority quota for a WVG with weights $w_1, w_2, \ldots, w_n$ is defined by

$$q = \left\lceil \frac{w_1 + w_2 + \ldots + w_n + 1}{2} \right\rceil.$$

A proper quota is one that is at least as big as the majority quota. An improper quota is one which is smaller than the majority quota. In real applications we usually consider only proper quotas, since improper quotas create that possibility that a bill and its negation could both pass. (See **[21]** pg. 59, however, for a real example of an improper quota.) A WVG with the majority quota will have the least probability of gridlock of WVGs with proper quotas. This is because for any coalition X, either X or X^c is winning, unless $w(X) = w(X^c)$. We now turn our attention to finding which hierarchies can be achieved with a majority quota, thus which hierarchies can be achieved with voting systems with as little gridlock as possible.

Returning to the case of Nassau County, the desired hierarchy was $=>>>=$. It turns out this hierarchy cannot be achieved by a majority quota (which is the quota that the designers of the system wanted) no matter what weights are assigned to the different districts.

To see that $=>>>=$ is impossible for a majority quota WVG, suppose $\{D_1, D_2, C, B, A_1, A_2\}$ induces the hierarchy $=>>>=$ with a majority quota and suppose the weights are ordered by $w(D_1) \geq w(D_2) > w(C) > w(B) > w(A_1) \geq w(A_2)$. (Remember if one player has more power than another, he/she must have more weight, but two players can have different weights and still have the same power.) Since $B > A_1$, there is a coalition $X \subseteq \{D_1, D_2, C, A_2\}$ such that $X \cup \{B\} \in \mathcal{W}$ but $X \cup \{A_1\} \notin \mathcal{W}$. Consider the coalition $\{D_1, D_2, A_1\}$ and its complement $\{C, B, A_2\}$. Since $w(\{D_1, D_2, A_1\}) > w(\{C, B, A_2\})$ and we have a majority quota, $\{D_1, D_2, A_1\}$ must be winning and $\{C, B, A_2\}$ losing. Thus X must contain exactly one of the D's.

Furthermore, since $w(\{D_1, C, A_1\}) > w(\{D_2, B, A_2\})$ it must be the case that $\{D_1, C, A_1\}$ must be winning and $\{D_2, B, A_2\}$ is losing. Since $D_1 = D_2$ in the sense of power, we can swap them without affecting whether or not the coalition is winning. Hence $\{D_2, C, A_1\}$ must be winning and $\{D_1, B, A_2\}$ must be losing. Thus X cannot contain C. Thus the largest X can be is $X = \{D_1, A_2\}$ and the largest $X \cup \{B\}$ can be is $\{D_1, B, A_2\}$, which is losing. So $=>>>=$ is impossible with a majority quota.

The following theorem, proved in [**3**], characterizes all hierarchies achievable by a WVG with a majority quota.

THEOREM 4.2. *All hierarchies can be achieved with a majority quota except:*

$$
\begin{gathered}
=^m>> \\
=^m>>> \\
=^m>>>> \\
=^{2m}>>= \\
=^{2m+1}>=>> \\
=^{2m+1}>>>=
\end{gathered}
$$

where $m \geq 0$.

We see it is impossible to achieve the hierarchy $=>>>=$ with a majority quota. It is possible, however, with a quota which is one more than a majority quota. $[9 : 4, 4, 3, 2, 1, 1]$ achieves this hierarchy with a quota of nine instead of the majority quota of eight. This is not a coincidence. It is shown in [**3**] that all hierarchies, except those in (1), which cannot be achieved by a majority quota, can be achieved by a quota which is one more than the majority quota.

5. Final Remarks

This work is in the spirit of Joseph Gallian's pioneering efforts in providing undergraduate research opportunities. As can be seen here, mathematics applied to political science and other social sciences yields interesting results that require little technical background. These questions are ideal for encouraging and inspiring young mathematicians. For example [**23**] formed the basis for an REU project at Valparaiso University(see [**6**]). At our own institution, students were motivated by [**11**] to undertake simulation studies comparing different voting schemes.

References

1. Kenneth J. Arrow, *A difficulty in the concept of social welfare*, The Journal of Political Economy **58** (1950), no. 4, 328–346.
2. John Banzhaf, *Weighted voting doesn't work: a mathematical analysis*, Rutgers Law Review **19** (1965), 317–343.
3. Dwight Bean, Jane Friedman, and Cameron Parker, *Simple majority achievable hierarchies*, Theory and Decision (2008), to appear.

4. Francesc Carreras and Josep Freixas, *Complete simple games*, Math. Social Sci. **32** (1996), no. 2, 139–155. MR MR1408858 (97d:90116)
5. COMAP, *For all practical purposes: Introduction to contemporary mathematics*, 3rd ed., W.H. Freeman and Company, New York, 1994.
6. A Cuttler, A De Guire, and S Rowell, *Using sets of winning coalitions to generate feasible banzhaf power distributions*, http://www.valpo.edu/mcs/pdf/finalbanzhafpaper.pdf, 2005.
7. Lawrence Diffo Lambo and Joël Moulen, *Ordinal equivalence of power notions in voting games*, Theory and Decision **53** (2002), no. 4, 313–325. MR MR1994396 (2004g:91007)
8. Dan S. Felsenthal and Moshé Machover, *Ternary voting games*, Internat. J. Game Theory **26** (1997), no. 3, 335–351. MR MR1467834 (98k:90005)
9. ______, *The measurement of voting power*, Edward Elgar Publishing Limited, Cheltenham, 1998. MR MR1761929 (2001h:91032)
10. ______, *Models and reality: the curious case of the absent abstention*, Power Indices and Coalition Formation, Kluwer, Dordrecht, 2001, pp. 87–104.
11. Peter C. Fishburn and Steven J. Brams, *Paradoxes of preferential voting*, Mathematics Magazine **56** (1983), no. 4, 207–214.
12. Josep Freixas and Montserrat Pons, *Hierarchies achievable in simple games*, Theory and Decision (2008), to appear.
13. Jane Friedman, Lynn McGrath, and Cameron Parker, *Achievable hierarchies in voting games*, Theory and Decision **61** (2006), no. 4, 305–318. MR MR2285280 (2007j:91029)
14. L.S. Penrose, *The elementary statistics of majority voting*, J. Roy. Statist. Soc. Ser. A **109** (1946), no. 1, 53–57. MR MR836481
15. Alvin E. Roth (ed.), *The shapley value*, Cambridge University Press, Cambridge, 1988, Essays in honor of Lloyd S. Shapley. MR MR989818 (89j:90274)
16. Lloyd S. Shapley and Martin Shubik, *A method for evaluating the distribution of power in a committee system*, American Political Science Review **48** (1954), 787–792.
17. Philip D. Straffin, Jr., *The Shapley-Shubik and Banzhaf power indices as probabilities*, The Shapley value, Cambridge Univ. Press, Cambridge, 1988, pp. 71–81. MR MR989823 (90a:90017)
18. Peter Tannenbaum, *Excursions in modern mathematics with mini-excursions*, 6th ed., Prentice Hall, New Jersey, 2007.
19. Alan Taylor and William Zwicker, *A characterization of weighted voting*, Proc. Amer. Math. Soc. **115** (1992), no. 4, 1089–1094. MR MR1092927 (92j:90021)
20. ______, *Simple games and magic squares*, J. Combin. Theory Ser. A **71** (1995), no. 1, 67–88. MR MR1335777 (96c:90152)
21. Alan D. Taylor, *Mathematics and politics*, Textbooks in Mathematical Sciences, Springer-Verlag, New York, 1995. MR MR1344686 (96j:90013)
22. Alan D. Taylor and William S. Zwicker, *Simple games*, Princeton University Press, Princeton, NJ, 1999. MR MR1714706 (2000k:91028)
23. John Tolle, *Power distribution in four-player weighted voting systems*, Mathematics Magazine **76** (2003), no. 1, 33–39. MR MR2084115
24. Yoshinori Tomiyama, *Simple game, voting representation and ordinal power equivalence*, International Journal on Policy and Information **11** (1987), no. 1, 67–75.

Department of Mathematics and Computer Science, University of San Diego, San Diego, California, 92110-2492 USA

E-mail address: janef@sandiego.edu

E-mail address: cparker@sandiego.edu

Contemporary Mathematics
Volume **479**, 2009

McKay's Canonical Graph Labeling Algorithm

Stephen G. Hartke and A. J. Radcliffe

ABSTRACT. The problem of deciding whether two graphs are isomorphic is fundamental in graph theory. Moreover, the flexibility with which other combinatorial objects can be modeled by graphs has meant that efficient programs for deciding whether graphs are isomorphic have also been used to study a variety of other combinatorial structures. Not only is the graph isomorphism problem a very practical one, it is also fascinating from a complexity-theoretic point of view. Graph isomorphism is one of the few problems that are clearly in NP but not known either to be solvable in polynomial time, or to be NP-complete.

Various people have worked to create algorithms for graph isomorphism which are "practical in practice". One of the most powerful and best known of these algorithms is due to Brendan McKay. It is known that his algorithm has exponential running time on some inputs, but it performs exceptionally well under most circumstances. In this article we aim to provide an introduction to the essential ideas of McKay's algorithm.

1. Introduction

The problem of deciding whether two graphs are isomorphic is fundamental in graph theory. Moreover, the flexibility with which other combinatorial objects can be modeled by graphs has meant that efficient programs for deciding whether graphs are isomorphic have also been used to study and enumerate (see [**16**]) a variety of other combinatorial structures, including distance-regular graphs [**5**], strongly regular graphs [**7**], block designs [**4**], Latin squares [**18**], partial geometries [**24**], and integer programs [**21**].

Not only is the graph isomorphism problem (which we abbreviate GI henceforth) a very practical one, it is also fascinating from a complexity-theoretic point of view. It is one of the few problems that are clearly in NP but not known either to be solvable in polynomial time, or to be NP-complete[1].

Key words and phrases. graph isomorphism, graph automorphism, canonical labeling, canonical isomorph, nauty.

The first author is partially supported by a Maude Hammond Fling Faculty Research Fellowship from the University of Nebraska Research Council.

[1]GI is in NP since an explicit isomorphism serves as a certificate that two graphs are isomorphic.

There is a large amount of heuristic evidence that GI is not NP-complete. For example GI behaves markedly differently from problems that are known to be NP-complete, in that seemingly mild constraints on the problem allow polynomial time algorithms. GI is polynomial time for: graphs of bounded degree [**13**], graphs of bounded genus [**8, 19**], graphs of bounded eigenvalue multiplicity [**2**], and graphs of bounded treewidth [**3**]. Contrast this behavior with that of, say, the graph coloring problem, which is NP-complete even for the special case of deciding 3-colorability of planar graphs of maximum degree at most four [**10**]. Even more strikingly, the problem of counting the number of isomorphisms between two given graphs is Turing reducible to GI itself [**14**].[2] Compare this to the problem of determining whether a bipartite graph has a perfect matching. This is a polynomial time decision problem, yet the corresponding counting problem is complete for #P (the class of counting problems corresponding to decision problems in NP) [**25**].

There is further evidence that the problem of graph isomorphism is not NP-complete, in that if it were the polynomial hierarchy (of complexity classes "above" NP and co-NP) would collapse[3]. For more details about the complexity issues around GI, see the lovely book by Köbler, Schöning, and Torán [**12**].

Despite this evidence that GI is in some ways "easy", there is no known polynomial time algorithm for it. GI is known to have time complexity at most $\exp\left(O(n^{2/3})\right)$ for graphs with n vertices [**1**]. Various people have worked to create general algorithms for GI which are "practical in practice". One of the most powerful and best known of these algorithms is due to Brendan McKay [**15**]. It is known that his algorithm has exponential running time on some inputs [**9, 20**], but in general it performs exceptionally well. McKay has implemented his algorithm in the software package `nauty` (No AUTomorphisms, Yes?), freely available at his website [**17**]. R. L. Miller has implemented the algorithm as the `NICE` (NICE Isomorphism Check Engine) module of the open-source SAGE mathematics software [**23**].

In this article we aim to provide an introduction to the essential ideas of McKay's algorithm. There are three main strands to the `nauty` algorithm:

- using, iteratively, degree information;
- building a search tree examining choices not determined by degree information; and
- using graph automorphisms, as they are found, to prune the search tree.

We introduce all three strands in Section 2 and discuss their connections. Later, in Sections 4, 5, and 6, we go into more detail about how they are incorporated into McKay's algorithm. In Section 7 we discuss additional aspects which, although useful in certain cases, do not affect the main thrust of the algorithm.

---

[2] I.e., if we are allowed to use GI as a constant time subroutine then the number of isomorphisms between two given graphs can be determined in polynomial time.

[3] The polynomial hierarchy consists of two inductively defined sequences of complexity classes Σ_k^P and Π_k^P, $k \geq 0$. (The superscript P stands for the complexity class P.) The sequences start with $\Sigma_0^P = \Pi_0^P =$ P. Then we say that a problem is in Σ_k^P if every instance (for which the answer is "yes") has some certificate (of length bounded by some polynomial in the length of the instance) such that the problem of checking the certificate is in Π_{k-1}^P. Then we let Π_k^P = co-Σ_k^P. For instance $\Sigma_1^P = NP$, Π_1^P =co-NP, and a problem is in Σ_2^P if it has a certification scheme for which the certificate-checking problem is in co-NP. It is widely believed, though not proven, that all these classes are distinct.

2. Motivation

We start by noting that we can assume, without loss of generality, that all the graphs we consider have vertex set $[n] = \{1, 2, \ldots, n\}$. We will only consider simple undirected graphs, so we can identify a graph with its edge set, considered as a subset of $\binom{[n]}{2}$, the set of all unordered pairs from $[n]$.

One natural approach to the graph isomorphism problem is the use of canonical isomorphs: picking a canonical representative from each isomorphism class of graphs on $[n]$. Then testing isomorphism between two graphs reduces to checking equality of their corresponding canonical isomorphs. Note that equality of graphs is easy to test; the real work is performed in computing the canonical isomorph.

DEFINITION 1. An *isomorph* of a graph G is a graph with vertex set $[n]$ that is isomorphic to G. A *canonical isomorph function* assigns to every graph G an isomorph $C(G)$ such that whenever H is isomorphic to G we have $C(H) = C(G)$. We call $C(G)$ the *canonical isomorph* of G.

In the rest of the literature, including the title of this article, $C(G)$ is referred to as a "canonical labeling". However, we prefer the term "canonical isomorph" since it emphasizes the fact that $C(G)$ is a graph and not a labeling of G.

One way to define a canonical isomorphism function is to specify a total order $\leq$ on graphs on $[n]$. Then we define $C_{\leq}(G)$ to be the $\leq$-largest graph in the isomorphism class of G. Clearly this satisfies the conditions for a canonical isomorph function. One such example is the canonical isomorph function $C_{\preceq}$ defined by Read [**22**] and independently by Faradžev [**6**]. The ordering $\preceq$ they use is a lexicographic total order induced by a fixed total ordering on $\binom{[n]}{2}$. Represent a graph G by the binary sequence $i(G)$ of length $\binom{n}{2}$ in which the j-th entry is 1 exactly if the j-th unordered pair (in our fixed ordering) is an edge in G. Then $G \preceq H$ if $i(G) \leq i(H)$ lexicographically.

Unfortunately $C_{\preceq}$ has the drawback of being difficult to compute. Typically there is no alternative to checking essentially every permutation of the vertices to see whether it produces the graph that is $\preceq$-greatest.

One reason for the weakness of this sort of approach is that it does not exploit any graph theoretical information. How might one use such information to pick out "landmarks" in the graph? One starting point might be to have vertices in the canonical isomorph appear in increasing order of degree. Clearly this is not sufficient to determine the canonical isomorph. However, this local information can then be propagated around the graph. For instance, if there is a unique vertex v of some particular degree, then the neighbors of v can be distinguished from the non-neighbors. Iterating this idea, the second neighborhood of v can then be distinguished, and so on. The first main strand of McKay's algorithm is to distinguish vertices according to degree, and then to propagate this local information using the process described above. We discuss this propagation scheme in Section 4.

Unfortunately, it is quite possible that the propagation of local information starting from degrees will stabilize without all of the vertices being distinguished. For instance, if G is regular, then the process does not even start. Using stronger local information than degrees might provide more information in certain cases, at the cost of more computation. However, in many instances no local information is useful. Indeed, in vertex-transitive graphs, *no* vertex can be distinguished by any information, local or global.

McKay's algorithm proceeds at this point by introducing artificial asymmetry. Faced with a set S of (to this point) indistinguishable vertices, we distinguish, in turn, each vertex of S. This information is then propagated to the rest of the graph. In this way, we build a search tree (examining all choices) whose leaves are isomorphs of G. McKay's canonical isomorph C_M is the $\preceq$-greatest of these isomorphs appearing at leaves. Thus, in comparison to $C_{\preceq}$, the algorithm uses graph theoretical information to reduce the number of candidate isomorphs from which the canonical isomorph is chosen. We discuss this aspect of McKay's algorithm in Section 5.

The final major aspect of McKay's algorithm is the use of automorphisms that are discovered while exploring the search tree to prune later computation. This allows us to compute C_M more efficiently. This topic is covered in Section 6.

Before we discuss the details of McKay's canonical isomorph function and its implementation in `nauty` we introduce the notation we will use for group actions and isomorphisms.

3. Group Actions and Isomorphism

In order to talk about isomorphism carefully it is helpful to briefly discuss group actions.

DEFINITION 2. An *action*[4] of a group Γ on a set X is a function from $X \times \Gamma$ to X, mapping (x, g) to x^g, that satisfies the following conditions:

- for all $x \in X$ and $g, h \in \Gamma$ we have $(x^g)^h = x^{(gh)}$; and
- If e is the identity of Γ then $x^e = x$ for all $x \in X$.

For us the most important examples are various actions of Σ_n, the group of all permutations of $[n]$. We have chosen to write our group actions on the right; as a consequence we will also choose to compose our permutations on the right: for σ, γ elements of Σ_n the group product $\sigma\gamma$ is defined to be the permutation obtained by first applying σ and then γ. With this convention Σ_n acts on $[n]$, since for $\sigma, \gamma \in \Sigma_n$, $v \in [n]$, we have $(v^\sigma)^\gamma = v^{(\sigma\gamma)}$ (where, of course, v^σ is the image of v under σ).

Whenever Γ acts on X it also acts on subsets of X simply by defining

$$A^g = \{x^g : x \in A\}.$$

Thus the natural action of Σ_n on $[n]$ induces an action on $\binom{[n]}{2}$. As we have identified simple, undirected graphs with their edge sets, this action on pairs induces in turn an action of Σ_n on the set of all graphs with vertex set $[n]$. This action defines the notion of isomorphism of graphs.

DEFINITION 3. Two graphs G and H with vertex set $[n]$ are *isomorphic* if there exists a permutation $\gamma \in \Sigma_n$ such that $H = G^\gamma$.

Similarly, if Γ acts on X, then it also acts on sequences $(x_i)_{i=1}^k$ from X by

$$\left((x_i)_1^k\right)^\gamma = (x_i^\gamma)_1^k.$$

[4]To be a little more careful, what we define here is called a *right action* of Γ on X.

4. Propagating Degree Information

The first aspect of McKay's algorithm we will focus on is the propagation of degree information. We describe a classification of the vertices of G using an ordered partition. From some such partitions we can deduce further distinctions; otherwise, we call the ordered partition *equitable.* We first give the appropriate definitions and then describe the procedure in McKay's algorithm which refines an ordered partition until it becomes equitable.

DEFINITION 4. An *ordered partition* π of $[n]$ is a sequence $(V_1, V_2, \dots, V_r)$ of nonempty subsets of $[n]$ such that $\{V_1, V_2, \dots, V_r\}$ is a partition of $[n]$. The subsets $V_1, V_2, \dots, V_r$ are called the *parts* of π. A *trivial part* is a part of size 1. A *discrete partition* only has trivial parts, while the *unit partition* μ only has one part, namely $[n]$. The *length* of an ordered partition π is the number of parts in π.

Ordered partitions come equipped with the natural partial order of refinement. Given two ordered partitions π_1 and π_2, we say that π_1 is *finer* than π_2 if:

- every part V_i of π_1 is contained in a part W_k of π_2, and
- earlier parts of π_1 are contained in earlier parts of π_2; *i.e.*, if V_i and V_j are parts of π_1 with $i \leq j$, and W_k and W_ℓ are parts of π_2 such that $V_i \subseteq W_k$ and $V_j \subseteq W_\ell$, then $k \leq \ell$.

If π_1 is finer than π_2, then π_2 is *coarser* than π_1. (Note that "finer than" is not strict: it includes the case of equality.) The set of ordered partitions of $[n]$ with the relation "finer than" forms a partially ordered set whose unique maximal element is the unit partition μ and whose minimal elements are the discrete ordered partitions of $[n]$.

Suppose now that the ordered partition π encodes a classification of the vertices. Vertices in different parts of π have already been distinguished from each other; vertices in the same part have not. If two vertices v, w belong to the same part of π but have different degrees into a part of π then we can distinguish v from w, refining our classification. Our next definition describes the situation in which no further propagation of degree information is possible.

DEFINITION 5. An ordered partition $\pi = (V_1, V_2, \dots, V_r)$ of $[n]$ is an *equitable ordered partition* (with respect to G) if, for all $1 \leq i, j \leq r$, $\deg(v, V_j) = \deg(w, V_j)$ for all $v, w \in V_i$. An equitable ordered partition τ is a *coarsest equitable refinement*[5] *of* π if τ is finer than π and there is no equitable ordered partition finer than π and strictly coarser than τ.

Thus, for example, in a regular graph the unit partition is equitable. Similarly, in a strongly regular graph the ordered partition

$$\left(\{v\}, N(v), [n] \setminus (\{v\} \cup N(v))\right)$$

(where $N(v)$ is the neighborhood of v) is equitable for any vertex v.

If π is an inequitable ordered partition we would like to extract, iteratively, all the consequences (and *only* the consequences) of the distinctions expressed by π. This amounts to finding a coarsest equitable refinement of π. In the remainder

[5]Despite the name, coarsest equitable refinements are not necessarily unique. A better name might be "maximal equitable refinements", but for reasons both of history and clarity we prefer McKay's terminology. However, coarsest equitable refinements *are* unique up to ordering of the parts; see Lemma 7.

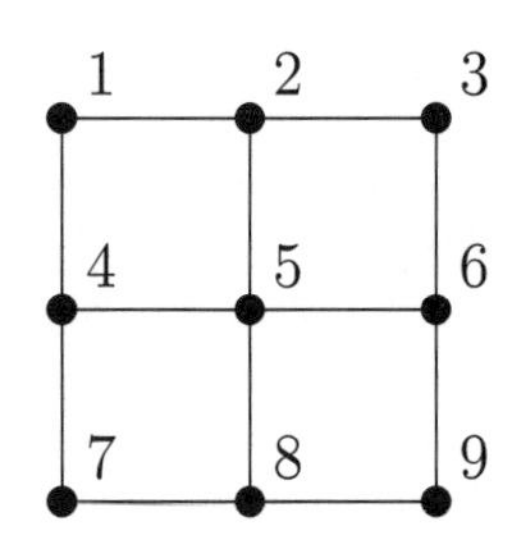

FIGURE 1. Example graph G on 9 vertices.

of this section we present the equitable refinement procedure used in McKay's algorithm[6]. The procedure accepts as input a graph G and an ordered partition π of $[n]$, and returns a coarsest equitable refinement $R(\pi)$ of π.

DEFINITION 6. Given an inequitable ordered partition $\pi = (V_1, V_2, \dots, V_r)$, we say that V_j *shatters* V_i if there exist two vertices $v, w \in V_i$ such that $\deg(v, V_j) \neq \deg(w, V_j)$. The *shattering of* V_i *by* V_j is the ordered partition $(X_1, X_2, \dots, X_t)$ of V_i such that if $v \in X_k$ and $w \in X_\ell$ then $k < \ell$ if and only if $\deg(v, V_j) < \deg(w, V_j)$. Thus, $(X_1, X_2, \dots, X_t)$ sorts the vertices of V_i by their degree to V_j.

Equitable Refinement Procedure.

Input: An unordered simple graph G with vertex set $[n]$, and an ordered partition π of $[n]$.

Output: An ordered partition $R(\pi)$.

Initialize: Let τ be the ordered partition π.

Iterate: If $\tau = (V_1, V_2, \dots, V_r)$, then let $B = \{(i, j) : V_j \text{ shatters } V_i\}$. If B is empty, then stop, reporting τ as the output $R(\pi)$. Otherwise, let (i, j) be the minimum element of B under the lexicographic order. Let $(X_1, X_2, \dots, X_t)$ be the shattering of V_i by V_j. Replace τ by the ordered partition where V_i is replaced by $X_1, X_2, \dots, X_t$; that is, replace $\tau = (V_1, V_2, \dots, V_r)$ with

$$(V_1, V_2, \dots, V_{i-1}, X_1, X_2, \dots, X_t, V_{i+1}, \dots, V_r).$$

EXAMPLE. Let G be the graph on 9 vertices shown in Figure 1. We show the execution of the equitable refinement procedure on two ordered partitions of [9], namely the unit partition μ and another partition.

π	B	V_i	V_j
$(1\ 2\ 3\ 4\ 5\ 6\ 7\ 8\ 9)$	$\{(1,1)\}$	(123456789)	(123456789)
$(1\ 3\ 7\ 9 \mid 2\ 4\ 6\ 8 \mid 5)$	$\emptyset$		

π	B	V_i	V_j
$(1 \mid 3\ 7\ 9 \mid 2\ 4\ 6\ 8 \mid 5)$	$\{(3,1),(3,2)\}$	(2468)	(1)
$(1 \mid 3\ 7\ 9 \mid 6\ 8 \mid 2\ 4 \mid 5)$	$\{(2,4)\}$	(379)	(24)
$(1 \mid 9 \mid 3\ 7 \mid 6\ 8 \mid 2\ 4 \mid 5)$	$\emptyset$		

[6]McKay's actual equitable refinement procedure differs slightly in the order in which parts are chosen to be shattered. However, only two properties of the procedure are necessary. Firstly, it must produce a coarsest equitable refinement of π. Secondly, the refinement $R(\pi)$ must satisfy $R(\pi^\gamma) = R(\pi)^\gamma$ for any $\gamma \in \Sigma_n$.

We prove, in Proposition 8 below, the correctness of the equitable refinement procedure. Our proof relies on the following lemma which is just a restatement of the fact (which is straightforward to prove) that in the lattice of *unordered* partitions, there is a unique unordered coarsest equitable refinement of any partition.

LEMMA 7. *Let π be an ordered partition of $[n]$. Suppose that ζ and ξ are coarsest equitable refinements of π. Then ζ and ξ have the same parts and differ only in the order of the parts.*

PROPOSITION 8. *The partition $R(\pi)$ returned by the equitable refinement procedure is a coarsest equitable refinement of π.*

PROOF. At each iteration of the procedure, some part V_i of τ is shattered, producing a new ordered partition whose length is greater than that of τ. Since every discrete partition is equitable, the length of τ is at most n. Hence the number of iterations is bounded, and the equitable refinement procedure terminates.

The algorithm halts only when an equitable ordered partition $R(\pi)$ is obtained. Since τ at each stage is finer than that of the previous iteration, and hence is finer than π, the partition $R(\pi)$ is also finer than π. That $R(\pi)$ is a coarsest equitable refinement follows by a straightforward argument using Lemma 7. □

5. The Search Tree

McKay's algorithm starts by forming the equitable refinement of the unit partition, thereby extracting all of the initial degree information. Having reached an equitable partition, we need to introduce artificial distinctions between vertices. However, we must be careful to examine all relevant choices. We systematically explore the space of equitable ordered partitions using a search tree. The next definition describes the way we make these artificial distinctions, forming children in the search tree.

DEFINITION 9. Let π be an equitable ordered partition of $[n]$ with a nontrivial part V_i, and let $u \in V_i$. The *splitting of π by u*, denoted by $\pi \perp u$, is the equitable refinement $R(\pi')$ of the ordered partition $\pi' = (V_1, V_2, \ldots, \{u\}, V_i \setminus \{u\}, V_{i+1}, \ldots, V_r)$. (Note that $\pi \perp u$ is strictly finer than π.)

Crucially, the children of an equitable ordered partition in the search tree do not correspond to all possible splittings of π. When we artificially distinguish a vertex u, the only alternatives we need to consider are vertices indistinguishable (so far) from u, i.e., vertices in the same part of π. At each stage we chose to split the first non-trivial part[7] of π.

We record in the search tree not only the current equitable ordered partition, but also the sequence of vertices used for splittings.

[7] To reduce the branching factor, McKay actually chooses the first *smallest* part of π. The method of choosing the part for splitting π is irrelevant as long as it is an isomorphism invariant of unordered partitions. That is to say, that if the i-th part of π is chosen, then also the i-th part of π^γ is chosen for any $\gamma \in \Sigma_n$.

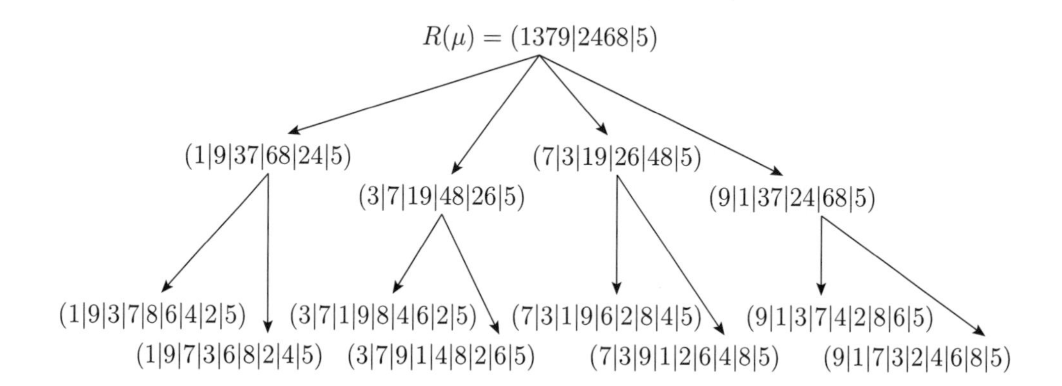

FIGURE 2. The search tree $T(G)$ for the graph G of Figure 1. Only the ordered partitions associated with each node of the search tree are shown.

DEFINITION 10. The *search tree* $T(G)$ is the rooted tree whose nodes

$$\Big\{(\pi;\underset{\sim}{u}) : \pi \text{ is an ordered partition of } [n];\ \underset{\sim}{u} = (u_1, u_2, \ldots, u_k) \text{ is a sequence of distinct vertices;}$$
$$\pi = (\ldots(R(\mu) \perp u_1) \perp u_2)\ldots) \perp u_k;$$
$$u_i \text{ is in the first non-trivial part of } (\ldots(R(\mu) \perp u_1) \perp u_2)\ldots) \perp u_{i-1} \text{ for each } i.\Big\}.$$

We allow the sequence $\underset{\sim}{u}$ to be empty, in which case $\pi = R(\mu)$. This is in fact the root node of the tree. There is an arc in $T(G)$ directed from the node $(\pi;\underset{\sim}{u})$ to the node $(\tau;\underset{\sim}{v})$ if $\underset{\sim}{u}$ is a prefix of $\underset{\sim}{v}$ and $\text{length}(\underset{\sim}{v}) = \text{length}(\underset{\sim}{u}) + 1$.

The *terminal nodes* of the search tree are nodes with no outgoing arcs. These terminal nodes correspond to discrete ordered partitions of $[n]$. Given a discrete ordered partition $\pi = (V_1, V_2, \ldots, V_n)$, we define $\sigma_\pi \in \Sigma_n$ to be the permutation such that $i^{(\sigma_\pi)} = j$ if V_j contains i. For example, if $\pi = (1|9|3|7|8|6|4|2|5)$, then σ_π maps $1 \mapsto 1$, $9 \mapsto 2$, $3 \mapsto 3$, $7 \mapsto 4$, $8 \mapsto 5$, and so on. To each terminal node $p = (\pi;\underset{\sim}{u})$ we associate the isomorph $G^{(\sigma_\pi)}$ of G.

We can now completely describe McKay's canonical isomorph function.

DEFINITION 11. McKay's canonical isomorph function $C_M(G)$ is defined to be

$$C_M(G) = \max_{\preceq} \Big\{G^{(\sigma_\pi)} : (\pi, \underset{\sim}{u}) \text{ is a leaf of } T(G)\Big\}.$$

We denote the set of isomorphs of G appearing on the right hand side by $L(G)$.

EXAMPLE. The search tree $T(G)$ for the graph G of Figure 1 is shown in Figure 2. For this example, all the isomorphs of G associated to the terminal nodes are the same; this isomorph is the canonical isomorph $C_M(G)$. (So in particular $L(G)$ has size 1.) Figure 3 shows this graph $C_M(G)$. Thinking of the various σ_π's as relabelings of G, we see many relabelings but only one isomorph. This is why we prefer the "canonical isomorph" terminology.

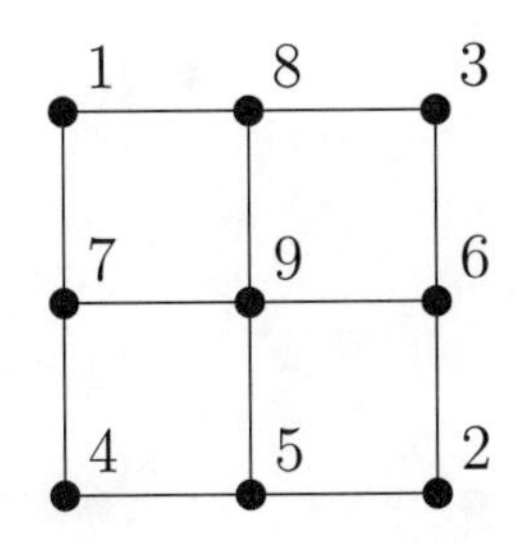

FIGURE 3. The canonical isomorph $C_M(G)$ for the graph G of Figure 1.

We have an action of Σ_n on sequences $\underset{\sim}{u}$ of vertices and on ordered partitions π. Thus we get an action on search trees, where the nodes of $T(G)^\gamma$ have the form $(\pi^\gamma; \underset{\sim}{u}^\gamma)$ for $(\pi; \underset{\sim}{u})$ in $T(G)$.

The following lemma contains the essence of the proof that $C_M(G)$ is a canonical isomorph function.

LEMMA 12. *If $H = G^\gamma$ for some $\gamma \in \Sigma_n$, then $T(H) = T(G)^\gamma$.*

PROOF. Note that the equitable refinement process respects the action of γ; that is, for an ordered partition π, $R(\pi^\gamma) = R(\pi)^\gamma$. Furthermore, if the first nontrivial part of π is V_i, then the first nontrivial part of π^γ is $(V_i)^\gamma$, which is the i-th part of π^γ. Therefore, the children of $(\pi; \underset{\sim}{u})^\gamma$ in $T(H)$ are the images under γ of the children of $(\pi; \underset{\sim}{u})$ in $T(G)$. The result follows by induction on length($\underset{\sim}{u}$). □

THEOREM 13. *$C_M(G)$ is a canonical isomorph function.*

PROOF. Note that in $T(G)$, the ordered partition π associated to a node m is strictly finer than the ordered partition associated to the parent of m. Hence, $T(G)$ is finite, $L(G)$ is finite, and the algorithm terminates. The output $C_M(G)$ is defined, by the algorithm, to be an isomorph of G.

By Lemma 12, for every terminal node $p = (\pi; \underset{\sim}{u})$ in $T(G)$, there exists a terminal node $p^\gamma = (\pi^\gamma; \underset{\sim}{u}^\gamma)$ in $T(H)$. From the action of γ on the partition π, we have $\sigma_{(\pi^\gamma)} = \gamma^{-1}\sigma_\pi$. Hence,

$$H^{\sigma_{(\pi^\gamma)}} = H^{\gamma^{-1}\sigma_\pi} = (G^\gamma)^{\gamma^{-1}\sigma_\pi} = G^{(\sigma_\pi)},$$

and so the sets $L(G)$ and $L(H)$ of graphs are the same. Thus, the maxima are the same, and so $C_M(G) = C_M(H)$. □

6. Pruning the Search Tree through the use of Automorphisms

One reason that the search tree $T(G)$ might be large is if G has a large automorphism group. In that case, the number of terminal nodes is at least the size of the automorphism group. However, the search tree is generated depth-first, and so automorphisms that are discovered during the search process can be used to *prune* the tree—discarding a section of the search tree not yet examined because it is known that no terminal node in that section will generate an isomorph of G that is better than the ones already discovered.

If there exist two terminal nodes $p = (\pi; \underset{\sim}{u})$ and $t = (\tau; \underset{\sim}{v})$ where $G^{(\sigma_\pi)} = G^{(\sigma_\tau)}$, then $\sigma_\pi(\sigma_\tau)^{-1}$ is an automorphism of G, since

$$G^{\sigma_\pi(\sigma_\tau)^{-1}} = (G^{\sigma_\tau})^{(\sigma_\tau)^{-1}} = G.$$

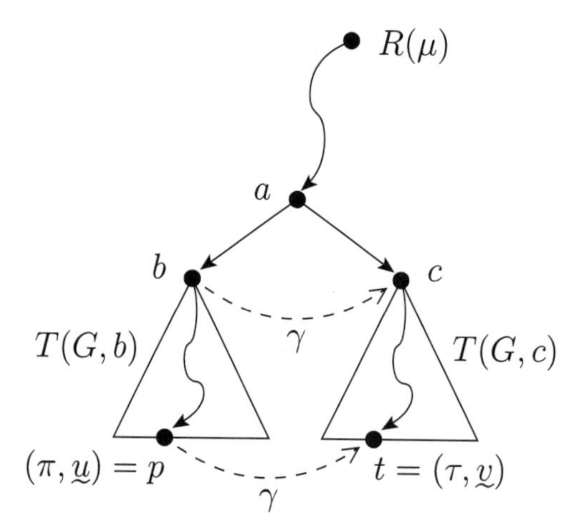

FIGURE 4. When an automorphism γ is discovered, certain subtrees of the search tree are known to be isomorphic.

Let a be the node in the search tree that is the deepest common ancestor of p and t, and let b be the child of a that is an ancestor of p, and c the child of a that is an ancestor of t. (See Figure 4 for an illustration.) The permutation $\gamma = \sigma_\pi(\sigma_\tau)^{-1}$ sends π to τ, fixes a, and sends b to c. For a node m of the search tree, let $T(G,m)$ denote the subtree of $T(G)$ rooted at m. Then $T(G,c)$ is isomorphic to $T(G,b)$ via γ, and so the set of graphs generated from terminal nodes of $T(G,c)$ is the same as the set of graphs generated from terminal nodes of $T(G,b)$. The search tree $T(G)$ is examined depth-first so all of $T(G,b)$ is examined before $T(G,c)$, and there is no reason to examine any more of $T(G,c)$; the search can continue at a.

Automorphisms discovered during the search can also be used to prune the search tree in another way. Let d be a node being re-visited by the depth-first search. Let Γ be the group generated by the automorphisms discovered thus far, and let Φ be the subgroup of Γ that fixes d. Suppose that b and c are children of d where some element of Φ maps b to c. If $T(G,b)$ has already been examined, then, as above, there is no need to examine $T(G,c)$. Hence $T(G,c)$ can be pruned from the tree[8].

EXAMPLE. The search tree $T(G)$ pruned through the use of automorphisms is shown in Figure 5. The search tree is examined depth-first by examining left children first. The automorphism γ_1 is discovered first, and at the root node, γ_1 has three orbits (1), (9), and (37) on the part (1379). Hence the child c of the root node must still be examined. When the automorphism γ_2 is discovered, the remainder of $T(G,c)$ is immediately discarded. At the root node, the subgroup of $\mathrm{Aut}(G)$ generated by γ_1 and γ_2 has one orbit in (1379), and hence no more children of the root need to be examined.

Note that for every automorphism $\gamma \in \mathrm{Aut}(G)$, there exist two terminal nodes with associated discrete ordered partitions π and τ such that $\gamma = \sigma_\pi(\sigma_\tau)^{-1}$. When

[8]In practice, only the generators of Γ are stored and only those which fix d are used to prune children of d. Moreover, the determination of which children to examine and which to prune is simplified by storing the orbits of this subgroup on $[n]$.

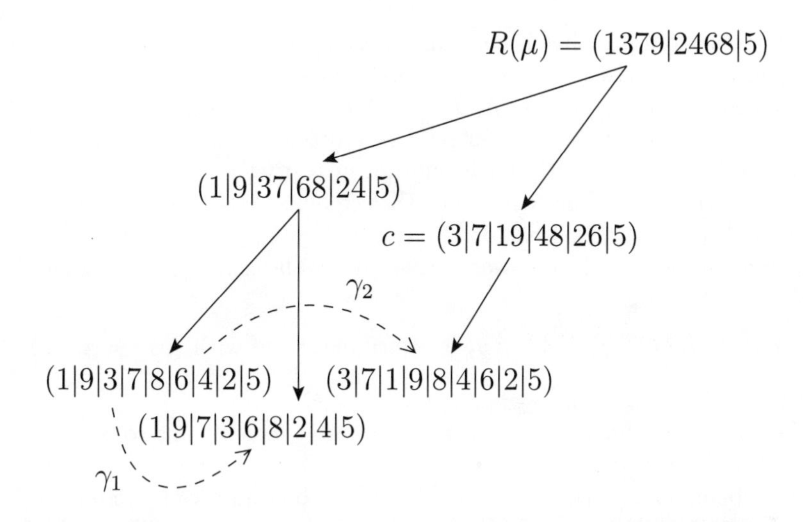

FIGURE 5. The search tree $T(G)$ pruned through the use of automorphisms.

pruning sections of the search tree, only the images of terminal nodes under the subgraph of $\mathrm{Aut}(G)$ generated thus far are removed. Thus, the canonical isomorph algorithm will find a complete set of generators for $\mathrm{Aut}(G)$.

EXAMPLE. The automorphism group of G from Figure 1 is the dihedral group on 8 elements, which is generated by γ_1 (a reflection) and γ_2 (a rotation).

At this point, we have completed a description of the essential elements of McKay's canonical isomorph algorithm.

7. Additional Aspects

In [**15**], McKay describes several improvements to the basic canonical isomorph algorithm. Many of these improvements are primarily implementation details and do not change the fundamental mathematical concepts. For instance, McKay chooses a different part to shatter in the equitable refinement procedure, and a different part to split when forming children in the search tree. We have omitted these details for the sake of clarity. However, we mention here just a few of the further improvements that are not purely details of a particular implementation.

IMPROVEMENT 1. Sometimes automorphisms of G can be determined at a nonterminal node $m = (\pi; \underset{\sim}{u})$ of the search tree. When this happens, they can be used to prune the search tree as before. For instance, this occurs when every nontrivial part of π has size 2 or when π has exactly one nontrivial part whose size is at most 5.

IMPROVEMENT 2. The search tree of the canonical isomorph algorithm can start with any equitable ordered partition, not just $R(\mu)$. With this modification, the canonical isomorph algorithm can be used to test for color-preserving isomorphisms of vertex-colored graphs.

IMPROVEMENT 3. It is possible to further restrict the number of candidates for the canonical isomorph by using a arbitrary function of pairs (G, π) (where π is an ordered partition of $[n]$) that is constant on isomorphism classes. McKay calls such isomorphism invariants *indicator functions*. Thus an indicator function $\Lambda(G, \pi)$ is a real-valued function such that $\Lambda(G, \pi) = \Lambda(G^\gamma, \pi^\gamma)$ for any $\gamma \in \Sigma_n$. (Note that Λ does not have to be different on different isomorphism classes of graphs; if it were, then it could be used directly for isomorphism testing!) Given a node $m = (\pi; \underset{\sim}{u})$ of the search tree, we define $\Lambda(m)$ to be the sequence $\Lambda(G, \pi_1), \ldots, \Lambda(G, \pi_k)$, where $\pi_0 = R(\mu)$ and $\pi_i = \pi_{i-1} \perp u_i$ for $1 \leq i \leq k$. Let $\Lambda_{\max}$ be the maximum of $\Lambda(p)$ for all terminal nodes p in the search tree, under the lexicographic order. We now set

$$L_\Lambda(G) = \left\{ G^{(\sigma_\pi)} : p = (\pi; \underset{\sim}{u}) \text{ is a terminal node with } \Lambda(p) = \Lambda_{\max} \right\},$$

and define $C_\Lambda(G)$ to be the $\preceq$-maximum of $L_\Lambda(G)$.

The advantage of using an indicator function is that it can also be used to prune the search tree. Suppose that a terminal node p has been found. If subsequently a node m is being examined where $\Lambda(m) < \Lambda(p)$, then no terminal node of $T(G, m)$ will be in $L_\Lambda(G)$ and hence $T(G, m)$ need not be examined. Note that when a nontrivial indicator function is used, the set $L_\Lambda(G)$ may be smaller than $L(G)$ and hence we may have $C_\Lambda(G) \neq C(G)$. However, $C_\Lambda(G)$ is still a canonical isomorph function, by an argument essentially identical to that of Theorem 13.

Acknowledgements

The authors wish to thank the anonymous referee and editor Tim Chow for comments that substantially improved the presentation of this article.

References

[1] L. Babai, Moderately exponential bound for graph isomorphism, in Fundamentals of computation theory (Szeged, 1981), Lecture Notes in Computer Science, Spinger, Berlin, 1981, v. 177, 34–50.

[2] L. Babai, D. Y. Grigoryev, and D. M. Mount, Isomorphism of graphs with bounded eigenvalue multiplicity, Proceedings of the 14th Annual ACM Symposium on Theory of Computing, 1982, 310–324.

[3] H. Bodlaender, Polynomial algorithms for graph isomorphism and chromatic index on partial k-trees, Journal of Algorithms 11 (1990), 631–643.

[4] D. Crnković and V. Mikulić, Block designs and strongly regular graphs constructed from the group $\mathrm{U}(3,4)$, *Glas. Mat. Ser. III*, 41(61) (2006), no. 2, 189–194.

[5] A. E. Brouwer, J. H. Koolen, and R. J. Riebeek, A new distance-regular graph associated to the Mathieu group M_{10}, J. Alg. Comb., 8 (1998), no. 2, 153–156.

[6] I. A. Faradžev, Generation of nonisomorphic graphs with a given distribution of the degrees of vertices, Algorithmic studies in combinatorics (Russian), "Nauka", Moscow, 1978, 11–19, 185.

[7] F. Fiedler, M. H. Klin, and M. Muzychuk, Small vertex-transitive directed strongly regular graphs, *Disc. Math.*, 255 (2002), no. 1–3, 87–115.

[8] I. S. Filotti and J. N. Mayer, A polynomial-time algorithm for determining the isomorphism of graphs of fixed genus, Proceedings of the 12th Annual ACM Symposium on Theory of Computing, 1980, 236–243.

[9] M. Fürer, A counterexample in graph isomorphism testing, Tech. Rep. CS-87-36, Department of Computer Science, The Pennsylvania State University, University Park, Penna., 1987.

[10] Michael R. Garey and David S. Johnson, Computers and Intractability: a guide to the theory of NP-completeness, W. H. Freeman and Co., San Francisco, CA, 1979.

[11] N. Giansiracusa, Determining graph isomorphism via canonical labeling, manuscript.

[12] J. Köbler, U. Schöning, and J. Torán, The graph isomorphism problem: its structural complexity. Progress in Theoretical Complexity, Birkhäuser Boston, Boston, MA, 1993.

[13] E. M. Luks, Isomorphism of graphs of bounded valence can be tested in polynomial time, Journal of Computer and System Sciences, 25 (1982), 42–65.
[14] Rudolf Mathon, A note on the graph isomorphism counting problem, Inform. Process. Lett. 8 (1979), no. 3, 131–132
[15] B. D. McKay, Practical graph isomorphism, Congr. Numer., 30 (1981), 45–87.
[16] B. D. McKay, Isomorph-free exhaustive generation, J. Algorithms, 26 (1998), no. 2, 306–324.
[17] B. D. McKay, `nauty` Software Program, Version 2.2, `http://cs.anu.edu.au/~bdm/nauty/`.
[18] B. D. McKay, and E. Rogoyski, Latin squares of order 10, Electron. J. Combin., 2 (1995), Note 3, 4 pp.
[19] G. Miller, Isomorphism testing for graphs of bounded genus, Proceedings of the 12th Annual ACM Symposium on Theory of Computing, 1980, 225–235.
[20] T. Miyazaki, The complexity of McKay's canonical labeling algorithm. in: Groups and computation, II (New Brunswick, NJ, 1995), DIMACS Ser. Discrete Math. Theoret. Comput. Sci., 28, Amer. Math. Soc., Providence, RI, 1997, 239–256.
[21] J. Ostrowski, J. Linderoth, F. Rossi, and S. Smirglio, Orbital branching, IPCO 2007: The Twelfth Conference on Integer Programming and Combinatorial Optimization, to appear, 2007.
[22] R. C. Read, Every one a winner or how to avoid isomorphism search when cataloguing combinatorial configurations, Algorithmic aspects of combinatorics (Conf., Vancouver Island, B.C., 1976), Ann. Discrete Math., 2 (1978), 107–120.
[23] SAGE Mathematics Software, Version 3.0, `http://www.sagemath.org/`.
[24] L. Soicher, Is there a McLaughlin geometry?, J. Algebra, 300 (2006), no. 1, 248–255.
[25] L. G. Valiant, The complexity of computing the permanent, Theoret. Comput. Sci., 8 (1979), no. 2, 189–201.

Department of Mathematics, University of Nebraska–Lincoln, Lincoln, NE 68588-0130 USA

E-mail address: {`shartke2,aradcliffe1`}`@math.unl.edu`

Contemporary Mathematics
Volume **479**, 2009

A multiplicative deformation of the Möbius function for the poset of partitions of a multiset

Patricia Hersh and Robert Kleinberg

Abstract. The Möbius function of a partially ordered set is a very convenient formalism for counting by inclusion-exclusion. An example of central importance is the partition lattice, namely the partial order by refinement on partitions of a set $\{1, \ldots, n\}$. It seems quite natural to generalize this to partitions of a multiset, i.e. to allow repetition of the letters. However, the Möbius function is not nearly so well-behaved. We introduce a multiplicative deformation, denoted μ', for the Möbius function of the poset of partitions of a multiset and show that it possesses much more elegant formulas than the usual Möbius function in this case.

1. Introduction

The partially ordered set (or "poset") of set partitions of $\{1, \ldots, n\}$ ordered by refinement, denoted Π_n, has a wealth of interesting properties. Among them are lexicographic shellability (a topological condition on an associated simplicial complex called the order complex) and a natural symmetric group action permuting elements in an order-preserving manner which induces an S_n representation on (top) homology of the order complex. These features lead to formulas for counting by inclusion-exclusion, growing out of the fact that a function called the Möbius function for the partition lattice has the elegant formula $\mu_{\Pi_n}(\hat{0}, \hat{1}) = (-1)^{n-1}(n-1)!$. This formula may be explained either topologically by counting descending chains in a lexicographic shelling (see [Bj]) or via representation theory (see [St1]), by interpreting this Möbius function as the dimension of a symmetric group representation on top homology of the order complex; up to a sign twist, this representation is an induced linear representation, induced from the cyclic group C_n up to the symmetric group S_n, hence having dimension $n!/n$.

It is natural to consider more generally the poset of partitions of a multiset $\{1^{\lambda_1}, \cdots, k^{\lambda_k}\}$, i.e. to allow repetition in values, or in other words to consider the quotient poset Π_n/S_λ for S_λ the Young subgroup $S_{\lambda_1} \times \cdots \times S_{\lambda_k}$ of the symmetric group. For example, the poset $\Pi_7/(S_3 \times S_2 \times S_2)$ consists of the partitions of the multiset $\{1^3, 2^2, 3^2\}$ into unordered blocks; a sample element would be the partition

Both authors were supported by Hertz Foundation Graduate Fellowships. The first author is now supported by NSF grant 0500638, and the second author is now supported by NSF CAREER award CCF-0643934 and NSF grant CCF-0729102.

made up of blocks $\{1,1,3\},\{1,2\},\{2\}$, and $\{3\}$, which is sometimes represented more compactly as $1^2,3|1,2|2|3$. We say that an element u is less than or equal to an element v, denoted $u \le v$, whenever the multiset partition u is a refinement of the partition v, or in other words each block of v is obtained by merging blocks of u. For example, if $u = 1|1$ and $v = 1^2$, then $u \le v$ because v is obtained from u by merging the two blocks of u. When $\lambda = (1,1,\dots,1)$, i.e. $\lambda_1 = \cdots = \lambda_k = 1$, then Π_n/S_λ is the usual partition lattice Π_n. On the other hand, the case $\lambda = (n)$ yields another important poset, namely the poset P_n of partitions of the positive integer n into smaller positive integers.

A Möbius function formula for the poset of partitions of any multiset would have a wide array of enumerative applications. A natural first question is whether this poset is shellable (or perhaps even has an interesting group representation on top homology of its order complex), since this could again give a convenient way of calculating the Möbius function. This turns out not to be the case – the example of Ziegler which we include in Section 2 demonstrates that this is not possible. This note therefore suggests a different approach.

Recall that the Möbius function of a finite partially ordered set P is a function μ_P on pairs $u,v \in P$ with $u \le v$. It is defined recursively as follows: $\mu_P(u,u) = 1$,

$$\mu_P(u,v) = -\sum_{u \le z < v} \mu_P(u,z)$$

for $u < v$, and $\mu_P(u,v) = 0$ otherwise. One may encode the Möbius function as an upper triangular matrix M by letting the rows and columns be indexed by the elements of P with $M_{u,v} = \mu_P(u,v)$, using a total order that is consistent with the partial order P to order the row and column indices. This is the inverse matrix to the "incidence matrix" I having $I_{u,v}$ equalling 1 for $u \le v$ and 0 otherwise. See [St2].

To see by way of an example how the Möbius function is useful for counting by inclusion-exclusion, consider the problem of counting lattice points in the portion of $\mathbb{R}^2$ consisting of points (x,y) with $|x| \le q$ and $|y| \le q$ such that (x,y) does not lie on any of the lines $x = 0, y = 0, x = y$. A natural approach is to count all lattice points $(x,y) \in \mathbb{R}^2$ with $|x| \le q$ and $|y| \le q$, subtract off those on each of the three lines, then add back 2 copies of the unique lattice point lying on all three lines. Thus, each allowable lattice point is counted once, and each forbidden point is counted 0 times altogether. Notice that the coefficients assigned to the various intersections of lines (i.e. coefficient 1 for the empty intersection, 2 for the triple intersection and -1 for each line by itself) are in fact the Möbius function values $\mu_P(\mathbb{R}^2, u)$ for the various elements u in the poset P of intersections (or "intersection poset") ordered by containment. The recursive definition of the Möbius function is designed exactly to count allowable points once and all others 0 times. Notice that the partition lattice may be viewed as another instance of an intersection poset, now taking the intersections of hyperplanes $x_i = x_j$ in $\mathbb{R}^n$ in which two coordinates are set equal.

A remarkably fruitful method for computing Möbius functions (as well as for finding beautiful formulas for them) grows out of the following topological interpretation for the Möbius function. Given a finite poset P, its order complex is the simplicial complex $\Delta(P)$ whose i-dimensional faces are the chains $u_0 < \cdots < u_i$ of $i+1$ comparable, distinct elements of P. It turns out that $\mu_P(u,v)$ equals the reduced

Euler charactistic for the order complex of the subposet of P comprised of exactly those elements z with $u < z < v$ (see e.g. [Ro]). Recall that the reduced Euler characteristic of a simplicial complex is $-1+f_0-f_1+f_2-\cdots = -1+\beta_0-\beta_1+\beta_2-\cdots$ where f_i is the number of i-dimensional faces in the complex and β_i is the rank of the i-th homology group of the complex.

Surprisingly many posets arising in combinatorics have a property called shellability (more specifically, EL-shellability) enabling easy calculation of the reduced Euler characteristic (and thus the Möbius function) by showing that the order complex is homotopy equivalent to a wedge of spheres and in fact giving a way to count these spheres by counting the so-called descending chains of the poset. See [Bj] to learn more about this technique. Shellability for a graded poset also implies an algebraic/topological property called the Cohen-Macaulay property. We will soon see that the poset of partitions of a multiset cannot be shellable in general, because it is not always Cohen-Macaulay.

A simplicial complex is Cohen-Macaulay if all its maximal faces have the same dimension and for each face σ, the subcomplex called the link of σ (which may be thought of as a neighborhood of σ in the original complex) also has all its maximal faces of equal dimension i and its reduced homology concentrated in this same dimension i. See [Mu], [St2] or [St3] and references therein for further background, including powerful bridges built largely by Stanley and Hochster in the 70's between combinatorics (e.g. counting faces) and commutative algebra (e.g. computing Hilbert series of rings), using the topology of simplicial complexes along the way. See [BH] for partial results and further open questions regarding the topology of the order complex for the poset of partitions of a multiset; in [BH], a relatively new technique called discrete Morse theory is used to analyze the topology of the order complex, since shellability is not an option in this case. Discrete Morse theory translates topological questions into combinatorial questions of finding nice graph matchings, and can be used in situations where other techniques such as shellability are not applicable.

Our main interest here is in the Möbius function, and in variations on it, for the poset of partitions of a multiset. While the special case of the partition lattice has been studied a great deal, little is known about the more general case. The point of this note is to suggest questions about this more general poset, in particular suggesting a variant on the Möbius function which we believe in this case may be much more useful to study than the usual Möbius function; we only scratch the surface here as far as carrying out this study, but we do prove two formulas demonstrating that our variation on the Möbius function is much more well-behaved than the usual Möbius function in these cases, and which seem like a good starting point for further study.

2. A cautionary example and a consequent deformation of the Möbius function

Ziegler showed in [Zi, p. 218] that Π_n/S_n is not Cohen-Macaulay for $n \geq 19$. The Möbius function for Π_n/S_λ also is not well-behaved, seemingly because poset intervals are no longer products of smaller intervals coming from single blocks. For instance, $\mu_{\Pi_4/S_4}(\hat{0}, 1^2|1^2) \neq [\mu_{\Pi_2/S_2}(\hat{0}, 1^2)]^2$. We introduce a variant of the Möbius function, denoted μ', which is defined to be multiplicative in the following sense: given a partition into blocks $B_1, \ldots, B_j$, let $\mu'(\hat{0}, B_1|\ldots|B_j) = \prod_{i=1}^{j} \mu'(\hat{0}, B_i)$. For each block B_i, let $\mu'(\hat{0}, B_i) = -\sum_{\hat{0} \le u < B_i} \mu'(\hat{0}, u)$, and let $\mu'(\hat{0}, \hat{0}) = 1$. In the case of the partition lattice, $\mu' = \mu$. The function μ' first arose in [He] as the inverse matrix to a weighted version of the incidence matrix used to count quasisymmetric functions by inclusion-exclusion by collecting them into groups comprising symmetric functions. See [Eh] and references therein for background on quasisymmetric functions and their connections to flag f-vectors, i.e. to vectors counting "flags" of faces $F_0 \subset F_1 \subset \cdots \subset F_r$ with each face contained in the next.

Ziegler gave the following example to show that Π_n/S_n is not Cohen-Macaulay for $n \ge 19$ [Zi, p. 218]. In that paper, he denotes Π_n/S_n by P_n, because he thinks of it as consisting of the partitions of an integer into smaller integers.

EXAMPLE 2.1 (Ziegler). *The open interval* $(1^6|1^5|1^3|1^2|1^2|1, 1^8|1^7|1^4)$ *prevents* Π_{19}/S_{19} *from being Cohen-Macaulay. Simply note that* $1^8|1^7|1^4$ *may be refined to* $1^6|1^5|1^3|1^2|1|1$ *in two different ways:*

$$8 = 6 + 2 = 5 + 3$$
$$7 = 5 + 2 = 6 + 1$$
$$4 = 3 + 1 = 2 + 2$$

Any saturated chain from $\hat{0}$ *to* $1^6|1^5|1^3|1^2|1^2|1$ *together with any saturated chain from* $1^8|1^7|1^4$ *to* $\hat{1}$ *gives a face* F *in the order complex* $\Delta(\Pi_{19}/S_{19})$ *such that* $lk(F)$ *is a disconnected graph, precluding the Cohen-Macaulay property.*

In [Eh], Richard Ehrenborg encoded the flag f-vector of any finite graded poset with unique minimal and maximal elements as a function called a quasisymmetric function. Quasisymmetric functions recently have been shown to play a key role in the theory of combinatorial Hopf algebras; they also may be regarded as a generalization of symmetric functions (a widely studied family of functions that arise in representation theory as a convenient way of encoding characters). An application to determining the flag f-vector for graded monoid posets in [He] (encoded as the quasisymmetric function F_P introduced in [Eh]) led us to examine μ' in the first place. We gave an inclusion-exclusion counting formula in [He] where we grouped collections of quasisymmetric functions together into symmetric functions, leading us to use a weighted version of the incidence matrix having μ' as its inverse matrix.

Thus, μ' is not just an abstraction admitting pleasant formulas, but does arise in counting by inclusion-exclusion, albeit in a way that is not yet well understood. The aim of this note is to take a first step towards understanding to what extent Π_n/S_λ shares the nice properties of Π_n. The fact that μ' satisfies much nicer formulas than the usual Möbius function suggests that the order complex may not be the best complex to associate to this poset, that rather one should study a complex with μ' as its reduced Euler characteristic. It seems that this should be a boolean cell complex, in the sense of [Bj2], where one poset chain may give rise to multiple cells in the complex, with weighted incidence numbers counting these.

3. Combinatorial formulas

Next we give formulas for μ' for two natural seeming classes of multisets. Let $n = \sum_{i=1}^{j} n_i$. Since $(-1)^{n-1} \frac{(n-1)!}{\prod_{i=1}^{j}(n_i)!}$ is not necessarily an integer unless some $n_i = 1$, it is too much to ask for the following to hold for all multisets.

THEOREM 3.1. *If P is the poset $\Pi_n/(S_{n_1} \times \cdots \times S_{n_k} \times S_1)$ of partitions of $\{1^{n_1}, \ldots, k^{n_k}, k+1\}$, so $n-1 = n_1 + \cdots + n_k$, then*

$$\mu'_P(\hat{0}, \hat{1}) = (-1)^{n-1} \frac{(n-1)!}{\prod_{i=1}^k (n_i)!}.$$

PROOF. Suppose by induction this is true for rank less than n. We decompose the set of partitions below the maximal element according to the content of the block containing the distinguished letter $k+1$. For any fixed block B such that $|B| < n-1$ and $k+1 \in B$, note that

$$\sum_{\{u | B \in u\}} \mu'_P(\hat{0}, \overline{u}) = 0$$

where $\overline{u}$ is obtained from u by deleting the block B. Since $\mu'_P(\hat{0}, u) = \mu'(\hat{0}, B)\mu'(\hat{0}, \overline{u})$, we have

$$\sum_{\{u|B\in u\}} \mu'(\hat{0}, u) = \mu'(\hat{0}, B) \sum_{\{u|B\in u\}} \mu'(\hat{0}, \overline{u}) = 0.$$

Therefore, we need only sum over all possible blocks B which contain $k+1$ and satisfy $|B| = n-1$. For each such B, we sum over partitions u containing any particular such block B, yielding

$$\begin{aligned}
\sum_B \sum_{\{u|B\in u\}} \mu'(\hat{0}, u) &= (-1)^{n-2} \sum_{j=1}^k \frac{(n-2)!}{(n_j - 1)! \prod_{i \neq j} (n_i)!} \\
&= (-1)^{n-2} \sum_{j=1}^k \frac{(n-2)!(n_j)}{\prod_{i=1}^k (n_i)!} \\
&= (-1)^{n-2} \frac{(n-2)!}{\prod_{i=1}^k (n_i)!} \sum_{j=1}^k n_j \\
&= (-1)^{n-2} \frac{(n-1)!}{\prod_{i=1}^k (n_i)!}.
\end{aligned}$$

This implies

$$\mu'(\hat{0}, \hat{1}) = -\sum_B \sum_{\{u|B\in u\}} \mu'(\hat{0}, u) = (-1)^{n-1} \frac{(n-1)!}{\prod_{i=1}^k (n_i)!}.$$

□

PROPOSITION 3.1. *Let P be the poset Π_n/S_n of partitions of $\{1^n\}$. Then $\mu'_P(\hat{0}, \hat{1}) = (-1)^{n-1}$ for n a power of 2 and $\mu'_P(\hat{0}, \hat{1}) = 0$ otherwise.*

PROOF. By induction, $\mu'_P(\hat{0}, 1^{n_1}|\ldots|1^{n_k}) = 0$ for $n_1 + \cdots n_k = n$ and $k > 1$ unless each n_i is a power of 2, in which case it equals $(-1)^{n-k}$. Hence, if we restrict to partitions where each block has size a power of 2 and give a bijection between such partitions of $\{1^n\}$ involving an even number of parts and those involving an odd number of parts, this will imply the result, since the partition with a single block of size n is paired with another partition by this bijection if and only if n is a power of 2. Such a correspondence is obtained by pairing a partition having a unique largest block with the partition obtained by splitting this block into two equal parts. □

In particular, the formula in Theorem 3.1 suggests the possibility of μ' recording the dimension of an induced representation on (top) homology of an associated cell complex, specifically a linear representation induced from the group $S_{n_1} \times S_{n_2} \times \cdots \times S_{n_k}$ to the full symmetric group S_{n-1}. See [He2] for a generalization of lexicographic shellability applicable to cell complexes having μ' as reduced Euler characteristic.

References

[BH] E. Babson and P. Hersh, *Discrete Morse functions from lexicographic orders*, Trans. Amer. Math. Soc. **357** (2005), 509–534.

[Bj] A. Björner, *Shellable and Cohen-Macaulay partially ordered sets*, Trans. Amer. Math. Soc. **260** (1980), no. 1, 159–183.

[Bj2] A. Björner, *Posets, regular CW complexes and Bruhat order*, European J. Combin. **5** (1984), no. 1, 7–16.

[Eh] R. Ehrenborg, *On posets and Hopf algebras*, Advances in Math. **119** (1996), 1–25.

[He] P. Hersh, *Chain decomposition and the flag f-vector*, J. Combin. Theory, Ser. A, **103** (2003), no. 1, 27–52.

[He2] P. Hersh, *Lexicographic shellability for balanced complexes*, J. Algebraic Combinatorics, **17** (2003), no. 3, 225–254.

[Mu] J. Munkres, Elements of algebraic topology, Addison-Wesley Publishing Company, Inc., 1984.

[Ro] G.-C. Rota, *On the foundations of combinatorial theory I: Theory of Möbius functions, Z. Wahrsch.* **2** (1964), 340–368.

[St1] R. Stanley, *Some aspects of groups acting on finite posets*, J. Combin. Theory, Ser. A, **32** (1982), no. 2, 132–161.

[St2] R. Stanley, Enumerative Combinatorics, vol. I. Wadsworth and Brooks/Cole, Pacific Grove, CA, 1986; second printing, Cambridge University Press, Cambridge/New York, 1997.

[St3] R. Stanley, Combinatorics and commutative algebra, Second edition. Progress in Mathematics, **41**, Birkhäuser Boston, Inc., Boston, MA, 1996. x+164 pp.

[Zi] G. Ziegler, *On the poset of partitions of an integer*, J. Combin. Theory Ser. A, **42 (2)** (1986), 215-222.

Department of Mathematics, Indiana University, Bloomington, IN 47405
E-mail address: phersh@indiana.edu

Department of Computer Science, Cornell University, Ithaca, NY 14853
E-mail address: rdk@cs.cornell.edu

Contemporary Mathematics
Volume **479**, 2009

Communicating, Mathematics, Communicating Mathematics — Joe Gallian style

Aparna Higgins

A tribute to Butch Cassidy from the Sundance Kid

1. Introduction

In the summer of 2007, a conference was held to celebrate communicating mathematics. This was an unusual theme — the conference would not focus on a specific area of mathematics, but on the idea of sharing mathematics between mathematicians who may not be in the same research area. The conference was held in Duluth, MN, which is home to one of the greatest communicators of mathematics of our times — Joe Gallian.

This article is an adaptation of a talk about the mathematical life of Joe Gallian given at the conference by Gallian's friend and collaborator, Aparna Higgins. The talk was accompanied by photographs of Gallian and the people and places important to him, with snippets of some music that is important to him. We will eschew those effects in this article, concentrating on Joe as a communicator and a mathematician.

The field of mathematics is fortunate to have several of its practitioners be simultaneously insightful researchers and excellent teachers who provide dedicated and tireless service to the profession. A few of these are even gifted expositors of mathematics, whether they are writing it or speaking it. Joe Gallian is one such. Gallian's work in mathematics and his work for the mathematical community have inspired his colleagues, his collaborators and his students.

2. Biographical Sketch

Joe Gallian was born on 5 January, 1942 to Joseph and Alvira Gallian. He grew up in Pennsylvania with a younger brother Gary. Although Joe claims that he was not nerdy as a child, he also revealed that in the 8th grade, his class held elections to have a student be a teacher for a day, and Joe was elected Teacher for the Day in mathematics.

2000 *Mathematics Subject Classification.* Primary 01A70.
Key words and phrases. Joseph Gallian, Communicating Mathematics.
The author is grateful for the referees' thoughtful and detailed comments, and for Dan Isaksen's guidance.

Joe attributes his affinity to mathematics to his boyhood interest in the various statistical measures associated with the performance of baseball players. This interest was demonstrated forty-five years later in his article *Who is the greatest hitter of them all?* [**1**]. Further, Joe says that, in an era when only one college football game per week was on television, every Notre Dame football game was broadcast on the radio. Hence it was easy to become a follower, not to mention a fan, of Notre Dame football. Joe, his father and his brother were big sports fans.

Years later in 1981, Joe wrote a paper entitled *An optimal football strategy* [**2**], which exemplifies his ability and willingness to talk about mathematics as it appears to non-mathematicians, and to demonstrate that mathematics is found all around us in many everyday situations. The article discusses two different strategies that a coach can pursue when his team is suffering a 14-point deficit late in a football game. The question arose while watching such a game with his brother and brother-in-law, with each stratagem having a supporter. Joe shows us which is the better stratagem mathematically-speaking — not because it has a larger probability of resulting in a win, but because it reduces the probability of a loss! It should be noted that when Joe wrote this article, it was still possible for a football game to end in a tie.

Joe graduated from high school in 1959, and went to work in a glass factory where his father had worked his entire life. Joe has also held janitorial jobs at a pool room, a night club and a country club. In the 1960s, Joe Gallian fell under the spell of Charlene Toy and the Beatles — in that order. Charlene and Joe were married in 1965, between Joe's junior and senior year in college. They have three children, Bill, Ronald and Kristin, and eight grandchildren.

Joe went to Slippery Rock State College in 1962, graduating with a Bachelor's degree in 1966, and then to the University of Kansas for a Master's program. He continued his education at the University of Notre Dame, earning his doctoral degree in mathematics in 1971, under the direction of Karl Kronstein. His thesis was entitled *Two-step centralizers in finite p-groups.* (Joe says he and his wife went to every home football game while they were at Notre Dame!) Joe stayed at Notre Dame as a Visiting Assistant Professor for a year after his PhD, and then came to the University of Minnesota Duluth, where he has been ever since. At Duluth, Joe has written about 100 papers, has generated about $3,000,000 in grants, has won every teaching award in the University of Minnesota system, and holds the title of Morse Alumni Distinguished University Teaching Professor.

3. Communicating Mathematics

Joe does mathematics, Joe teaches mathematics, Joe teaches how to create mathematics, Joe gives talks about mathematics, Joe helps to disseminate mathematics.

Joe has written approximately one hundred articles. Joe himself categorizes the papers in his vita thus: "Publications not related to Teaching", which number about 30, with the rest under the heading "Publications Related to Teaching".

In 1976, Joe discovered one of his talents — that of being an organizer. He organized an NSF-CBMS (National Science Foundation-Conference Board of Mathematical Sciences) conference on finite groups. Since then, Joe has helped to organize and plan many mathematical get-togethers, including the Summer Seminars of the North Central Section of the Mathematical Association of America (MAA) in 1987 and 1999, and two National Security Agency (NSA)-funded, American Mathematical Society (AMS)-hosted conferences on undergraduate research in mathematics in 1999 and 2006.

4. Communicating Mathematics Through Writing

Joe writes well. He allows himself time for several re-writes so that the piece reads well. Joe works to provide context. He takes care of the little things that take a paper from being merely readable to being enjoyable. Often his writing is targeted to undergraduates or to teachers of undergraduates.

An example of such writing is in his paper *The search for finite simple groups* [**3**], which appeared in 1976. This paper begins, "At present, simple group theory is the most active and glamorous area of research in the theory of groups and it seems certain that this will remain the case for many years to come. Roughly speaking, the central problem is to find some reasonable description of all finite simple groups." He continues, "A number of expository papers ... detailing progress on this problem have been written for professional group theorists, but very little has appeared which is accessible to undergraduates." Gallian writes that he hopes to emphasize that "problems of one generation have deep roots in the work of previous generations," and that there is frequently "a large temporal gap between certain results and their subsequent improvements." The paper is organized into sixteen sections in more or less chronological order according to theme. An interesting aspect of the article is Joe's inclusion of several tables that help the reader to consolidate this history in a comprehensible way. He uses a table to provide a chronological history of highlights, such as "1963 — Feit-Thompson — Proved simple groups have even order." In another table he lists each known finite simple group, along with its name, discoverer, type, notation and order. Not surprisingly, this article won one of the two very first Carl B. Allendoerfer awards given by the MAA for expository excellence in 1977. As a follow-up to the story of this paper, Gallian wrote an article in the first-ever issue of FOCUS (MAA's newsletter at that time) in 1981 announcing that "The classification of finite simple groups is complete" [**4**].

In a biographical note that Joe wrote for *The mathematics of identification numbers* [**5**], Joe acknowledges that although he writes the typical research papers that are of interest to only a few, he prefers to write articles that can be used in the classroom. A beautiful paper by Gallian and David Moulton (a former student in the Duluth REU) illustrates this sentiment well. The paper is entitled *On groups of order pq* [**6**]. In just one page, Gallian and Moulton provide a proof of the result that a group of order pq is cyclic, where p and q are primes with $q < p$, and q not dividing $p-1$. The advantage of their proof is that it avoids using Sylow theorems. As a result, this pretty theorem is now accessible to students who have taken only a single semester of abstract algebra which may not have included coverage of the Sylow theorems. Gallian and Moulton use only Lagrange's Theorem and sophisticated counting to prove this theorem. A quick outline of this proof by contradiction is as follows: Suppose that every non-identity element of G has order either p or q. The centralizer of a non-identity element cannot have order pq, so it has order either p or q. The number of elements of order p must be a multiple of q and of $p-1$, hence a multiple of $q(p-1)$, and similarly, the number of elements of order q must be a multiple of $p(q-1)$. Since neither $q(p-1)$ nor $p(q-1)$ divides

$pq - 1$, not all the nonidentity elements of G have the same order, and the number of nonidentity elements exceeds pq, which is impossible.

Another of Joe's great writing ventures has been his book *Contemporary Abstract Algebra* [**7**], now in its sixth edition and used in undergraduate abstract algebra courses at scores of universities. Joe has also contributed to the writing of the popular liberal arts text *For all Practical Purposes* [**8**] on topics such as zip codes, bar codes and identification numbers.

Joe's writing is not always on mathematical topics. As a good communicator of mathematics, he often writes about the interplay of mathematics and other areas of life. He has co-authored reviews of films with mathematical themes. In addition to writing a couple of articles on Putnam trivia in the American Mathematical Monthly, he has written a lengthy history of the Putnam Competition [**9**] that he updates every year.

It is not surprising to see why Joe has won both the Trevor Evans (1996) and Carl B. Allendoerfer awards for mathematical exposition from the MAA.

5. Communicating by Driver's Licenses(!)

Joe's analysis of driver's license numbers surely counts as some of his best-known work! Joe reverse-engineered the Minnesota Driver's License algorithm by poring over lists of driver's license numbers with attributes of the owner to see which characteristics (such as birth date, last name, first name) may be encoded and by what part of the number. This is a story that could be read in *Assigning driver's license numbers* [**10**], but it is a story that must be heard! And heard from Joe himself! There is just something about the way that he shows examples of finding the sequences that correspond to the names, his delight in finding enough evidence to conjecture a pattern, his disappointment in seeing the gaps in the supposed pattern, and then the "aha" when he figured out what purpose the gaps serve. You can read that "The three most common methods [of assigning driver's license numbers], a sequential number, the social security number, and a computer-generated number, are uninteresting mathematically." But to watch Joe's expression as he says this is a real treat! He makes you see how exciting it could be to code and decode numbers that are individually tailored, and why the excitement is in the mathematics, and not in the numbers. And so we come to Joe's talks!

6. Communicating by talks

Joe prepares his talks meticulously, and delivers them with an unmatched sense of humor and timing. The incredible infusion of energy in his talks is what makes them so memorable. Joe has given over two hundred and fifty talks at colleges and universities around America. As of this conference, he had given talks at forty MAA Section meetings in twenty-four of MAA's twenty-nine Sections.

In the late 1980s, the MAA Committee on Student Chapters was created to found MAA Student Chapters and help to sustain them. The Committee also put into place MAA sessions directed towards undergraduates at the national meetings, one of which was the MAA Student Lecture. The Committee worked on creating a list of potential presenters. In early 1993, Don Kreider, the then MAA President, called Higgins, who chaired the MAA Student Chapter Committee, and explained that it was his duty to find a speaker to address the United States of America Mathematical Olympiad (USAMO) winners. It was a formal occasion, he said, and a difficult crowd, because there would be mathematically super-smart high school students, along with their parents, who may or may not be mathematically inclined. Kreider asked for recommendations for a speaker. Higgins offered a few names, including Joe Gallian's. Kreider had been told that he would probably get a recommendation of Gallian from Higgins, and he proceeded to ask questions about Gallian to determine whether Gallian would be able to deal with students from places like Berkeley and Harvard. Higgins remembers telling Kreider about the institutions from which the Duluth REU students come, and saying something like, "Joe has those guys eating out of his hands. They all love him. They are all charmed by him." Although Higgins did not think that she had convinced Kreider that Joe would be an appropriate speaker for the Awards ceremony, Joe was invited to present a talk to the USAMO winners in 1993, and of course, more recently, he has presided over the ceremonies as MAA President.

Joe has the unique honor of presenting the only MAA-AMS-PME (Pi Mu Epsilon) Invited Lecture ever at Boulder in the summer of 1989 at one of the last AMS-MAA Summer Joint Meetings.

Anyone who has heard Joe give a talk is not surprised that Joe has won every teaching award in the University of Minnesota system. Joe was a recipient of the MAA's highest teaching honor, the Deborah and Tepper Haimo Award for Distinguished University or College Teaching in Mathematics, in 1993 — the year of the inception of this award. The MAA chose Joe Gallian to be the MAA Polya Lecturer from 1999 to 2001.

One of Joe's more popular articles is *How to give a good talk* [**11**], which he modified to PowerPoint presentations in *Advice on giving a good PowerPoint presentation* [**12**]. Joe tells the story of giving a talk at an institution where all the students in the audience had a printed page on which they occasionally checked off something. Joe was well into the talk when he realized that each student had a copy of his article and was checking off items as Joe practiced what he had preached in the article! Joe's upgraded PowerPoint talks remain strong because of Joe's delivery, but one must comment nostalgically on the loss of the folksy handwritten transparencies in various colors with page numbers for multiple presentations in the top right corner.

7. Communicating to the next generation(s)

Joe Gallian is one of the mathematical community's most vigorous proponents of undergraduate research. Joe directs undergraduate research and promotes it in two distinct, but related, ways: through the Duluth REU and by doing academic-year research with students at his own institution. For over thirty years, Joe has been helping to create mathematics research experiences for undergraduates that are as authentic as possible to those experienced by professional mathematicians. The most striking undergraduate research that Joe directs is the Duluth REU. This program has served over one hundred fifty undergraduates, producing over one hundred thirty publications in mainstream research journals, eight Morgan Prize winners, and eleven Alice Schafer Prize winners. Joe has taught these students to create mathematics, but he has also taught them to communicate it. Many of his REU students teach, many continue with professional research in mathematics, many direct students in undergraduate research, and at least one has written articles communicating the mathematics of knitting, crocheting and parking lot patterns! A vital and exciting feature of the program is the return of alums as visitors to the program for days and even weeks, communicating with the current participants. Details of the Duluth REU program, including some history and a listing of participants and papers produced can be found in Joe's articles on his webpage [**13**] and in the Proceedings of two conferences funded by the NSA and hosted by the AMS, Summer Undergraduate Mathematics Research Programs (1999) and Promoting Undergraduate Research in Mathematics (2007). Both conference proceedings are available free on the web [**14**] [**15**]. Joe initiated both these conferences as invitation-only conferences, designed for various practitioners of undergraduate research to come together and share best practices, successes and concerns. Joe sought funding for these conferences and edited both the Proceedings.

8. Communicating by collaborating

Joe Gallian is a masterful collaborator. He says he knows his strengths and recognizes complementary strengths in others. He claims that his collaborative ventures resulted in much better outcomes than if he had gone solo. He has co-authored about thirty papers. He has invited people who have used computer algebra systems in abstract algebra courses to share their information on his book's website.

For the last ten years, Gallian and Higgins have co-presented an MAA minicourse intended for faculty on getting students involved in undergraduate research. The minicourse is always well-attended, attesting to the need for such a course. Gallian and Higgins promote academic year research at one's own institution. Higgins presented such a minicourse at the Allegheny Section meeting in 1996, where Gallian was an invited speaker. Joe attended the course and asked her if she would be interested in teaming up with him to present a minicourse at the Joint Mathematics Meetings. The premise of this course is that undergraduate research is an introduction to our profession. Research in mathematics develops mathematics ability in students differently from course work. Gallian and Higgins advocate tailoring interesting and new problems to each individual student's talents and abilities. Extending their collaboration, Gallian and Higgins had their unique common student present a portion of the minicourse from the student's point of view. Stephen Hartke was an undergraduate at the University of Dayton (1995-1999), where he wrote an Honors thesis under the direction of Higgins. He participated in the Duluth REU in 1998, and later returned to serve as a research advisor for two years. His doctoral studies were at Rutgers University. While Hartke was an undergraduate and a beginning graduate student, Gallian and Higgins invited him to present a comparison of his four distinct research experiences — his academic-year undergraduate Honors thesis and his summer REUs at Lafayette College, Duluth, and at the NSA's Director's Summer Program — in this minicourse.

Besides promoting undergraduate research in mathematics, Joe has also helped to establish ways of integrating undergraduates into our profession by supporting opportunities for students to present their work at a national forum where students can interact with many professional mathematicians. In the early 1990s, Joe supported the creation of student paper sessions at the winter meetings (Pi Mu Epsilon had successfully established undergraduate paper sessions at the summer meetings already), and a poster session at the winter meetings. The first poster session started with thirteen posters, and has grown to over one hundred and fifty posters, with the number of posters limited only by the size of the room. Joe helped establish judging of the posters and arranged for prizes. Although giving prizes for posters is sometimes controversial, the act of judging ensures that many professional mathematicians see some of the students' work in detail. This may renew their enthusiasm and broaden their horizons regarding undergraduate research.

9. Communicating with the Project NExT Fellows

Joe Gallian has been involved with the MAA's Project NExT (New Experiences in Teaching) since its inception fourteen years ago in 1994. Project NExT is a professional development program for new and recent PhDs in mathematics and it addresses all aspects of an academic career: improving the teaching and learning of undergraduate mathematics, maintaining research and scholarship, and participating in professional activities. Project NExT's co-founders and directors were T. Christine Stevens and the late James R. C. Leitzel. Each year, about seventy Project NExT Fellows are selected from applicants around the country. Selection criteria include being in the first or second year of full-time post-PhD teaching at a college or university. The Fellows meet at three consecutive national meetings of the MAA in their Fellowship year and at a two-and-a-half day workshop at the first of these meetings.

After Jim Leitzel died in 1998, Chris Stevens invited Gallian and Higgins to join the Project NExT team as Co-Directors. Gavin Larose and Judith Covington, both of whom were in the first cohort of Project NExT Fellows, were already serving on the team as Associate Co-Directors. The nature of Project NExT compels each of the five team members to do something for it almost every day. Most of the work is done in collaboration. Joe's many contributions to the program include reading applications and helping to select the Fellows, providing a Project NExT course on getting your research off to a good start, organizing a panel at the winter meetings, helping to set policy on various parts of the program, and matching Fellows with more experienced consultants. But Joe's enduring, and endearing, contribution to Project NExT is the address "Finding Your Niche in the Profession," dubbed by the Fellows as the "Just say Yes!" talk. This is the closing address of the first workshop that the Fellows attend. This inspiring talk is legendary, and it has now achieved almost-cult status, with people listing the number of times they have heard the talk, and some boasting of being able to lip-synch it! In the talk, which loses much in this paraphrasing, Joe encourages each Project NExT Fellow to find her/his niche in our profession by availing of opportunities present at the Fellows' own institution and trying different ideas of service to students, institution and the profession. He suggests ways to be useful, even irreplaceable, and he tells great stories of trying to make a difference at the institution where you are, instead of pining to be at a place you are not. This is the one session in the Project NExT program that meets with almost unanimous acclamation. But more importantly, Project NExT Fellows send email messages years after they have been inspired by Joe's talk saying that they accepted assignments they would not have taken on and enjoyed the work, sometimes finding new avenues for their creativity, and always finding satisfaction in the service they have performed.

10. Conclusion

For all his talents and many activities, Joe Gallian is a simple man. He is open with his thoughts and he enjoys what is around him — his family, his flowers, the snow! This conference was a celebration of mathematics and of communicating mathematics, at the same time paying tribute to, and thanking, an inspiring communicator of mathematics — Joe Gallian!

References

[1] J. A. Gallian, *Who is the greatest hitter of them all?*, Math Horizons, September 2002, 13-16.

[2] J. A. Gallian, *An optimal football strategy*, The Two-Year College Mathematics Journal **12** (1981) 330-331.

[3] J. A. Gallian, *The search for finite simple groups*, Mathematics Magazine **49** (1976) 163-179.

[4] J. A. Gallian, *The classification of finite simple groups is complete*, FOCUS, 1981.

[5] J. A. Gallian, *The mathematics of identification numbers*, The College Mathematics Journal **22** (1991) 194-202.

[6] J. A. Gallian and D. Moulton, *On groups of order pq*, Mathematics Magazine **68** (1995) 287-288.

[7] J. A. Gallian, *Contemporary Abstract Algebra*, Houghton Mifflin, 6th edition, 2006.

[8] J. A. Gallian, *The digital revolution*, in For All Practical Purposes, W. H. Freeman, 7th edition, 2005.

[9] J. A. Gallian, *History of the Putnam Competition*, `http://www.d.umn.edu/~jgallian`

[10] J. A. Gallian, *Assigning driver's license numbers*, Mathematics Magazine **64** (1991) 13-22.

[11] J. A. Gallian, *How to give a good talk*, Math Horizons **5** (April 1998) 29-30.

[12] J. A. Gallian, *Advice on giving a good PowerPoint presentation*, Math Horizons **13** (April 2006) 25-27.

[13] J. A. Gallian, The Duluth Undergraduate Research Program, `http://www.d.umn.edu/~jgallian/progdesc.html`

[14] Proceedings of the Conference on Summer Undergraduate Mathematics Research Programs, edited by Joseph A. Gallian, American Mathematical Society, 1999. `http://www.ams.org/outreach/REUproceedings.pdf`

[15] Proceedings of the Conference on Promoting Undergraduate Research in Mathematics, edited by Joseph A. Gallian, American Mathematical Society, 2007. `http://www.ams.org/outreach/PURMproceedings.pdf`

Department of Mathematics, University of Dayton, Dayton, OH 45469

Contemporary Mathematics
Volume **479**, 2009

Fair Allocation Methods for Coalition Games

David Housman

Abstract. A coalition game is a mathematical model of situations in which players can make enforceable agreements to cooperate. For each set of players, there is a numerical gain that can be distributed should the players agree to cooperate. This paper describes some interesting classes of coalition games, methods for allocating the gains among the players, and properties that formalize intuitive notions of fairness. We determine on which class of games each method satisfies each property. Some methods are characterized via properties.

Game theory uses mathematical models to explore situations in which two or more decision makers, the players, have an effect on outcomes that each player may value differently. A coalition game is an austere model of situations in which players can make enforceable agreements to cooperate. If some players agree to cooperate, they must know, at minimum, what can be accomplished by their cooperation. A coalition game uses numbers to describe what all sets of players can accomplish through cooperation. These numbers can be thought of as money, utility, value, or gain that can be distributed among the cooperating players.

Given a coalition game, we can ask what *will* happen? Presumably, players will negotiate a distribution of the obtainable gains that is acceptable to each of the cooperating players. We can also ask what *should* happen? For example, an external arbitrator may be asked to impose a distribution of the obtainable gains. Fairness will be part of the answer to either question. An arbitrator is supposed to impose a fair distribution, and in a negotiation, a player is unlikely to agree to a distribution that does not seem fair.

This article will describe coalition games, methods to distribute obtainable gains, and method properties that formalize intuitive notions of fairness. We will see which methods satisfy which properties and then characterize some methods with properties. One goal is to provide an introduction to, not a complete survey of, the literature. Some of the results are new, but most of the results are in the literature. A second goal is to suggest directions for future research that could be accomplished by undergraduate or graduate students of mathematics. Good

2000 *Mathematics Subject Classification.* Primary 91A12, 91A06.
Key words and phrases. Cooperative game, allocation method, axiomatics.
The author thanks Joe Gallian for giving him his first taste of research during the summer of 1977, Bill Lucas for instilling in him an enthusiasm for game theory, and an anonymous referee for prodding him to look more carefully at simple games.

sources to obtain further pointers into the appropriate literature are [**1**, chapters 13, 17, and 18], [**2**, chapters 34 and 36], [**3**, chapters 53-56], [**8**], [**21**], [**22**, chapter 7], [**23**, chapter 5], [**25**], [**32**], and [**37**].

Section 1 defines coalition game and allocation, describes a few examples and classes of coalition games, and describes a few fairness properties for allocations. Section 2 describes several fairness properties for allocation methods and illustrates these properties with a very simple allocation method. Sections 3, 4, and 5 describe many reasonable allocation methods and determine which properties each method satisfies. Section 6 states a mutual incompatibility of three properties and three characterizations of allocation methods via properties. Section 7 suggests directions for future research.

1. Coalition Games

This section defines coalition game and allocation, describes a few examples and classes of coalition games (the containment relationships among these are summarized in Figure 1), and describes a few fairness properties for allocations.

DEFINITION 1.1 (Coalition Game). A *coalition game* consists of a finite set N and a real-valued function w from the subsets of N that satisfies $w(\varnothing) = 0$. An element i in N is called a *player*, and a nonempty subset S of N is called a *coalition*.

The real number $w(S)$ is interpreted as the worth, value, utility, or gain of the coalition S, that is, the amount available to distribute among the players in S if that coalition forms.

EXAMPLE 1.2 (Savings). The government has mandated improvements in the sewage treatment facilities in the cities of Avon, Barport, Claron, and Delmont. Each city could work separately, but \$140 million would be saved by all four working together. If one of the cities was unwilling to cooperate, some triples of cities could also save money: without Delmont's cooperation, Avon, Barport, and Claron could save \$108 million; without Claron's cooperation, Avon, Barport, and Delmont could save \$96 million; without Belmont's cooperation, Avon, Claron, and Delmont could save \$84 million; and without Claron's or Delmont's cooperation, Avon and Barport could save \$24 million. No other subset of the cities could save money over completing the projects individually. In particular, Barport, Claron, and Delmont cannot save any money without the assistance of Avon. This situation can be modeled with a coalition game involving the players $N = \{A, B, C, D\}$, where we represent each city with the first letter of its name, and the function w defined by $w(N) = 140$, $w(\{A, B, C\}) = 108$, $w(\{A, B, D\}) = 96$, $w(\{A, C, D\}) = 84$, $w(\{A, B\}) = 24$, and $w(S) = 0$ for all other coalitions S. In a standard abuse of notation, we usually remove the set brackets and commas in examples. For example, instead of $w(\{A, C, D\}) = 84$, we write $w(ACD) = 84$.

REMARK 1.3. The set of games w with a fixed player set N can be viewed as vector in a $2^{|N|} - 1$ dimensional convex subset of $\mathbb{R}^{2^N}$, where 2^N is the set of subsets of N. So, it makes mathematical sense to add games using vector addition and multiply games by scalars.

Given a coalition game, our goal is to determine a fair distribution of the possible gains. Our presumption will be that all players either choose or are forced (say by government decree) to cooperate.

DEFINITION 1.4 (Allocation). Given a coalition game (N, w), an *allocation* is a real-valued vector x indexed by the players that satisfies the *efficiency condition* $\sum_{i \in N} x_i = w(N)$. The number x_i is called player i's *payoff*.

For the savings coalition game, $(20, 30, 40, 50)$ and $(100, 25, 25, -10)$ are allocations, although neither seems the least bit fair.

EXERCISE 1. *This is a good place for the reader to stop for a few minutes and do one or both of the following. First, find three other people and play the savings coalition game, that is, have each person be a representative for one of the four cities and then negotiate to an acceptable allocation. Second, act as an arbitrator and choose what you think is the most fair allocation.*

In this paper, we will typically consider games in which there is no disincentive for players to cooperate with other players.

DEFINITION 1.5 (Superadditive Game). A coalition game (N, w) is *superadditive* if $w(S) + w(T) \leq w(S \cup T)$ for all coalitions S and T satisfying $S \cap T = \varnothing$.

The savings game is superadditive. Indeed, if $S \cap T = \varnothing$, then at most one of S and T contains player A, and so $w(S) = 0$ or $w(T) = 0$. Since worths are non-decreasing in the cardinality of the coalition, $w(S) + w(T) = \max\{w(S), w(T)\} \leq w(S \cup T)$.

Player D would object to the fairness of $(100, 25, 25, -10)$ because player D could obtain a payoff of 0 without cooperating with anyone else. We posit that players will agree to an allocation only if each player receives at least as much as that player could receive on its own.

DEFINITION 1.6 (Player-Rational Allocation). Suppose (N, w) is a coalition game. The allocation x is *player-rational* if $x_i \geq w(\{i\})$ for all $i \in N$.

The savings game allocation $(20, 30, 40, 50)$ is player-rational, but players A, B, and C may object because they only receive $20+30+40 = 90$ but as a coalition they could obtain $w(ABC) = 108$. A coalition-rational allocation avoids such objections.

DEFINITION 1.7 (Coalition-Rational Allocation). Suppose (N, w) is a coalition game. The allocation x is *coalition-rational* if $\sum_{i \in S} x_i \geq w(S)$ for all $S \subseteq N$.

Clearly, a coalition-rational allocation is player-rational. The savings game allocations $(140, 0, 0, 0)$ and $(8, 56, 44, 32)$ are coalition-rational, but few would argue either is fair. Hence, more intuitions about fairness need to be formalized. At the same time, insisting on coalition-rational allocations may be too restrictive because some games have no coalition-rational allocation.

EXAMPLE 1.8 (Simple Majority). There are at least three players and

$$w(S) = \begin{cases} 1, & \text{if } |S| > |N|/2 \\ 0, & \text{otherwise} \end{cases}$$

This game has no coalition-rational allocations. Indeed, if x were a coalition-rational allocation, then $\sum_{i \in N \setminus \{j\}} x_i \geq 1$ for all $j \in N$. Summing these inequalities, we obtain $(n-1) \sum_{i \in N} x_i \geq n$, contradicting the efficiency condition $\sum_{i \in N} x_i = 1$.

DEFINITION 1.9 (Balanced Game). A coalition game (N, w) is *balanced* if there are coalition-rational allocations.

Since no coalition-rational allocation exists for a simple majority game, we cannot expect a fair allocation to be coalition-rational. In a negotiation, perhaps a coalition of just over half of the players will form and evenly split the gain while the other players are left with payoffs of zero. In an arbitration, there is no fair way to distinguish among the players, and so the same amount should be given to each player. We now formalize this intuition.

DEFINITION 1.10 (Unbiased Allocation). Suppose (N, w) is a coalition game. Players i and j are *indistinguishable* if $w(S \cup \{i\}) = w(S \cup \{j\})$ for all $S \subseteq N \backslash \{i, j\}$. The allocation x is *unbiased* if $x_i = x_j$ for all indistinguishable players i and j.

The unique unbiased allocation for a simple majority game gives a payoff of $1/|N|$ to each player. However, the unbiased property does not directly eliminate any allocations to the savings game. So, instead of only considering players that are indistinguishable, we strengthen the property by considering players that are clearly distinguishable. too.

DEFINITION 1.11 (Strongly Unbiased Allocation). Suppose (N, w) is a coalition game. Player i is called *weaker than* player j if $w(S \cup \{i\}) \leq w(S \cup \{j\})$ for all $S \subseteq N \backslash \{i, j\}$. The allocation x is *strongly unbiased* if $x_i \leq x_j$ whenever player i is weaker than player j.

In the savings game, player D is weaker than player C is weaker than player B is weaker than player A. So, the allocation $(50, 35, 35, 20)$ is strongly unbiased, but the allocation $(8, 56, 44, 32)$ is not strongly unbiased.

The superadditivity criterion can be strengthened to obtain games with very strong incentives for cooperation.

DEFINITION 1.12 (Convex Game). A coalition game (N, w) is *convex* if $w(S) + w(T) \leq w(S \cup T) + w(S \cap T)$ for all $S, T \subseteq N$.

The simple majority game is a game that has no coalition-rational allocation. On the other hand, convex games have "large" sets of coalition-rational allocations [**30**]. Games that are not convex may have coalition-rational allocations. For example, the savings game has the coalition-rational allocation $(140, 0, 0, 0)$, but it is not convex because $w(ABC) + w(ABD) \not\leq w(ABCD) + w(AB)$.

DEFINITION 1.13 (Simple Game). A coalition game (N, w) is *simple* if $w(N) = 1$ and $w(S) \in \{0, 1\}$ for all $S \subseteq N$. If $w(S) = 1$, then S is called *winning*, and if $w(S) = 0$, then S is called *losing*.

Simple games model voting systems, and a fair allocation is interpreted as the voting powers of the voters. The simple majority games are examples of simple games. Here is a more complex example of a simple game.

EXAMPLE 1.14 (Federal Law). The players are the 435 Representatives, 100 Senators, and President of the United States of America. We think of the Vice President, who can break ties among Senators, as a proxy for the President. In order for a proposal to become federal law, it must be approved by the President and simple majorities of the Representatives and Senators, or without approval of the President, it must be approved by two-thirds majorities of the Representatives and Senators. Hence, a coalition is winning if and only if it contains (1) at least 218 Representatives, at least 50 Senators, and the President; or (2) at least 290 of the

Representatives and at least 67 of the Senators. Intuitively, the President has more voting power than a Senator who has more voting power than a Representative.

A superadditive simple game corresponds to voting situations which satisfy the following natural conditions: (1) supersets of winning coalitions are winning, and (2) the complement of a winning coalition must be losing. Convex simple games are the unanimity games.

DEFINITION 1.15 (Unanimity Game). Suppose $T \subseteq N$. The *unanimity game* on T is the game (N, u^T) where

$$u^T(S) = \begin{cases} 1, \text{ if } T \subseteq S \\ 0, \text{ otherwise} \end{cases}$$

In words, a coalition is winning if and only if the coalition contains T. If $T = \{i\}$, then we call player i a *dictator*.

In the unanimity game (N, u^T), two players in T are indistinguishable, and two players in $N \backslash T$ are indistinguishable. Hence, an unbiased allocation gives the same payoff to each player in T and gives the same payoff to each player in $N \backslash T$. Since the players in $N \backslash T$ neither contribute nor detract from gains obtained by any coalition, it seems only fair that players in $N \backslash T$ receive nothing and pay nothing.

DEFINITION 1.16 (Subsidy-Free Allocation). Suppose (N, w) is a coalition game. Player i is called a *dummy* if $w(S) = w(S \backslash \{i\}) + w(\{i\})$ for all coalitions S containing i. Note that we are using the convention $w(\varnothing) = 0$. The allocation x is *subsidy-free* if $x_i = 0$ for all dummy players i.

The unique unbiased and subsidy-free allocation for the unanimity game (N, u^T) is x satisfying $x_i = 1/|T|$ if $i \in T$ and $x_i = 0$ otherwise. The same allocation is the unique unbiased and coalition-rational allocation.

Besides having uniquely determined fair allocations, the unanimity games on n-players also form a basis for the space of games on n-players. Indeed, the number of unanimity games, $2^n - 1$, is the same as the dimension of the space of games, and the space of games is spanned by the unanimity games: $w = \sum_{T \subseteq N} d_T u^T$ where $d_T = \sum_{S \subseteq T} (-1)^{|T|-|S|} w(S)$ [**29**]. The game w is convex if $d_T \geq 0$ for all coalitions T satisfying $|T| \geq 2$

The subsidy-free property is also useful in selecting the most reasonable allocation in another class of games.

DEFINITION 1.17 (Additive Game). If there exists a real-valued vector $z \in \mathbb{R}^N$ for which $w(S) = \sum_{i \in S} z_i$ for all $S \subseteq N$, then the coalition game (N, w) is *additive*.

The unique unbiased and subsidy-free allocation for the additive game (N, w) is x satisfying $x_i = w(\{i\})$. The same allocation is the unique player-rational allocation.

It is sometimes useful to compare allocations in different, but related, games.

EXAMPLE 1.18 (Veto Power). Suppose $N = \{1, 2, \ldots, n\}$, $k \in N$, and $2 \leq r \leq n - 1$. Define

$$v^{k,r}(S) = \begin{cases} 1, \text{ if } k \in S \text{ and } |S| \geq r \\ 0, \text{ otherwise} \end{cases}$$

If x is coalition-rational, then for each player $i \neq k$, $0 \leq x_i = 1 - \sum_{j \in N \backslash \{i\}} x_i \leq 0$, which implies $x_i = 0$, and so $x_k = 1$. Thus, the kth unit vector χ^k is the unique coalition-rational allocation.

The allocation $\chi^1 = (1, 0, \ldots, 0)$ is the unique coalition-rational allocation in both the dictator game $u^{\{1\}}$ and the veto power games $v^{1,r}$. In the dictator game $u^{\{1\}}$, player 1 can obtain 1 without cooperating with the other players. In the veto power games $v^{1,r}$, player 1 needs to cooperate with at least $r - 1$ other players to obtain 1. Restricting ourselves to coalition-rational allocations will not reflect the apparent differences in power player 1 has in these games.

EXAMPLE 1.19 (Cost Overrun). The savings game is based on cost estimates before the improvements have been started. Suppose that the four cities decide to cooperate and a \$20 million cost overrun occurs. Although a cost overrun might have occurred had a different coalition decided to cooperate, for simplicity, we will assume that is not the case. This situation can be modeled with the coalition game (N, w) where $N = \{A, B, C, D\}$, $w(N) = 120$, $w(ABC) = 108$, $w(ABD) = 96$, $w(ACD) = 84$, $w(AB) = 24$, and $w(S) = 0$ otherwise.

It seems reasonable that whatever the fair allocation is for the savings game, each player should receive less in the cost overrun game.

DEFINITION 1.20 (Zero-Normalized Game). The coalition game (N, w) is *zero-normalized* if $w(\{i\}) = 0$ for all $i \in N$.

Except for the unanimity games $u^{\{i\}}$, all of our examples have been zero-normalized. Another non-zero-normalized game could have been defined had we described the savings game differently. Presumably for each coalition S, there is a cost $c(S)$ for the cities in S to jointly improve their sewage treatment facilities. The savings game would have been obtained by computing $w(S) = \sum_{i \in S} c(\{i\}) - c(S)$. It would have been reasonable to instead define the game $v(S) = -c(S)$. This later game would not be zero-normalized. Nonetheless, it would seem that the allocation to player i in w should be $c(\{i\})$ plus the allocation to player i in v.

As suggested by our comparison of the savings and cost overrun games and our comparison of two ways of representing joint costs as a coalition game, a desire for consistent treatment of players across different games motivates the definition of allocation *methods*, the topic of the next section.

In this section, we have described two economics examples (savings and cost overrun), three political science examples (federal law, simple majority, and veto power), and seven classes of games (zero-normalized, superadditive, balanced, convex, simple, unanimity, and additive). The containment relationship among these is illustrated in Figure 1.

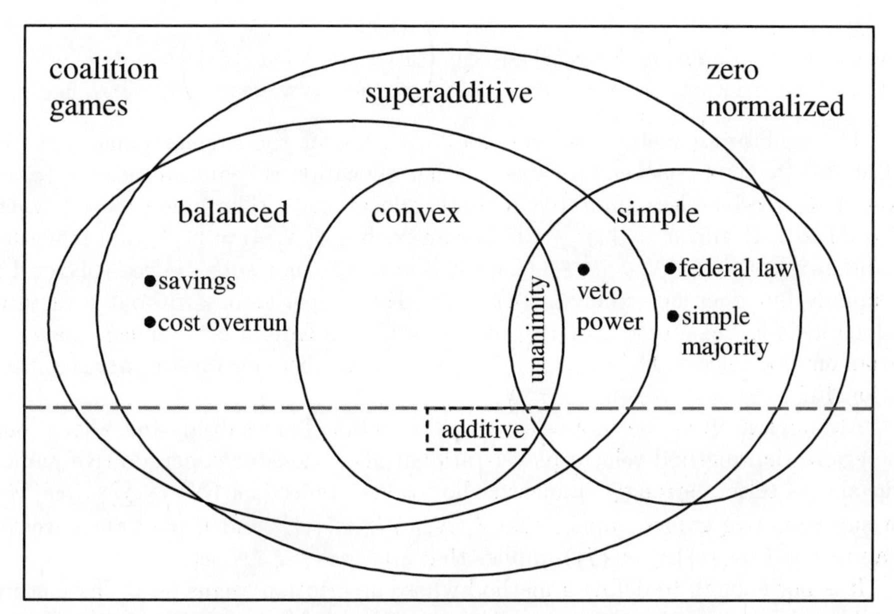

Figure 1. Containment relationships among the classes and examples of coalition games.

There are many other situations that can be modeled as coalition games. Lucas [**18**] describes several voting bodies including the United Nations Security Council, the Israeli Knesset, the United States Electoral College, and New York State county boards. Lucas and Billera [**19**] describe several economic games including airport landing fees, WATS telephone lines, waterway pollution abatement costs, and the sharing of the cost of a communication satellite. Moulin [**22**] describes coalition game models for bilateral assignment markets, marriage markets, output-sharing of production, exchange economies, and fair division with money. Young [**36**] describes a cost allocation problem among six municipalities in Sweden and airport landing fees. Moulin [**23**] models mail distribution and access to a network as coalition games. Driessen [**8**] models water resource development in Japan and bankruptcy as coalition games. A primary focus of Taylor and Zwicker [**32**] is the relationship between weighted voting games and simple games.

2. Allocation Methods and Properties

An allocation can seem fair if it is selected by a method whose description seems fair.

DEFINITION 2.1 (Allocation Method). An *allocation method* or *value* is a function from coalition games to allocations.

Here is a simple example of an allocation method.

DEFINITION 2.2 (Egalitarian Method). The *egalitarian method* gives player i her individual gain, $w(\{i\})$, and an equal share of any possible additional gain to be had by all players cooperating, $w(N) - \sum_{j \in N} w(\{j\})$. An explicit formula is

$$\xi_i(N,w) = w(\{i\}) + \frac{1}{|N|}\left(w(N) - \sum_{j \in N} w(\{j\})\right)$$

The egalitarian method selects $(35, 35, 35, 35)$ for the savings game and $(30, 30, 30, 30)$ for the cost overrun game; neither allocation is coalition-rational (coalition ABC receives less than 108 in both allocations). The game (N,w) where $N = \{1,2,3,4\}$ and $w = 3u^{12} + u^{13}$ is convex, but $\xi(N,w) = (1,1,1,1)$ is neither coalition-rational ($\xi_1(N,w) + \xi_2(N,w) < 3 = w(12)$) nor subsidy-free (player 4 is a dummy but does not receive $w(4) = 0$). For simple games without a dictator (which include the simple majority, veto power, and federal law games), the egalitarian method selects $(1/|N|, \ldots, 1/|N|)$, which is neither subsidy-free nor coalition-rational if $w(S) = 1$ for some $S \neq N$.

Although it often does not select coalition-rational nor subsidy-free allocations, the egalitarian method selects player-rational allocations at superadditive games, and always selects strongly unbiased allocations. Indeed, $w(N) \geq \sum_{j \in N} w(\{j\})$ for superadditive games implies that $\xi_i(N,w) \geq w(\{i\})$, and if player i is weaker than player j, $w(\{i\}) \leq w(\{j\})$ implies that $\xi_i(N,w) \leq \xi_j(N,w)$.

It is not enough to define a method whose description seems fair. The method should satisfy properties that formalize our notions of fairness. The fairness properties described in the previous section for allocations can be turned into fairness properties for methods.

DEFINITION 2.3 (Unbiased, Strongly Unbiased, Subsidy-Free, Player-Rational, and Coalition-Rational Properties). An allocation method α is *unbiased, strongly unbiased, subsidy-free, player-rational,* or *coalition-rational* on a set Γ of coalition games if $\alpha(N,w)$ is unbiased, strongly unbiased, subsidy-free, player-rational, or coalition-rational, respectively, for all $(N,w) \in \Gamma$.

The egalitarian method is player-rational on superadditive games, and unbiased and strongly unbiased on all games. The egalitarian method is neither subsidy-free nor coalition-rational on superadditive, balanced, convex, or simple games. We would like to describe allocation methods that satisfy fairness properties on as large of a class of games as possible.

The goal of the following sections is to describe several allocation methods, apply them to the previously described examples, and explore their properties. In this section, we describe the fairness properties we will consider.

If we give the players different names, the economic situation is left unchanged and so the allocation should be unchanged except for the renaming. Such an allocation method is anonymous, and an anonymous method is unbiased.

DEFINITION 2.4 (Anonymous Property). If (N,w) is a game, $x \in \mathbb{R}^N$, and $\pi : N \to M$ is a bijection, then the game $(\pi(N), \pi w)$ is defined by $(\pi w)(\pi(S)) = w(S)$ and the vector $\pi(x)$ is defined by $\pi(x)_{\pi(i)} = x_i$. An allocation method α is *anonymous* on a set Γ of coalition games if $\alpha(\pi(N), \pi w) = \pi(\alpha(N,w))$ whenever $(N,w) \in \Gamma$, $(\pi(N), \pi w) \in \Gamma$, and $\pi : N \to M$ is a bijection.

A change in units (e.g., euros instead of dollars) in the data should result in only a change of units in the allocations.

DEFINITION 2.5 (Proportionate Property). An allocation method α is *proportionate* on a set Γ of coalition games if $\alpha(N, \lambda w) = \lambda\alpha(N, w)$ for all $(N, \lambda w) \in \Gamma$, $(N, w) \in \Gamma$, and $\lambda > 0$.

If a game is changed by adding 1 to every coalition containing player i, then the allocation should be unchanged except for giving an additional 1 to player i.

DEFINITION 2.6 (Player-Separable Property). An allocation method α is *player-separable* on a set Γ of coalition games if $\alpha_i(N, w + v) = \alpha_i(N, w) + v(\{i\})$ for all $(N, w + v) \in \Gamma$, $(N, w) \in \Gamma$, additive games (N, v), and $i \in N$.

The egalitarian method is clearly anonymous, proportionate, and player-separable.

We can now present our first allocation method characterization theorem. Notice that the characterization holds on four different classes of games. While we want allocation methods that will satisfy fairness properties on as large of a class of games as possible, when we are characterizing allocation methods, the smaller the class of games, the stronger the result.

THEOREM 2.7. *If α is unbiased and player-separable on (all, superadditive, balanced, or convex) 2-player games, then α is the egalitarian method on (all, superadditive, balanced, or convex) 2-player games.*

PROOF. Suppose (AB, u) is a 2-player game. Define (AB, w) by $w(AB) = u(AB) - u(A) - u(B)$ and $w(A) = w(B) = 0$. If u is superadditive, $w(AB) \geq 0$ and so w is convex. Since α is unbiased, $\alpha_A(AB, w) = \alpha_B(AB, w) = \frac{1}{2}w(AB)$. Let (AB, v) be the additive game defined by $v(AB) = u(A) + u(B)$, $v(A) = u(A)$, and $v(B) = u(B)$. Since $u = w + v$ and α is player-separable, $\alpha_i(AB, u) = \alpha_A(AB, w) + v(i) = \frac{1}{2}(u(AB) - u(A) - u(B)) + u(i) = \xi_i(AB, u)$ for $i = A, B$. $\square$

In comparing the savings and cost overrun games, we suggested that each player should be allocated at least as much in the savings game as in the cost overrun game. This notion is captured by the following property:

DEFINITION 2.8 (Aggregate-Monotone Property). An allocation method α is *aggregate-monotone* on a set Γ of coalition games if $\alpha_i(N, v) \leq \alpha_i(N, w)$ for all $(N, v) \in \Gamma$, $(N, w) \in \Gamma$, and $i \in N$ satisfying $v(N) \leq w(N)$ and $v(S) = w(S)$ for all $S \subseteq N$ but $S \neq N$.

In addition to player payoffs increasing with increasing $w(N)$, we may wish to require the payoffs to players in any coalition S to increase with increasing $w(S)$. This notion is captured by the following property:

DEFINITION 2.9 (Coalition-Monotone Property). An allocation method α is *coalition-monotone* on a set Γ of coalition games if $\alpha_i(N, v) \leq \alpha_i(N, w)$ for all $(N, v) \in \Gamma$, $(N, w) \in \Gamma$, and $i \in N$ satisfying $v(S) \leq w(S)$ for all $S \subseteq N$ and $v(S) = w(S)$ for all $S \subseteq N \backslash \{i\}$.

Clearly, a coalition-monotone method is aggregate-monotone. The egalitarian method is coalition-monotone. Indeed, $\partial\xi_i(N, w)/\partial w(S) \geq 0$ for all coalitions S containing i.

Felsenthal and Machover [**11**] argue that measuring voting power in superadditive simple games should satisfy their transfer property, which is a strengthening of the coalition-monotone property. Allocation methods that are not coalition-monotone are subject to serious paradoxes when used as measures of *a priori* relative voting power.

This completes our introduction to some fairness properties for allocation methods and our analysis of the egalitarian method. The next three sections describe and analyze many allocation methods. A table summarizing our analysis can be found in section 6. The reader may find it helpful to refer to that table now for the egalitarian method and then while reading the following sections.

3. Weighted Contribution Methods

The egalitarian method is almost never subsidy-free. To define a subsidy-free allocation method, we need to give a payoff of $w(\{i\})$ to a player i whenever $w(S) = w(S\backslash\{i\})+w(\{i\})$ for all coalitions S that contain i. This suggests that after assigning each player their individual gain, we split any additional gains proportional to quantities involving $w(S)-w(S\backslash\{i\})-w(\{i\})$, which can be interpreted as the gain player i brings to coalition S in addition to what player i can gain as an individual.

DEFINITION 3.1 (Marginal and Synergy Contribution). Suppose (N,w) is a game. Player i's *individual gain* is $w(\{i\})$. Player i's *marginal contribution* to the coalition S is the quantity $w(S) - w(S\backslash\{i\})$. Player i's *synergy contribution* to the coalition S is the quantity $w(S) - w(S\backslash\{i\}) - w(\{i\})$.

DEFINITION 3.2 (Synergy, Banzhaf, and Shapley Values). Suppose $a = (a_2, a_3, \ldots, a_n)$ is a non-zero vector of nonnegative real numbers. Player i's a-weighted sum of synergy contributions is

$$\bar{\sigma}_i^a(N,w) = \sum_{S\ni i} a_{|S|}\left(w(S) - w(S\backslash\{i\}) - w(\{i\})\right) \tag{3.1}$$

where the notation $S \ni i$ means to sum over all $S \subseteq N$ satisfying $i \in S$. If w is additive, then $\bar{\sigma}_i^a(N,w) = 0$ for all $i \in N$. If w is superadditive, $\bar{\sigma}_i^a(N,w) \geq 0$ for all $i \in N$. The *a-synergy value* σ^a gives player i his individual gain, $w(\{i\})$, plus the remaining available non-individual gains, $w(N) - \sum_{j\in N} w(\{j\})$, in proportion to $\bar{\sigma}_i^a(N,w)$:

$$\sigma_i^a(N,w) = w(\{i\}) + \frac{\bar{\sigma}_i^a(N,w)}{\sum\limits_{j\in N}\bar{\sigma}_j^a(N,w)}\left(w(N) - \sum_{j\in N} w(\{j\})\right) \tag{3.2}$$

if $\sum\limits_{j\in N}\bar{\sigma}_j^a(N,w) \neq 0$ and

$$\sigma_i^a(N,w) = \xi_i(N,w)$$

otherwise. Note that for any $\lambda > 0$, the (λa)-synergy value is the same as the a-synergy value. If $a_s = 1$ for all s, then the a-synergy value is denoted by β and called the (normalized) *Banzhaf value*. If $a_s = (s-1)!(n-s)!$ for all s, then the a-synergy value is denoted by φ and called the *Shapley value*.

The Shapley value was defined in [**29**].and has been studied extensively (e.g., [**3**, chapters 53-54] and [**27**]). The Banzhaf value as defined here was mentioned in [**10**]. The Banzhaf value was originally defined as the average sum of marginals in [**4**] on a subclass of simple games called weighted voting games. This corresponds to (3.1) with $a_s = 1/2^{|N|-1}$ for all s and the synergy contributions replaced with marginal contributions. Of course, such functions on games need not yield payoff vectors that satisfy the efficiency condition. Dubey, Neyman, and Weber [**9**] characterized such functions that are linear, anonymous, and player-separable, and so introduced the idea of weighted sums of marginal values. See Malawski [**20**] for some characterizations of the absolute Banzhaf value and references to earlier characterizations. van den Brink and van der Laan [**33**] characterized the "normalized" Banzhaf value, which yields allocations proportional to the sum of marginals. The normalized Banzhaf value and the Banzhaf value as defined here are identical on simple games and on zero-normalized games. The advantage of the a-synergy values are that they are subsidy-free while the normalized Banzhaf value is not.

The computation of the Banzhaf value for the savings game is summarized in the table. Each upper cell contains the synergy contribution for the player named in the column header to the coalition named in the row header. The last two cells record the sum of the synergy contributions and the Banzhaf value.

	Synergy Contribution			
Player	A	B	C	D
ABCD	140	56	44	32
ABC	108	108	84	
ABD	96	96		72
ACD	84		84	84
AB	24	24		
Sum	452	284	212	188
β_i	55.7	35.0	26.1	23.2

For the cost overrun game, each number in the "ABCD" row and the "Sum" row is reduced by 20, and so the Banzhaf value selects $(49.1, 30.0, 21.8, 19.1)$.

For the Shapley value, before summing the synergy contributions, those in the first row are multiplied by $a_4 = (4-1)!(4-4)! = 6$, those in the second through fourth rows are multiplied by $a_3 = (3-1)!(4-3)! = 2$, and those in the fifth row are multiplied by $a_2 = (2-1)!(4-2)! = 2$. The weighted sums are $(1464, 792, 600, 504)$ and the Shapley value selects $(61, 33, 25, 21)$. For the cost overrun game, each weighted sum is reduced by 120, and so the Shapley value selects $(56, 28, 20, 16)$. Since $56+28+20 < 108$, this example shows that the Shapley value is not coalition-rational on balanced games.

Other a-synergy values can be computed in a similar manner. The next theorem allows us to determine for a game the set of all a-synergy values simultaneously.

THEOREM 3.3. *If (N, w) is superadditive, then the set of $\sigma^a(N, w)$ for all non-zero vectors of nonnegative real numbers $a = (a_2, a_3, \ldots, a_n)$ is the convex hull of $\sigma^{\chi^i}(N, w)$ for $i = 2, 3, \ldots, n$, where χ^i is the ith unit vector*

PROOF. If w is additive, then $\sigma^a(N, w) = \xi(N, w)$ for all a, and so the theorem follows. Now suppose w is not additive. Let

$$d_{i,s} = \sum_{S:i\in S,|S|=s} (w(S) - w(S\backslash\{i\}) - w(\{i\}))$$

and

$$A = w(N) - \sum_{j\in N} w(\{j\}).$$

By the superadditivity condition, $d_{i,s} \geq 0$. Since w is not additive, $d_{i,s} > 0$ for some i and s. We can now let

$$\lambda_s = a_s \left(\sum_{j=1}^{n} d_{j,s}\right) / \left(\sum_{r=2}^{n} a_r \left(\sum_{j=1}^{n} d_{j,r}\right)\right).$$

Clearly, $\lambda_s \geq 0$ and $\sum_{s=2}^{n} \lambda_s = 1$. Now

$$\begin{aligned}
\sigma_i^a(N,w) &= w(i) + A \left(\sum_{s=2}^{n} a_s d_{i,s}\right) / \left(\sum_{j=1}^{n} \sum_{r=2}^{n} a_r d_{j,r}\right) \\
&= \sum_{s=2}^{n} \lambda_s w(i) + \sum_{s=2}^{n} a_s A d_{i,s} / \left(\sum_{r=2}^{n} a_r \left(\sum_{j=1}^{n} d_{j,r}\right)\right) \\
&= \sum_{s=2}^{n} \lambda_s \left(w(i) + A d_{i,s} / \sum_{j=1}^{n} d_{j,s}\right) \\
&= \sum_{s=2}^{n} \lambda_s \sigma_i^{\chi^s}(N,w)
\end{aligned}$$

and the theorem follows. □

For the savings game,

$$\begin{aligned}
\bar{\sigma}^{\chi^2}(N,w) &= (24, 24, 0, 0), \\
\bar{\sigma}^{\chi^3}(N,w) &= (288, 204, 168, 156), \\
\bar{\sigma}^{\chi^4}(N,w) &= (140, 56, 44, 32).
\end{aligned}$$

Hence any a-synergy allocation is a convex combination of $\sigma^{\chi^2}(N,w) = (70, 70, 0, 0)$, $\sigma^{\chi^3}(N,w) \approx (49.4, 35.0, 28.8, 26.8)$, and $\sigma^{\chi^4}(N,w) \approx (72.1, 28.8, 22.6, 16.5)$.

We now examine some properties of a-synergy values. It is easy to verify that a-synergy values are player-rational on superadditive games and are unbiased, anonymous, proportionate, player-separable, and subsidy-free on all games.

The a-synergy values are not coalition-rational on balanced games. Indeed, for an a-synergy value, $a_s > 0$ for some s satisfying $2 \leq s \leq n$. If $a_s > 0$ for some s satisfying $2 \leq s \leq n-1$, then for a veto power game $\varepsilon = \sigma_n^a(N, v^{1,s}) > 0$ which implies that $\sum_{i=1}^{n-1} \sigma_1^a(N, v^{1,s}) = 1 - \sigma_n^a(N, v^{1,s}) < 1 = v^{1,s}(N \backslash \{n\})$, and so σ^a is not coalition-rational on the balanced game $(N, v^{1,s})$. If $a_s = 0$ for all s satisfying $2 \leq s \leq n-1$, then $a_n > 0$. Let $w(1234) = 8$, $w(123) = w(124) = w(134) = w(234) = w(12) = 5$, and $w(S) = 0$ otherwise. Then $(3,3,1,1)$ is coalition-rational and $x = \sigma^a(N,w) = (2,2,2,2)$. Hence, $x_1 + x_2 = 4 < w(12)$, and so $\sigma^a(N,w)$ is not coalition-rational on the balanced game (N,w).

On convex games, Shapley [**30**] showed that the Shapley value is coalition-rational and is in the interior of the coalition-rational allocations. This implies that for weights a sufficiently close to the Shapley weights, the a-synergy value is also coalition-rational on convex games.

On the other hand, the Banzhaf value is not coalition-rational on convex games. Indeed, if $w = 31u^{12} + 93u^{23456}$, then w is convex and

$$\begin{aligned}
\bar{\beta}(N,w) &= 31\bar{\beta}(N, u^{12}) + 93\bar{\beta}(N, u^{23456}) \\
&= 31(2^4, 2^4, 0, 0, 0, 0) + 93(0, 2, 2, 2, 2, 2) \\
&= (496, 682, 186, 186, 186, 186),
\end{aligned}$$

and so $\beta(N,w) = (32, 44, 12, 12, 12, 12)$, which implies that $\sum_{i=2}^{6} \beta_i(N,w) = 92 < 93 = w(23456)$.

Not all a-synergy values are aggregate-monotone. Indeed, let $a = \chi^n$, and consider increasing the worth of the grand coalition from a veto power game, $w = v^{1,n-1} + \varepsilon u^N$. Then $\bar{\sigma}^a(N,w) = (1+\varepsilon, \varepsilon, \ldots, \varepsilon)$, and $\sigma_1^a(N,w) = (1+\varepsilon)^2/(1+n\varepsilon)$. Hence, $(\partial\sigma_1^a/\partial\varepsilon)_{\varepsilon=0} = 2 - n < 0$, and so σ^a is not aggregate-monotone.

The Banzhaf value is aggregate-monotone on superadditive games. Because the Banzhaf value is player-separable, it is sufficient to show $\partial\beta_i/\partial w(N)$ at superadditive games (N,w) satisfying $w(S) = 0$ if $|S| = 1$. This simplifies

$$\bar{\beta}_i(N,w) = \sum_{S \ni i} (w(S) - w(S \backslash \{i\})) = \sum_{S \ni i} w(S) - \sum_{S \not\ni i} w(S)$$

(by separating the individual terms into separate sums),

$$\sum_{j \in N} \bar{\beta}_j(N,w) = \sum_{S \subseteq N} (2\,|S| - |N|) w(S)$$

(by interchanging summations over j and S), and

$$\beta_i(N,w) = \frac{\bar{\beta}_i(N,w)}{\sum_{j \in N} \bar{\beta}_j(N,w)} w(N).$$

Using the quotient and product derivative rules, the numerator of $\partial\beta_i/\partial w(N)$ is

$$\begin{aligned}
&\sum_{S \subseteq N} (2\,|S| - |N|) w(S) \left(w(N) + \bar{\beta}_i(N,w) \right) - \bar{\beta}_i(N,w) w(N)\,|N| \\
&= w(N) \sum_{S \subseteq N} (2\,|S| - |N|) w(S) + \bar{\beta}_i(N,w) \sum_{S \neq N} (2\,|S| - |N|) w(S) \\
&= w(N) \sum_{j \in N} \bar{\beta}_j(N,w) + \bar{\beta}_i(N,w) \sum_{s=1}^{\lfloor |N|/2 \rfloor} (|N| - 2s) \left(\sum_{\substack{S \subseteq N: \\ |S| = n-s}} w(S) - \sum_{\substack{S \subseteq N: \\ |S| = s}} w(S) \right)
\end{aligned}$$

which is nonnegative because (N,w) is superadditive.

The Banzhaf value is neither coalition-monotone nor strongly unbiased on convex games. Indeed, let $v(1234) = 32$, $v(S) = 8$ if $|S| = 3$, $v(S) = 4$ if $|S| = 2$, and $v(S) = 0$ if $|S| = 1$. Let $w(1) = 4$ and $w(S) = v(S)$ otherwise. It is easily verified that v and w are convex, $\beta(v) = (8, 8, 8, 8)$, and $\beta(w) = (\frac{146}{19}, \frac{154}{19}, \frac{154}{19}, \frac{154}{19})$. Since $\beta_1(v) > \beta_1(w)$, the Banzhaf value is not coalition-monotone. Since $w(R \cup \{1\}) \geq w(R \cup \{2\})$ for all $R \subseteq \{3,4\}$ and $\beta_1(w) < \beta_2(w)$, the Banzhaf value is not strongly unbiased.

The Banzhaf value is not coalition-monotone on simple games. Felsenthal and Machover [**11**] provided this counterexample: let $N = \{1,2,3,4,5\}$, $v = u^{12}$, and w be the same as v except that $w(1345) = 1$. Then $\beta(N,v) = (\frac{1}{2}, \frac{1}{2}, 0, 0, 0)$ and $\beta(N,w) = (\frac{9}{19}, \frac{7}{19}, \frac{1}{19}, \frac{1}{19}, \frac{1}{19})$. The non-monotoncity is provided by $\beta_1(N,v) > \beta_1(N,w)$.

All a-synergy values are strongly unbiased on zero-normalized games and on superadditive simple games. Indeed, suppose $w(S \cup \{i\}) \leq w(S \cup \{j\})$ for all $S \subseteq$

$N\backslash\{i,j\}$. If (N,w) is zero-normalized, then

$$\begin{aligned}\bar{\sigma}_i^a(N,w) &= \sum\nolimits_{R\subseteq N\backslash\{i,j\}} a_{|R|+1}(w(R\cup\{i\})-w(R)) \\ &\quad+\sum\nolimits_{R\subseteq N\backslash\{i,j\}} a_{|R|+2}(w(R\cup\{i,j\})-w(R\cup\{j\})) \\ &\leq \sum\nolimits_{R\subseteq N\backslash\{i,j\}} a_{|R|+1}(w(R\cup\{j\})-w(R)) \\ &\quad+\sum\nolimits_{R\subseteq N\backslash\{i,j\}} a_{|R|+2}(w(R\cup\{i,j\})-w(R\cup\{i\})) \\ &= \bar{\sigma}_j^a(N,w).\end{aligned}$$

If (N,w) is superadditive and simple but not zero-normalized, then $w=u^{\{k\}}$ for some $k\in N\backslash\{i\}$, and $\bar{\sigma}^a(N,w)=\chi^k$, which implies $\bar{\sigma}_i^a(N,w)\leq\bar{\sigma}_j^a(N,w)$.

In a superadditive simple game, we say that a player i *swings* for coalition S if S is winning and $S\backslash\{i\}$ is losing. For a superadditive simple game without a dictator, the sum in the absolute Banzhaf value formula is the number of times the player swings. For the federal law game, a Representative swings whenever in a coalition of (1) 217 other Representatives, at least 50 Senators,and the President, or (2) 289 other Representatives and at least 67 other Senators. So, a Representative has the following number of swings:

$$r=\binom{434}{217}\sum_{s=50}^{100}\binom{100}{s}+\binom{434}{289}\sum_{s=67}^{100}\binom{100}{s}$$

A Senator swings whenever in a coalition of (1) at least 218 Representatives, 49 other Senators, and the President, or (2) at least 290 Representatives and 66 other Senators. So, a Senator has the following number of swings:

$$s=\binom{99}{49}\sum_{r=218}^{435}\binom{435}{r}+\binom{99}{66}\sum_{r=290}^{435}\binom{435}{r}$$

The President swings whenever in a coalition of (1) at least 218 Representatives and between 50 and 66 Senators, or (2) between 218 and 289 Representatives and at least 67 Senators. So, the President has the following number of swings:

$$p=\left(\sum_{r=218}^{435}\binom{435}{r}\right)\left(\sum_{s=50}^{66}\binom{100}{s}\right)+\left(\sum_{r=218}^{289}\binom{435}{r}\right)\left(\sum_{s=67}^{100}\binom{100}{s}\right)$$

The Banzhaf payoffs can be obtained by dividing the above quantities by $435r+100s+p$. The Banzhaf payoff for a Representative, Senator, and President are 0.00153, 0.00295, and 0.03996, respectively.

4. Shapley Value

In the previous section, we defined the Shapley value φ on n-player games as the a-synergy value with $a_s=(s-1)!(n-s)!$. An equivalent description is that the Shapley value gives each player that player's marginal contribution to each coalition, averaged over all player orders.

THEOREM 4.1. *The* Shapley value φ *is given by*

$$\varphi_i(N,w)=\frac{1}{|N|!}\sum_{\pi}\left(w\left(S^{\pi,i}\right)-w\left(S^{\pi,i}\backslash\{i\}\right)\right) \tag{4.1}$$

where the sum is over all player orders, i.e., one-to-one and onto functions $\pi : N \to \{1, 2, \ldots, |N|\}$, *and* $S^{\pi,i} = \{j \in N : \pi(j) \leq \pi(i)\}$ *is the coalition of player* i *and the players that come before* i *in the order* π. *An equivalent formula is*

$$\varphi_i(N, w) = \sum_{S \subseteq N} \frac{(|S|-1)!\,(|N|-|S|)!}{|N|!} \left(w(S) - w(S \setminus \{i\})\right). \tag{4.2}$$

PROOF. By definition (3.1),

$$\bar{\varphi}_i(N, w) = \sum_{S \ni i} (|S|-1)!(n-|S|)!\,(w(S) - w(S\setminus\{i\}) - w(\{i\}))$$

where $n = |N|$. Since

$$\begin{aligned}\sum_{S : i \in S} (|S|-1)!(|N|-|S|)! &= \sum_{s=1}^{n} \binom{n-1}{s-1}(s-1)!(n-s)! \\ &= \sum_{s=1}^{n} (n-1)! = n!,\end{aligned}$$

it follows that

$$\bar{\varphi}_i(N, w) = \sum_{S \ni i} (|S|-1)!(n-|S|)!\,(w(S) - w(S\setminus\{i\})) - n!w(\{i\}).$$

By interchanging the summations over i and S, we obtain

$$\sum_{i \in N} \bar{\varphi}_i(N, w) = \sum_{S \subseteq N} \sum_{i \in S} (|S|-1)!(n-|S|)!\,(w(S) - w(S\setminus\{i\})) - n! \sum_{i \in N} w(\{i\})$$

For the double sum and a fixed coalition S, the worth $w(S)$ appears as a positive term with coefficient $(|S|-1)!(n-|S|)!$ for each $i \in S$, and the worth $w(S)$ appears as a negative term if $S \neq N$ with coefficient $|S|!(n-|S|-1)!$ for each $i \in N\setminus S$. After simplifying the double sum, the coefficient for $w(N)$ is

$$(n-1)!(n-n)!n = n!,$$

and if $S \neq N$, the coefficient for $w(S)$ is

$$(|S|-1)!(n-|S|)!\,|S| - |S|!(n-|S|-1)!(n-|S|) = 0.$$

Hence,

$$\sum_{i \in N} \bar{\varphi}_i(N, w) = n! \left(w(N) - \sum_{i \in N} w(\{i\}) \right)$$

Plugging into (3.2) and simplifying, we obtain (4.2). We can also obtain (4.2) from (4.1) by counting the number of orders in which player i comes after the players in $S\setminus\{i\}$ and before the players in $N\setminus S$. □

The computation of the Shapley value for the savings game is summarized in the table. Each row corresponds to one of the 24 possible player orders. In the eighth row (in boldface), player B decides to cooperate first and is given the marginal contribution $w(B) - w(\varnothing) = 0 - 0 = 0$, player A decides to cooperate second and is given the marginal contribution $w(BA) - w(B) = 24 - 0 = 24$, player D decides to cooperate third and is given the marginal contribution $w(BAD) - w(BA) = 96 - 24 = 72$, and player C decides to cooperate fourth and is given the marginal contribution $w(BADC) - w(BAD) = 140 - 96 = 44$. The other rows are calculated in an analogous fashion. Finally, the marginal contributions are averaged to obtain the Shapley allocation (61, 33, 25, 21).

Order				Marginal Contribution			
$\pi(A)$	$\pi(B)$	$\pi(C)$	$\pi(D)$	A	B	C	D
1	2	3	4	0	24	84	32
1	2	4	3	0	24	44	72
1	3	2	4	0	108	0	32
1	4	2	3	0	56	0	84
1	3	4	2	0	96	44	0
1	4	3	2	0	56	84	0
2	1	3	4	24	0	84	32
2	**1**	**4**	**3**	**24**	**0**	**44**	**72**
3	1	2	4	108	0	0	32
4	1	2	3	140	0	0	0
3	1	4	2	96	0	44	0
4	1	3	2	140	0	0	0
2	3	1	4	0	108	0	32
2	4	1	3	0	56	0	84
3	2	1	4	108	0	0	32
4	2	1	3	140	0	0	0
3	4	1	2	84	56	0	0
4	3	1	2	140	0	0	0
2	3	4	1	0	96	44	0
2	4	3	1	0	56	84	0
3	2	4	1	96	0	44	0
4	2	3	1	140	0	0	0
3	4	2	1	84	56	0	0
4	3	2	1	140	0	0	0
				61	33	25	21

For the cost overrun game, the marginal contributions in the table stay the same except that for each order, the fourth player's marginal contribution is reduced by 20. Since each player is fourth in one-fourth of the orders, each player's average marginal contribution is reduced by $20/4 = 5$. Hence, the Shapley allocation for the cost overrun game is $(56, 28, 20, 16)$, which is not coalition-rational (consider coalition ABC).

Using (4.2), it is easy to verify that the Shapley value is strongly unbiased and coalition-monotone on all games.

For a superadditive simple game and each player order, there is exactly one marginal contribution of 1, corresponding to the *pivotal* player whose addition changes a losing coalition to a winning coalition, and all other marginal contributions are 0. Hence for superadditive simple games, the Shapley payoff to a player is the fraction of times that player is pivotal among all player orders.

For the federal law game, a Representative is pivotal whenever in a coalition of (1) 217 other Representatives, at least 50 Senators, and the President, or (2) 289 other Representatives and at least 67 other Senators. So, the Shapley value for a

Representative is

$$\frac{1}{536!}\left(\binom{434}{217}\sum_{s=50}^{100}\binom{100}{s}(218+s)!(317-s)!+\right.$$
$$\left.\binom{434}{289}\sum_{s=67}^{100}\binom{100}{s}(289+s)!(246-s)!\right)\approx 0.0010069$$

A Senator is pivotal whenever in a coalition of (1) at least 218 Representatives, 49 other Senators, and the President, or (2) at least 290 Representatives and 66 other Senators. So, the Shapley value for a Senator is

$$\frac{1}{536!}\left(\binom{99}{49}\sum_{r=218}^{435}\binom{435}{r}(50+r)!(485-r)!+\right.$$
$$\left.\binom{99}{66}\sum_{r=290}^{435}\binom{435}{r}(66+r)!(469-r)!\right)\approx 0.0039658$$

Since the Shapley value selects allocations, the Shapley value for the President is

$$1-(435)(0.0010069)-(100)(0.0039658)\approx 0.16542.$$

It is interesting to speculate whether the President has roughly 4% of the voting power related to the passage of federal laws, as suggested by the Banzhaf value, or more than 16%, as suggested by the Shapley value. Brams, Affuso, and Kilgour [**6**] argue that informal analysis and empirical data suggest that the President's power should be comparable to the power of at least one half of the Representatives or Senators combined. As this is more than suggested by either the Banzhaf or Shapley values, Brams, Affuso, and Kilgour advocate the use of a method introduced by Johnston [**16**], which suggests that the President has 77% of the power. On the other hand, Johnston's method is not coalition-monotone, and Felsenthal and Machover [**11**] argue that any allocation method for voting power must be coalition-monotone. There are a-synergy values that assign the President large powers, e.g., the χ^{356}-synergy value assigns the President 99.6% of the power. There is some regularity to Presidential power assigned by χ^r-synergy values, e.g., President power monotonically increases from 9.4% to 99.6% with r increasing from 280 to 356 (one less than needed for a veto override). It is unclear why it would be appropriate to choose the a_r weights with r near but below 356 to be greatest. Perhaps further investigation of which a-synergy values are coalition-monotone would be helpful.

For the veto power games, it is straightforward to calculate

$$\bar{\beta}_i(N,v^{1,r})=\begin{cases}\sum_{s=r}^{n}\binom{n-1}{s-1}, & \text{if } i=1\\ \binom{n-2}{r-2}, & \text{if } i>1\end{cases}$$

and

$$\varphi_i(N,v^{1,r})=\begin{cases}1-\frac{r-1}{n}, & \text{if } i=1\\ \frac{r-1}{n(n-1)}, & \text{if } i>1\end{cases}$$

The table provides a numerical comparison among the Banzhaf, Shapley, and coalition-rational values on small veto power games. The key observation is that a-synergy values detect differences in the powers of the players in different veto power games while a coalition-rational allocation method would not.

n	3	4	4	5	5	5
r	2	3	2	4	3	2
$\beta_1(N, v^{1,r})$	0.60	0.40	0.70	0.29	0.48	0.79
$\varphi_1(N, v^{1,r})$	0.67	0.50	0.75	0.40	0.60	0.80
coalition-rational	1.00	1.00	1.00	1.00	1.00	1.00

5. Weighted Nucleoli

A coalition-rational allocation x for the game (N, w) satisfies $\sum_{i \in S} x_i \geq w(S)$ for all $S \subseteq N$. This motivates us to find an allocation that minimizes the quantities $w(S) - \sum_{i \in S} x_i$. Of course, we cannot simultaneously minimize all of these quantities. Instead, we will minimize the maximum of these quantities after multiplying each quantity by some weighting factor.

DEFINITION 5.1 (Nucleolus). Suppose (N, w) is a coalition game and $a = (a_1, a_2, \ldots)$ is a sequence of positive real numbers. The *a-excess* of coalition S at an allocation x is the quantity

$$e^a(S, x) = a_{|S|}\left(w(S) - \sum_{i \in S} x_i\right).$$

The *a-excess vector* at an allocation x, denoted $e^a(x)$, is the vector of numbers $e^a(S, x)$ for $S \subseteq N$ ordered from largest to smallest. We order excess vectors lexicographically, that is, $e(x) <_{lex} e(y)$ if there is a positive integer k for which $e_j(x) = e_j(y)$ for $j < k$, and $e_k(x) < e_k(y)$. The *a-nucleolus* $\nu^a(N, w)$ for (N, w) is the player-rational allocation whose a-excess vector is the lexicographic minimum. The *a-prenucleolus* for (N, w) is the allocation (not necessarily player-rational) whose a-excess vector is the lexicographic minimum. If a is a constant sequence, we obtain the *nucleolus* $\nu = \nu^a$. If $a_k = 1/k$, we obtain the *per capita nucleolus* $\nu^{PC} = \nu^a$.

The nucleolus was defined by Schmedler [**28**], and the per capita nucleolus was defined by Grotte [**12**]. Wallmeier [**34**] defined weighted versions but limited his investigation to nonincreasing sequences a. Derks and Haller [**7**] considered a more general class of weighted nucleoli in which the weights depend on n and S, instead of only the cardinality of S. Our smaller class was chosen so that the values would be anonymous.

The excess $e^a(S, x)$ is a measure of coalition S's dissatisfaction with the allocation x. The allocation x is coalition-rational if and only if all excesses at x are nonpositive. By minimizing the maximum excess, the a-nucleolus will be coalition-rational if coalition-rational allocations exist. Since the veto power game $(N, v^{i,r})$ has the unique coalition-rational allocation χ^i, it follows that $\nu^a(N, v^{i,r}) = \chi^i$ for arbitrary a.

It is not immediately clear from its definition whether an a-nucleolus exists and is unique. Proofs appear in [**7**], [**28**], and [**34**]. The approach for the existence proof is to note that the set X_0 of player-rational allocations is nonempty and compact. We now inductively construct subsets of X_0 and subcollections of $\mathcal{C}_0$, the collection of all coalitions: Given a nonempty and compact set of allocations X_k

and a non-empty collection $\mathcal{C}_k$ of subsets of N, the function

$$f_k(x) = \max_{S \in \mathcal{C}_k} e^a(S, x)$$

is a continuous function on X_k, and so

$$b_{k+1} = \min_{x \in X_k} f_k(x)$$

exists, the set

$$X_{k+1} = \{x \in X_k : f_k(x) = b_k\}$$

is nonempty and compact,

$$\mathcal{B}_{k+1} = \{S \in \mathcal{C}_k : e^a(S, x) = b_k \text{ for all } x \in X_{k+1}\}$$

is nonempty, and the collection

$$\mathcal{C}_{k+1} = \mathcal{C}_k \backslash \mathcal{B}_{k+1}$$

is a strict subset of $\mathcal{C}_k$. Eventually, $\mathcal{C}_{k+1} = \varnothing$, and the elements of X_{k+1} are allocations whose a-excess vector are the lexicographic minimums among allocations in X_0. Notice that each b_k and X_k can be found by solving a linear program.

By definition, every a-nucleolus is player-rational on all games. It is easy to verify that every a-nucleolus is unbiased, anonymous, proportionate, and player-separable on all games. Before we consider other properties, we will determine the nucleolus and per capita nucleolus for our examples. Instead of solving linear programs, we will propose an allocation and provide a verification that the proposal is the a-nucleolus.

For the savings game, the table shows the excesses for the allocations $x = (74, 28, 22, 16)$, $y = (84, 20, 20, 16)$, and $z = (68, 36, 20, 16)$.

S	$e(S,x)$	$e(S,y)$	$e(S,z)$
ABC	-16	-16	-16
ABD	-22	-24	-24
ACD	-28	-20	-36
BCD	-66	-72	-56
AB	-78	-80	-80
AC	-96	-88	-104
AD	-90	-84	-100
BC	-50	-56	-40
BD	-44	-52	-36
CD	-38	-36	-36
A	-74	-68	-84
B	-28	-36	-20
C	-22	-20	-20
D	-16	-16	-16

The first three components of $e(x)$ are -16, -16, and -22, and the first three components of $e(y)$ are -16, -16, and -20; hence, $e(x) <_{lex} e(y)$. The vectors $e(y)$ and $e(z)$ agree on the first 7 components and $e_8(y) = -52 < -36 = e_8(z)$; hence, $e(y) <_{lex} e(z)$. This shows that neither y nor z is the nucleolus.

The allocation $x = (74, 28, 22, 16)$ is the nucleolus for the savings game. Indeed, suppose p is an allocation satisfying $e(p) \leq_{lex} e(x)$. First, since $e_1(x) = -16$, it follows that $e(S,p) \leq -16$ for all $S \neq \varnothing, N$. In particular,

$$-16 \geq e(ABC,p) = 108 - p_A - p_B - p_C$$
$$-16 \geq e(D,p) = -p_D.$$

Summing these two inequalities and using the efficiency condition, we obtain

$$\begin{aligned} -32 &\geq 108 - p_A - p_B - p_C - p_D \\ &= 108 - 140 = -32. \end{aligned}$$

So, the inequalities must be equalities, and we have $p_D = 16$ and $e(ABC,p) = e(D,p) = -16$. Second, since $e_i(p) = e_i(x)$ for $i \leq 2$, and $e(p) \leq_{lex} e(x)$, it follows that $e(S,p) \leq e_3(x) = -22$ for all $S \neq \varnothing, N, ABC, D$. In particular,

$$-22 \geq e(ABD,p) = 96 - p_A - p_B - p_D$$
$$-22 \geq e(C,p) = -p_C.$$

Summing these two inequalities and using the efficiency condition, we obtain

$$\begin{aligned} -44 &\geq 96 - p_A - p_B - p_C - p_D \\ &= 96 - 140 = -44. \end{aligned}$$

So, the inequalities must be equalities, and we have $p_C = 22$ and $e(ABD,p) = e(C,p) = -22$. Third, since $e_i(p) = e_i(x)$ for $i \leq 4$, and $e(p) \leq_{lex} e(x)$, it follows that $e(S,p) \leq e_5(x) = -28$ for all $S \neq \varnothing, N, ABC, D, ABD, C$. In particular,

$$-28 \geq e(ACD,p) = 84 - p_A - p_B - p_D$$
$$-28 \geq e(B,p) = -p_B.$$

Summing these two inequalities and using the efficiency condition, we obtain

$$\begin{aligned} -56 &\geq 84 - p_A - p_B - p_C - p_D \\ &= 84 - 140 = -56. \end{aligned}$$

So, the inequalities must be equalities, and we have $p_B = 28$. Fourth, using the efficiency condition, $p_A = 140 - p_B - p_C - p_D = 74$, and $p = x$.

The allocation $x = (84, 18, 12, 6)$ is the nucleolus for the cost overrun game. Indeed, the reader can readily verify that the first six components of $e(x)$ are again the six excesses $e(ABC,x) = e(D,x)$, $e(ABD,x) = e(C,x)$, and $e(ACD,x) = e(B,x)$. The argument of the previous paragraph, with $w(ABCD)$ changed from 140 to 120, shows that x is the nucleolus for the cost overrun game.

Since $\nu_1(w^{\text{savings}}) < \nu_1(w^{\text{cost overrun}})$, the nucleolus is not aggregate-monotone on balanced games.

The per capita nucleolus for the savings game is $x = (101.6, 17.6, 12.8, 8)$. Indeed, the first seven components of $e^{PC}(x)$ are

$$e^{PC}(ABC,x) = e^{PC}(D,x) = -8$$
$$e^{PC}(ABD,x) = e^{PC}(CD,x) = -10.4$$
$$e^{PC}(ACD,x) = e^{PC}(BD,x) = e^{PC}(C,x) = -12.8.$$

Suppose p is an allocation satisfying $e^{PC}(p) \leq_{lex} e^{PC}(x)$. First, since $e_1^{PC}(x) = -8$, it follows that $e(S,p) \leq -8$ for all $S \neq \varnothing, N$. In particular,

$$-8 \geq e^{PC}(ABC,p) = (108 - p_A - p_B - p_C)/3$$
$$-8 \geq e^{PC}(D,p) = -p_D.$$

Adding three times the first inequality to the second inequality and using the efficiency condition, we obtain

$$-32 \geq 108 - p_A - p_B - p_C - p_D = -32.$$

So, the inequalities must be equalities, and we have $p_D = 8$ and $e^{PC}(ABC, p) = e^{PC}(D, p) = -8$. Second, since $e_i^{PC}(p) = e_i^{PC}(x)$ for $i \leq 2$, and $e^{PC}(p) \leq_{lex} e^{PC}(x)$, it follows that $e^{PC}(S, p) \leq e_3^{PC}(x) = -10.4$ for all $S \neq \varnothing, N, ABC, D$. In particular,

$$-10.4 \geq e^{PC}(ABD, p) = (96 - p_A - p_B - p_D)/3$$
$$-10.4 \geq e^{PC}(CD, p) = (-p_C - p_D)/2.$$

Adding three times the first inequality to two times the second inequality and using the efficiency condition and $p_D = 8$, we obtain

$$\begin{aligned} -52 &\geq 96 - p_A - p_B - p_C - 2p_D \\ &= 96 - 140 - 8 = -52. \end{aligned}$$

So, the inequalities must be equalities, and we have $p_C = 12.8$ and $e^{PC}(ABD, p) = e^{PC}(C, p) = -10.4$. Third, since $e_i^{PC}(p) = e_i^{PC}(x)$ for $i \leq 4$, and $e^{PC}(p) \leq_{lex} e^{PC}(x)$, it follows that $e^{PC}(S, p) \leq e_5^{PC}(x) = -12.8$ for all $S \neq \varnothing, N, ABC, D, ABD, CD$. In particular,

$$-12.8 \geq e^{PC}(ACD, p) = (84 - p_A - p_C - p_D)/3$$
$$-12.8 \geq e^{PC}(BD, p) = (-p_B - p_D)/2.$$

Adding three times the first inequality to two times the second inequality and using the efficiency condition and $p_D = 8$, we obtain

$$\begin{aligned} -64 &\geq 84 - p_A - p_B - p_C - 2p_D \\ &= 84 - 140 - 8 = -64. \end{aligned}$$

So, the inequalities must be equalities, and we have $p_B = 17.6$. Fourth, using the efficiency condition, $p_A = 140 - p_B - p_C - p_D = 101.6$, and $p = x$.

The per capita nucleolus for the savings game is $x = (96.6, 12.6, 7.8, 3)$. Indeed, the first six components of $e^{PC}(x)$ are the six excesses $e^{PC}(ABC, x) = e^{PC}(D, x)$, $e^{PC}(ABD, x) = e^{PC}(CD, x)$, and $e^{PC}(ACD, x) = e^{PC}(BD, x)$. The argument of the previous paragraph, with $w(ABCD)$ changed from 140 to 120, shows that x is the per capita nucleolus for the cost overrun game.

The approach we have been using to verify that a proposed allocation is the a-nucleolus for a game was first generalized by Kohlberg [**17**] to the nucleolus and then by Potters and Tijs [**26**] for the a-nucleolus. They showed that an allocation is the a-nucleolus if and only if each collection of coalitions whose excess is greater than some number satisfies a combinatorial balancing condition.

We now compute the nucleolus and per capita nucleolus for the federal law game. Since the a-nucleolus is unbiased, the payoff to each representative is the same number r, the payoff to each senator is the same number s, and the payoff to the president is p. Using the efficiency condition, $435r + 100s + p = 1$. Since the a-nucleolus is player-rational, the winning coalition excesses are nonnegative and the losing coalition excesses are nonpositive. If the weights a are nonincreasing (as is the case for the nucleolus and per capita nucleolus), it follows that the maximum excesses will occur with the minimal winning coalitions..The excess for the regular minimal winning coalitions is $a_{269}(1 - 218r - 50s - p) = a_{269}(217r + 50s)$, and the excess for the veto override minimal winning coalitions is $a_{357}(1 - 290r - 67s)$. In order to minimize the maximum excess, we must solve

$$\begin{array}{ll} \min & z \\ \text{s.t.} & z \geq a_{269}(217r + 50s) \\ & z \geq a_{357}(1 - 290r - 67s) \\ & 435r + 100s + p = 1 \\ & r \geq 0, s \geq 0, p \geq 0 \end{array}$$

For the weights a of interest,

$$\begin{aligned} r &= 0 \\ s &= a_{357}/(50a_{269} + 67a_{357}) \\ p &= (50a_{269} - 33a_{357})/(50a_{269} + 67a_{357}). \end{aligned}$$

For the nucleolus, $s = 1/117 \approx 0.0085$ and $p = 17/117 \approx 0.1453$. For the per capita nucleolus, $s \approx 0.0075$ and $p \approx 0.2501$. Just as for the veto power games, the a-nucleoli allocate more to the strongest player in the federal law game than the a-synergy values. It is interesting to note that for large a_{269} in comparison to a_{357} the a-nucleolus allocates almost all power to the President. This suggests that if larger coalitions are considered far less important than smaller coalitions (perhaps because smaller coalitions are far more likely to form), then Presidential power is greater.

Every a-nucleolus is subsidy-free on all games. Indeed, suppose (N, w) is a game in which player k is a dummy and x is a player-rational allocation for which $x_k \neq w(\{k\})$. Since x is player-rational, $x_k > w(\{k\})$. We will show that x is not the a-nucleolus. Since k is a dummy, for all coalitions S containing k, it follows that

$$\begin{aligned} e^a(S\backslash\{k\}, x) - e^a(S, x) &= -w(S\backslash\{k\}) - w(S) + x_k \\ &> -w(S\backslash\{k\}) - w(S) + w(\{k\}) \\ &= 0. \end{aligned}$$

By taking a small amount from k and giving it to other players, we can obtain a new player-rational allocation y satisfying $e_1^a(y) < e_1^a(x)$, and so x is not the a-nucleolus.

Every a-nucleolus is strongly unbiased on all games. Indeed, suppose (N, w) is a game in which player i is weaker than player j, and x is a player-rational allocation for which $x_i > x_j$. We will show that x is not the a-nucleolus. For all $R \subseteq N\backslash\{i, j\}$, our supposition implies

$$\begin{aligned} e^a(R \cup \{j\}, x) - e^a(R \cup \{i\}, x) &= w(R \cup \{j\}) - w(R \cup \{i\}) - x_j + x_i \\ &> 0. \end{aligned}$$

Hence, by taking a small amount from i and giving it to j, we can obtain a new player-rational allocation y satisfying $e^a(R \cup \{j\}, x) > e^a(R \cup \{j\}, y) > e^a(R \cup \{i\}, y) > e^a(R \cup \{i\}, x)$ for all $R \subseteq N\backslash\{i, j\}$ and $e^a(S, x) = e^a(S, x)$ for all $S \subseteq N$ satisfying $S \cap \{i, j\} = \{i, j\}$ or $\varnothing$. Thus, $e^a(y) <_{lex} e^a(x)$, and so x is not the a-nucleolus.

The nucleolus is not aggregate-monotone on convex games. Hokari [**13**] provided the following counterexample. Let $N = \{1, 2, 3, 4\}$ and $v(1234) = 10$,

$v(123) = 4$, $v(124) = v(134) = v(234) = 6$, $v(12) = v(14) = v(23) = v(24) = v(34) = 2$, and $v(S) = 0$ otherwise. It can be verified that (N, v) and $(N, v+2u^{1234})$ are convex games, $\nu(N, v) = (2, 2, 2, 4)$, and $\nu(N, v + 2u^{1234}) = (3, 3, 3, 3)$.

The per capita nucleolus is aggregate-monotone on all games. Young, Okada, and Hashimoto [**38**] proved this, and we sketch the proof for the per capita *pre*nucleolus. Suppose (N, v) and (N, w) are games satisfying $v(N) \leq w(N)$ and $v(S) = w(S)$ for all $S \neq N$. Let f be defined by $f_i(x) = x_i + \varepsilon$, where $\varepsilon = (w(N) - v(N))/|N|$. Clearly, f is a bijection between the allocations of (N, v) and the allocations of (N, w). Also,

$$\begin{aligned}(v(S) - \sum_{i \in S} x_i)/|S| &= (w(S) - \sum_{i \in S}(f_i(x) - \varepsilon))/|S| \\ &= (w(S) - \sum_{i \in S} f_i(x))/|S| + \varepsilon,\end{aligned}$$

that is, in going from x to $f(x)$, the excesses all increase by ε. Thus, if x is the $\leq_{lex}$ minimum on the set of allocations of (N, v), then $f(x)$ is the $\leq_{lex}$ minimum on the set of allocations of (N, w). That is, if x is the per capita prenucleolus, then $f(x)$ is the per capita prenucleolus.

Theorem 6.1 will show that no a-nucleolus is coalition-monotone on balanced games.

If the weights a are nonincreasing (which includes the nucleolus and per capita nucleolus), then the a-nucleolus is not coalition-monotone on simple games. Indeed, let (N, v) be the five-player superadditive simple game with minimal winning coalitions 134, 135, 145, 234, 235, and 245. Then $\nu^a(N, v) = (0, 0, \frac{1}{3}, \frac{1}{3}, \frac{1}{3})$ because the corresponding excess for each minimal winning coalitions is $\frac{1}{3}a_3$, the only other positive excess is $\frac{1}{3}a_4 \leq \frac{1}{3}a_3$ for 1234, 1235, and 1245, and if x is any other player-rational allocation satisfying $e^a(x) \leq_{\text{lex}} e^a(0, 0, \frac{1}{3}, \frac{1}{3}, \frac{1}{3})$, then $e^a(S, x) \leq \frac{1}{3}a_3$ for all minimal winning coalitions S, $a_3 x_1 \geq 0$, and $a_3 x_2 \geq 0$, which when summed yield $2a_3 \leq 2a_3$ implying that these inequalities hold with equality yielding $x = (0, 0, \frac{1}{3}, \frac{1}{3}, \frac{1}{3})$. Let (N, w) be the five-player superadditive simple game with minimal winning coalitions 134, 135, 145, 234, 235, 245, and 123. Then $\nu^a(N, w) = (\frac{1}{5}, \frac{1}{5}, \frac{1}{5}, \frac{1}{5}, \frac{1}{5})$ because the corresponding excess for each minimal winning coalitions is $\frac{2}{5}a_3$, the only other positive excess is $\frac{1}{5}a_4 < \frac{2}{5}a_3$ for the four-player coalitions, and if x is any other player-rational allocation satisfying $e^a(x) \leq_{\text{lex}} e^a(\frac{1}{5}, \frac{1}{5}, \frac{1}{5}, \frac{1}{5}, \frac{1}{5})$, then $e^a(S, x) \leq \frac{2}{5}a_3$ for $S =$ 134, 145, 235, 245, and 123, which when summed yield $2a_3 \leq 2a_3$ implying that these inequalities hold with equality yielding $x = (\frac{1}{5}, \frac{1}{5}, \frac{1}{5}, \frac{1}{5}, \frac{1}{5})$. Since $\nu^a_3(N, v) > \nu^a_3(N, w)$, ν^a is not coalition-monotone.

It is an open question whether some a-nucleoli are coalition-monotone on convex or simple games. It would also be interesting to characterize which a-nucleoli are aggregate-monotone on all games.

6. Method Characterizations

In the following table, we state the maximal class(es) of games (among all, superadditive, balanced, convex, zero-normalized, and/or simple) on which each method satisfies each property. There are question marks when it is known that the property does not hold on any larger classes but it is not known whether the property holds on the stated class. All of these methods are unbiased, anonymous, proportionate, and player-separable on all games.

	ξ	β	φ	v	ν^{PC}
Subsidy-Free	None	All	All	All	All
Strongly Unbiased	All	Zero $\cup$ Super Simple	All	All	All
Player-Rational	Super	Super	Super	All	All
Coalition-Rational	None	None	Convex	Balanced	Balanced
Aggregate-Monotone	All	Super	All	None	All
Coalition-Monotone	All	Zero $\cap$ Convex?	All	None	Convex?

Each method has its positive and negative features. In particular, the Shapley value and the per capita nucleolus satisfy most of the properties. However, the Shapley value is not coalition-rational on balanced games and the per capita nucleolus is not coalition-monotone on balanced games. Since both properties are desirable, it is natural to ask whether a method exists that has both properties.

THEOREM 6.1. *There is no coalition-rational and coalition-monotone allocation method for balanced games with four or more players .*

This impossibility was shown for five or more players by Young [**35**]. The proof below is by Housman and Clark [**15**].

PROOF. Suppose α is a coalition-rational and coalition-monotone allocation method for balanced games. Let $N = \{1, 2, 3, 4\}$ and $w(N) = 2$, $w(123) = w(124) = w(134) = w(234) = w(13) = w(14) = w(23) = w(24) = 1$, and $w(S) = 0$ otherwise. Clearly, $(\frac{1}{2}, \frac{1}{2}, \frac{1}{2}, \frac{1}{2})$ is coalition-rational, and so (N, w) is a balanced game. Let w^1, w^2, w^3, and w^4 be the same as w except that $w^1(134) = 2$, $w^2(234) = 2$, $w^3(123) = 2$, and $w^4(124) = 2$. Suppose x is a coalition-rational allocation for (N, w^1). Then $0 = w^1(2) \leq x_2 = w^1(N) - x_1 - x_3 - x_4 \leq w^1(N) - w^1(134) = 0$, and so $x_2 = 0$. Furthermore, $x_3 = x_2 + x_3 \geq w^1(23) = 1$ and $x_4 = x_2 + x_4 \geq w^1(24) = 1$. It now follows from efficiency and $x_1 \geq 0$ that $x_3 = x_4 = 1$. Hence, $(0, 0, 1, 1)$ is the only possible coalition-rational allocation, and it is easily verified that $(0, 0, 1, 1)$ is coalition-rational. Since α is coalition-rational, $\alpha(w^1) = (0, 0, 1, 1)$. Analogous arguments imply $\alpha(w^2) = (0, 0, 1, 1)$, $\alpha(w^3) = (1, 1, 0, 0)$, and $\alpha(w^4) = (1, 1, 0, 0)$.

Notice that $w(134) < w^1(134)$ and $w(S) = w^1(S)$ for all $S \neq 134$. Since α is coalition-monotone, $\alpha_1(w) \leq \alpha_1(w^1) = 0$. Analogous arguments imply $\alpha_i(w) \leq \alpha_i(w^i) = 0$ for $i = 1, 2, 3, 4$. But this violates the efficiency of $\alpha(w)$. This contradiction proves the theorem. □

Theorem 6.1 shows that there are limits to the number and strength of the fairness properties we can impose. If players are able to choose whether or not they will cooperate in a joint economic venture, the use of a coalition-rational method, such as the weighted nucleoli, is indicated, and if changes in coalition worths are unlikely, then the coalition-monotone property is not crucial. As argued by Felsenthal and Machover [**11**], the measurement of voting power should use a coalition-monotone method such as the Shapley value and other, as of yet unidentified, weighted contribution values; the loss of the coalition-rational property is bolstered by our earlier remark that coalition-rational methods do not recognize the power differences among the dictator and veto power games.

Theorem 6.1 also suggests that there may be greater possibilities if we can restrict ourselves to smaller classes of games. For example, we have already noted that the Shapley value is coalition-rational and coalition-monotone on convex games. Housman and Clark [**15**] showed that there are many coalition-rational and coalition-monotone allocation methods (the nucleolus and per capita nucleolus being two of them) when restricted to three-player balanced games.

Some collections of properties characterize a single allocation method. Although we want methods to satisfy properties on large classes of games, it is useful to characterize methods on the smallest class of games possible.

player-separable requires that the allocation of a sum of two games, where one game is additive, is the sum of the allocations for the two games. Mathematically, it would perhaps be elegant to strengthen this property by removing the restriction that one game be additive. This has a real-world fairness interpretation when a coalition game could be considered the sum of separate games (e.g., the savings game may be the sum of savings from land acquisition, materials purchasing, and labor). The allocation from the sum game and the sum of the allocations from the separate summand games should be the same. Otherwise, the method of accounting has an effect on the allocation.

DEFINITION 6.2 (Additive Property). An allocation method α is *additive* on a set Γ of coalition games if $\alpha(N, v+w) = \alpha(N, v) + \alpha(N, w)$ for all $(N, v) \in \Gamma$, $(N, w) \in \Gamma$, and $(N, v+w) \in \Gamma$.

The egalitarian method and Shapley value are additive. No other a-synergy value is additive. The a-nucleoli are piecewise additive, that is, an a-nucleolus is additive on each element of some partition (the partitioning depending on a) of the space of all games into convex sets. The following theorem was first proved by Shapley [**29**].

THEOREM 6.3. *If an allocation method α is unbiased, player-separable, and additive on convex games, then α is the Shapley value.*

PROOF. The Shapley value clearly satisfies the four properties. We showed in section 1 that given any game (N, w), we can write $w = \sum_{T \subseteq N} d_T u^T$, where u^T is the unanimity game on T and d_T is a number dependent on w. Since α is unbiased and player-separable, $\alpha(N, d_T u^T) = (d_T / |T|)\chi^T$. Since α is additive, $\alpha(N, w) = \sum_{T \subseteq N} \alpha(N, d_T u^T)$. This shows that the method is uniquely determined at each game. We can limit our characterizing class to convex games because the unanimity games are convex and any convex game can be written as a linear combination of unanimity games. □

We can strengthen aggregate and coalition monotonicity one step further.

DEFINITION 6.4 (Strongly Monotone Property). An allocation method α is *strongly monotone* on a set Γ of coalition games if $\alpha_i(N, v) \leq \alpha_i(N, w)$ for all $(N, v) \in \Gamma$, $(N, w) \in \Gamma$, and $i \in N$ satisfying $v(S) - v(S\backslash\{i\}) \leq w(S) - w(S\backslash\{i\})$ for all $S \subseteq N$ containing i.

Young [**35**] proved the following theorem by inducting on the number of non-zero terms in the sum $w = \sum_{T \subseteq N} d_T u^T$.

THEOREM 6.5. *If an allocation method* α *is unbiased and strongly monotone on superadditive games, then* α *is the Shapley value.*

We have suggested what a fair allocation method should select when games are added (player-separable and additive) and when some of the coalition worths are changed (aggregate, coalition, and strongly monotone). The last property suggests what a fair allocation method should select if some players want to renegotiate their payoffs amongst themselves. Suppose an allocation x has been proposed, the players in a coalition T wish to renegotiate amongst themselves, and the remaining players are satisfied with their payoffs. To determine its worth in the renegotiation, a coalition S of T should be able to join with a coalition Q outside of T as long as the players in Q are compensated in accordance with x. Presumably, S will choose Q to maximize its worth.

DEFINITION 6.6 (Reduced Game). Let (N, w) be a game, T a coalition, and x an allocation. The *reduced game* with respect to T and x is the game $(T, w^{T,x})$ defined by

$$w^{T,x}(S) = \begin{cases} 0 & \text{if } S = \varnothing \\ \sum_{i \in T} x_i & \text{if } S = T \\ \max_{Q \subseteq N \setminus T} \left(w(S \cup Q) - \sum_{i \in Q} x_i \right) & \text{otherwise} \end{cases}$$

For the savings game (N, w), the coalition $T = \{A, B\}$, and the egalitarian allocation $x = (35, 35, 35, 35)$, the reduced game $(T, w^{T,x})$ is defined by $w^{T,x}(AB) = x_A + x_B = 70$, $w^{T,x}(A) = \max\{0, 0 - 35, 0 - 35, 84 - 70\} = 14$, and $w^{T,x}(B) = \max\{0, 0-35, 0-35, 0-70\} = 0$. Notice that $\xi_A(T, w^{T,x}) = 42 \neq 35 = \xi_A(N, w)$. If the egalitarian method is used to allocate, renegotiations among smaller coalitions may result in inconsistencies. Formally, the egalitarian method is not reduced game consistent.

DEFINITION 6.7 (Reduced Game Consistent Property). An allocation method α is *reduced game consistent* on a set Γ of coalition games if $(N, w) \in \Gamma$, $T \subseteq N$, and $T \neq \varnothing$ implies $(T, w^{T,x}) \in \Gamma$ and $\alpha_i(T, w^{T,x}) = \alpha_i(N, w)$ for all $i \in T$.

Notice that the reduced game consistent property insists that the reduced games created are in the focal class of games. This may force us to expand the class of games under consideration. For example, the 3-player veto power game is superadditive and simple, but the reduced game with respect to any pair of players and the egalitarian allocation is neither superadditive nor simple. So, the egalitarian method is not reduced game consistent on superadditive or simple games.

Suppose α is a player-separable, strongly unbiased, and reduced game consistent allocation method. Suppose $x = \alpha(N, w)$ for the savings game (N, w). Since x is strongly unbiased, $84 - x_C - x_D > 0$, and so $w^{AB,x}(AB) = x_A + x_B$, $w^{AB,x}(A) = 84 - x_C - x_D$, and $w^{AB,x}(B) = 0$. By Theorem 2.7,

$$\begin{aligned} \alpha_B(AB, w^{AB,x}) &= \xi_B(AB, w^{AB,x}) \\ &= \frac{1}{2}(x_A + x_B - (84 - x_C - x_D)) \\ &= \frac{1}{2}(140 - 84) = 28. \end{aligned}$$

Since α is reduced game consistent, $x_B = 28$. Similar arguments using reduced games on AC and AD require that $x_C = 22$ and $x_D = 16$. By the efficiency condition, $x = (74, 28, 22, 16)$, which is the (pre)nucleolus.

THEOREM 6.8. *If an allocation method α is anonymous, player-separable, proportionate, and reduced game consistent on all games, then α is the prenucleolus.*

Sobolev [**31**] proved this theorem. Unlike the motivating savings game example, the proof requires the focal game to be embedded in multiple ways inside an enormously larger game. Orshan [**24**] was able to obtain the same conclusion using unbiased instead of anonymous.

7. Summary

We have described several classes of coalition games, allocation methods, and fairness properties. We have determined which methods satisfy which properties on which classes of coalition games. We have shown that a set of fairness properties can be mutually inconsistent on a class of coalition games. We have also shown that some sets of fairness properties can uniquely characterize an allocation method. No method appears to be perfectly fair for all circumstances, and so it is important to identify the fairness properties most appropriate for any given class of models. For example, the coalition-monotone property is particularly compelling in measuring voting power while the coalition-rational property is particularly compelling when economic agents can choose whether or not to cooperate. A careful examination of particular situations are likely to lead to different, and perhaps new, fairness properties.

This leaves many questions unanswered or even as yet unasked. There are other classes of games, other allocation methods and set-valued solution concepts, and other fairness properties either described in the literature (see the references cited in the introduction) or yet to be discovered. In addition to coalition games, there are other mathematical models of cooperation such as nontransferable utility games [**3**, chapter 55], partition function form games [**5**], and partially defined games [**14**].

For the reader who would like to make a contribution, choose a situation and a corresponding model. Examples would be airport landing fees modeled as convex coalition games or Presidential power in legislation modeled as a simple coalition game. Intuit what seems correct. Perhaps it is that each airplane landing should pay an equal share of the cost for a runway only as long as was needed or that the President should have at least as much power as all of the Senators combined. Apply known allocation methods such as the Shapley value and nucleolus. Think more deeply about the properties that an allocation method should have in your context and characterize which allocation methods satisfy the properties you have defined. Perhaps you will characterize a method that confirms your intuition or find that your proof contradicts your intuition. Either way, you will have learned something and made a contribution. So, ask your own questions and find the answers!

References

[1] Aumann, R. J. and Hart, S., *Handbook of Game Theory with Economic Applications*, Volume 1, Elsevier Science Publishers, 1992.

[2] Aumann, R. J. and Hart, S., *Handbook of Game Theory with Economic Applications*, Volume 2, Elsevier Science Publishers, 1994.

[3] Aumann, R. J. and Hart, S., *Handbook of Game Theory with Economic Applications*, Volume 3, Elsevier Science Publishers, 2002.
[4] Banzhaf, J. F., "Weighted voting doesn't work: a mathematical analysis", *Rutgers Law Review*, 1965, **19**, 317-343.
[5] Bolger, E. M., "A set of axioms for a value for partition function games", *International Journal of Game Theory*, 1989, **18**, 37–44.
[6] Brams, S. J., Affuso, P. J., and Kilgour, D. M., "Presidential power: a game-theoretic analysis", in: P. Brace, C. B. Harrington, and G. King, eds., *The Presidency in American Politics*, New York University Press, 55-74.
[7] Derks, J. and Haller, H., "Weighted nucleoli", *International Journal of Game Theory*, 1999, **28**, 173-187.
[8] Driessen, T., *Cooperative Games, Solutions and Applications*, Kluwer Academic, 1998.
[9] Dubey, P., Neyman, A., and Weber, R. J., "Value theory without efficiency", Mathematics of Operations Research, 1981, 6, 122–128.
[10] Dubey, P. and Shapley, L., "Mathematical properties of the Banzhaf power index", *Mathematics of Operations Research*, 1979, **4**, 99-131.
[11] Felsenthal, D. and Machover, M., "Postulates and paradoxes of relative voting power–a critical re-appraisal", *Theory and Decision*, 1995, **38**, 195-229.
[12] Grotte, J. H., "Computation of and observations on the nucleolus, the normalized nucleolus and the central games", M. S. Thesis, Field of Applied Math., Cornell University, Ithaca, New York, 1970.
[13] Hokari, T., "The nucleolus is not aggregate-monotonic on the domain of convex games", *International Journal of Game Theory*, 2000, **29**, 133–137.
[14] Housman, D., "Linear and symmetric allocation methods for partially defined cooperative games", *International Journal of Game Theory*, 2001, **30**, 377–404
[15] Housman, D.; Clark, L., "Core and monotonic allocation methods", *International Journal of Game Theory*, 1998, **27**, 611–616.
[16] Johnston, R. J., "On the measurement of power: some reactions to Laver", *Environment and Planning A*, **10**, 907-914.
[17] Kohlberg, E, "On the nucleolus of a characteristic function game", *SIAM J. Appl. Math.*, 1971, **20**, 62–66.
[18] Lucas, W. F., "Measuring power in weighted voting systems", in: S. J. Brams, W. F. Lucas, and P. D. Straffin, Jr., eds., Political and Related Models (Modules in Applied Mathematics 2), Springer-Verlag, 1983, 183-238.
[19] Lucas, W. F. and Billera, L. J., "Modeling coalitional values", in: S. J. Brams, W. F. Lucas, and P. D. Straffin, Jr., eds., Political and Related Models (Modules in Applied Mathematics 2), Springer-Verlag, 1983, 66-97.
[20] Malawski, M, "Equal treatment, symmetry and Banzhaf value axiomatizations", *International Journal of Game Theory*, 2002, **31**, 47–67.
[21] Moulin, H., *Axioms of Cooperative Decision Making*, Cambridge University Press, 1988.
[22] Moulin, H., *Cooperative Microeconomics*, Princeton University Press, 1995.
[23] Moulin, H., *Fair Division and Collective Welfare*, The MIT Press, 2003.
[24] Orshan, G., "The prenucleolus and the reduced game property: Equal treatment replaces anonymity", *International Journal of Game Theory*, **22**, 241-248.
[25] Peleg, B. and Sudhölter, P., *Introduction to the Theory of Cooperative Games*, 2nd Edition, Springer-Verlag, 2007.
[26] Potters, J. A. M. and Tijs, S. H., "The nucleolus of a matrix game and other nucleoli", *Math. Oper. Res.*, 1992, **17**, 164–174.
[27] Roth, A. E., *The Shapley Value: Essays in Honor of Lloyd S. Shapley*, Cambridge University Press, 1988.
[28] Schmeidler, D., "The nucleolus of a characteristic function game", *SIAM J. Appl. Math.*, 1969, **17**, 1163–1170.
[29] Shapley, L. S., "A value for n-person games", in: H. W. Kuhn and A. W. Tucker, eds., *Contributions to the Theory of Games II (Annals of Mathematics Studies 28)*, Princeton University Press, 307-317.
[30] Shapley, L. S., "Cores of convex games", *International Journal of Game Theory*, 1971/72, **1**, 11–26; errata, ibid. (1971/72), 1, 199.

[31] Sobolev, A. I., "The characterization of optimality principles in cooperative games by functional equations", in: N. N. Vorobjev, ed., *Matematischeskie metody v socialnix naukakh*, Proceedings of the Seminar, Issue 6, Vilnius: Institute of Physics and Mathematics, Academy of Sciences of the Lithuanian SSR, 94-151.
[32] Taylor, A. D. and Zwicker, W. S., *Simple Games*, Princeton University Press, 1999.
[33] van den Brink, R. and van der Laan, G., "Axiomatizations of the normalized Banzhaf value and the Shapley value", *Social Choice and Welfare*, 1998, **15**, 567–582.
[34] Wallmeier, E., "Der f-Nukleolus als Lösungskonzept für n-Personenspiele in Funktionsform", Institut für Mathematische Statistik der Universität Münster, Working paper WO1, 1980.
[35] Young, H. P., "Monotonic solutions of cooperative games", *International Journal of Game Theory*, 1985, **14**, 65-72.
[36] Young, H. P., "Cost allocation", in: H. P. Young, ed., *Fair Allocation*, American Mathematical Society, 1985, 69-94.
[37] Young, H. P., *Equity in Theory and Practice*, Princeton University Press, 1994.
[38] Young, H. P., Okada, N., and Hashimoto, T., "Cost allocation in water resources development", *Water Resources Research*, 1982, **18**, 463-475.

Department of Mathematics, Goshen College
E-mail address: dhousman@goshen.edu
URL: http://www.goshen.edu/~dhousman

Contemporary Mathematics
Volume **479**, 2009

Sums-of-squares formulas

Daniel C. Isaksen

ABSTRACT. We discuss a certain kind of polynomial identity called a sums-of-squares formula. One motivation for studying such formulas is the construction of normed algebras. We describe how topological methods of cohomology can be applied to this purely algebraic problem.

1. Introduction

This article is concerned with the existence of certain kinds of polynomial identities called sums-of-squares formulas. Such identities date back at least to the seventh century. Many famous mathematicians, including Adams [**A**], Atiyah [**At**], Hopf [**Ho**], Hurwitz [**Hu1**] [**Hu2**], and Pfister [**P**], have worked on the problem or closely related topics.

The question of when such formulas exist has been extensively studied. The depth and breadth of the subject are exhibited in the survey articles [**L**] and [**Sh1**] and the detailed sourcebook [**Sh2**].

The overall theme of the article is to use topologically inspired methods to prove theorems in algebra. Although sums-of-squares formulas are easy to understand at a naive level, it turns out that the proofs of some theorems require deep modern cohomological methods. Thus, the subject is useful as a demonstration of the power of cohomology.

Sums-of-squares formulas are an important part of the subject of quadratic forms. They are also important because of their relationship to other mathematical areas. Below, we will motivate the search for sums-of-squares formulas by showing how they are related to the existence of normed algebras and division algebras. From this perspective, sums-of-squares formulas are closely linked to the four classical division algebras: the real numbers $\mathbb{R}$, the complex numbers $\mathbb{C}$, the quaternions $\mathbb{H}$, and the octonions $\mathbb{O}$. The book [**CS**] is a pleasant introduction to these important algebraic structures. The article [**B**] is a thorough survey of the role of the octonions in algebra, geometry, and physics.

Sums-of-squares formulas are also closely linked to many problems in topology, but we will not discuss these links in detail. For example, consider the fact that the projective plane cannot be embedded into 3-dimensional Euclidean space. This

1991 *Mathematics Subject Classification.* Primary 11E25; Secondary 55Nxx, 17A35.
Key words and phrases. Sums-of-squares formula, division algebra, cohomology.

fact inspires us to consider embeddings of higher dimensional projective spaces into higher dimensional Euclidean spaces. This purely topological question is essentially equivalent to the search for sums-of-squares formulas [**S**]. Another example of a related topological problem is the construction of linearly independent vector fields on spheres [**A**].

The goal of the article is to give a flavor for the results and techniques in the subject. For the most part, we will avoid the technical details. The exception is Section 6, which is more difficult than the rest of the article. This section explores the methods of proof more deeply. Ultimately, it is the methods that are most interesting because they teach us how to solve other kinds of problems also.

For the most part, we will work only over the field $\mathbb{R}$ of real numbers. In Section 9, we describe some very recent work that generalizes classical results over $\mathbb{R}$ to arbitrary fields [**DI1**] [**DI2**].

In 2005, a student in Joe Gallian's REU at the University of Minnesota – Duluth worked on a problem about sums-of-squares formulas. This work is described in Section 7.

2. Normed algebras

The four classical normed $\mathbb{R}$-algebras are the real numbers $\mathbb{R}$, the complex numbers $\mathbb{C}$, the quaternions $\mathbb{H}$, and the octonions $\mathbb{O}$ [**CS**]. An old and obvious question is to consider whether there exist any other similar algebraic structures. The following definition makes explicit what we are looking for.

Definition 2.1. A **normed $\mathbb{R}$-algebra** is a finite-dimensional $\mathbb{R}$-vector space A equipped with a positive-definite inner product $\langle -, - \rangle$ and a bilinear map $\mu : A \times A \to A$ that respects the inner product.

We think of μ as a multiplication rule on A. Therefore, we usually suppress the notation μ and just write xy for $\mu(x, y)$.

The bilinearity of μ expresses that multiplication respects scalars in $\mathbb{R}$ and also that multiplication satisfies both the left and right distributive laws. The multiplication is not required to be commutative; in fact, $\mathbb{H}$ and $\mathbb{O}$ are non-commutative. The multiplication is not even required to be associative; in fact, $\mathbb{O}$ is non-associative.

Let us explain what it means for μ to respect the inner product. For any vectors x and y in A, we require that

$$\langle xy, xy \rangle = \langle x, x \rangle \langle y, y \rangle .$$

After taking square roots, this is equivalent to the equation $|xy| = |x| \cdot |y|$.

The following classical theorem says that normed $\mathbb{R}$-algebras are very rare.

Theorem 2.2 (Hurwitz, 1898). *If A is a normed $\mathbb{R}$-algebra, then A is isomorphic to $\mathbb{R}$, $\mathbb{C}$, $\mathbb{H}$, or $\mathbb{O}$.*

The original reference for this result is [**Hu1**]. Hurwitz's proof relies on matrix algebra. A modern treatment of a more general result appears in [**Sh1**]. Another proof appears in [**CS**].

3. Sums-of-squares formulas

Consider the normed $\mathbb{R}$-algebra $\mathbb{C}$. If i is the usual square root of -1, we know that

$$|(x_1 + ix_2)(y_1 + iy_2)| = |x_1 + ix_2|\,|y_1 + iy_2|$$

for all real numbers x_1, x_2, y_1, and y_2. After squaring both sides, we obtain a polynomial identity

$$(x_1^2 + x_2^2)(y_1^2 + y_2^2) = (x_1y_1 - x_2y_2)^2 + (x_1y_2 + x_2y_1)^2$$

in four variables x_1, x_2, y_1, and y_2. This formula was known to the Indian mathematician Brahmagupta in the seventh century [**Br**]. The book [**Pr**] explains how to convert Brahmagupta's text into modern mathematical notation. Brahmagupta actually studied solutions to Pell's equation; the above polynomial identity is a special case of his work.

Now consider the normed $\mathbb{R}$-algebra $\mathbb{H}$. The equation $|xy| = |x|\,|y|$ leads to another polynomial identity

$$(x_1^2 + x_2^2 + x_3^2 + x_4^2)(y_1^2 + y_2^2 + y_3^2 + y_4^2) =$$
$$(x_1y_1 - x_2y_2 - x_3y_3 - x_4y_4)^2 + (x_1y_2 + x_2y_1 + x_3y_4 - x_4y_3)^2 +$$
$$(x_1y_3 + x_3y_1 + x_4y_2 - x_2y_4)^2 + (x_1y_4 - x_4y_1 + x_2y_3 - x_3y_2)^2$$

in eight variables x_1, x_2, x_3, x_4, y_1, y_2, y_3, and y_4. This formula, known sometimes as the four-square identity, was known to Euler in the eighteenth century [**E**].

Similarly, a careful examination of the multiplication in the octonions $\mathbb{O}$ leads to a polynomial identity involving sixteen variables $x_1, \ldots, x_8$ and $y_1, \ldots, y_8$. This formula, sometimes known as the eight-square identity, was known to several mathematicians in the nineteenth century.

We generalize these three classical examples to the following definition.

DEFINITION 3.1. A **sums-of-squares formula** of type $[r, s, t]$ over $\mathbb{R}$ is a polynomial identity of the form

$$\left(x_1^2 + \cdots + x_r^2\right)\left(y_1^2 + \cdots + y_s^2\right) = z_1^2 + \cdots + z_t^2,$$

where the x's and y's are indeterminate variables and the z_i's are bilinear expressions in the x's and y's over $\mathbb{R}$.

Let us explain what it means for each z_i to be bilinear in the x's and y's over $\mathbb{R}$. For $1 \leq i \leq t$, there exist scalars α_{ijk} in $\mathbb{R}$ such that

$$z_i = \sum_{j=1}^{r}\sum_{k=1}^{s} \alpha_{ijk} x_i y_j.$$

In other words, each z_i is a sum of monomials of the form $\alpha x_i y_j$, where α belongs to $\mathbb{R}$.

The examples described earlier in this section are sums-of-squares formulas of type $[2, 2, 2]$, $[4, 4, 4]$, and $[8, 8, 8]$ over $\mathbb{R}$. We have already seen from these examples that sums-of-squares formulas are linked to normed algebras. Now we make this relationship formal.

PROPOSITION 3.2. *A sums-of-squares formula of type $[n, n, n]$ over $\mathbb{R}$ exists if and only if a normed $\mathbb{R}$-algebra of dimension n exists.*

PROOF. First suppose that a sums-of-squares formula of type $[n, n, n]$ exists. Let A be the $\mathbb{R}$-vector space $\mathbb{R}^n$, equipped with the standard positive-definite inner product. Define $\mu : A \times A \to A$ by the formula

$$\mu\left((x_1, \ldots, x_n), (y_1, \ldots, y_n)\right) = (z_1, \ldots, z_n).$$

The sums-of-squares formula implies that A is a normed $\mathbb{R}$-algebra.

Conversely, suppose that A is a normed $\mathbb{R}$-algebra. Up to isomorphism, the inner product space A is equal to $\mathbb{R}^n$ equipped with its standard positive-definite inner product. The multiplication on A gives a bilinear map $\mu : \mathbb{R}^n \times \mathbb{R}^n \to \mathbb{R}^n$. Define z_i to be the ith coordinate of μ. A sums-of-squares formula follows from the fact that μ respects the inner product. □

Hurwitz's original classification of normed $\mathbb{R}$-algebras (see Theorem 2.2) involved an analysis of sums-of-squares formulas of type $[n, n, n]$, as suggested by Proposition 3.2. See [**CS**] for a distinct approach.

4. Existence of Sums-of-squares-formulas

We begin by observing that if a formula of type $[r, s, t]$ exists, then a formula of type $[r, s, t+1]$ also exists. To see why, just take $z_{t+1} = 0$. Therefore, we have the following natural question.

QUESTION 4.1. *Given r and s, find the smallest t such that a sums-of-squares formula of type $[r, s, t]$ exists over $\mathbb{R}$.*

Question 4.1 turns out to be extremely hard. We don't expect ever to find a complete answer. There are many partial results, some of which we will discuss below.

In one special case, the complete answer to Question 4.1 is known.

DEFINITION 4.2. The **Hurwitz-Radon function $\rho(n)$** is defined as follows. Write $n = 2^{c+4d}a$, where a is odd and $0 \leq c \leq 3$. Then $\rho(n) = 2^c + 8d$.

EXAMPLE 4.3. Here is a table of some values of $\rho(n)$.

n	2	4	8	16	32	64	128	256
$\rho(n)$	2	4	8	8	10	12	16	16

THEOREM 4.4 (Hurwitz, Radon, 1922). *Let $n \geq 1$. There exists a sums-of-squares formula over $\mathbb{R}$ of type $[n, m, m]$ if and only if $n \leq \rho(m)$.*

The original references for this result are [**Hu2**] and [**R**]. The standard proof of the Hurwitz-Radon theorem is straightforward. It involves manipulations with square matrices that satisfy certain equations. A modern treatment appears in [**Sh1**].

If we take Theorem 4.4 for granted, then there is a simple proof of Theorem 2.2. Just look for all n such that $\rho(n) = n$.

Despite the long history of the subject, we have made relatively little progress on constructions of sums-of-squares formulas. There are the Hurwitz-Radon constructions of Theorem 4.4. Another example is a family of sums-of-squares formulas over $\mathbb{R}$ constructed by Adem [**Ad**]. The first two formulas in Adem's family are of type $[10, 10, 16]$ and $[12, 12, 26]$.

5. The Hopf-Stiefel condition

A major breakthrough on Question 4.1 occurred with the nearly simultaneous work of Hopf, Stiefel, and Behrend around 1940. Rather than construct examples, they proved a non-existence result.

THEOREM 5.1 (Hopf, Stiefel, Behrend). *If a sums-of-squares formula of type $[r, s, t]$ exists over $\mathbb{R}$, then the binomial coefficients $\binom{t}{i}$ are even for $t - r < i < s$.*

We will refer to Theorem 5.1 as the **Hopf-Stiefel condition**. The original references for this result are [**Be**], [**Ho**], and [**St**]. We will describe Hopf's proof below in Section 6, which uses cohomology rings. Theorem 5.1 convinced many mathematicians of the fundamental importance of cohomological methods, which were still new at the time.

EXAMPLE 5.2. Does a formula of type $[6, 10, 13]$ exist over $\mathbb{R}$? The Hopf-Stiefel condition tells us to inspect the binomial coefficients $\binom{13}{4}$ and $\binom{13}{5}$. Notice that $\binom{13}{4}$ is odd. Therefore, the answer is no; a formula of type $[6, 10, 13]$ does not exist over $\mathbb{R}$.

EXERCISE 5.3. Does a formula of type $[6, 10, 14]$ exist over $\mathbb{R}$?

Given r and s, let $t(r, s)$ be the smallest integer t such that the Hopf-Stiefel condition is satisfied for $[r, s, t]$. Theorem 5.1 says that a sums-of-squares formula of type $[r, s, n]$ over $\mathbb{R}$ does *not* exist for $n < t(r, s)$. However, a sums-of-squares formula of type $[r, s, t(r, s)]$ over $\mathbb{R}$ might exist.

The following two results describe simple and practical methods for computing $t(r, s)$.

PROPOSITION 5.4. *For any r and s, $t(r, s)$ equals*

$$\min_{k \geq 0} \left\{ 2^k \left(\left\lceil \frac{r}{2^k} \right\rceil + \left\lceil \frac{s}{2^k} \right\rceil - 1 \right) \right\},$$

where $\lceil x \rceil$ indicates the smallest integer greater than or equal to x.

The article [**Ka**] contains a short and elementary proof of Proposition 5.4.

PROPOSITION 5.5. *The values of $t(r, s)$ can be computed recursively using the following three rules.*

(1) $t(r, s) = t(s, r)$.
(2) $t(r, s) = 2^n$ *if* $2^{n-1} < r, s \leq 2^n$.
(3) $t(r, s) = 2^n + t(r, s - 2^n)$ *if* $r \leq 2^n < s$.

Proposition 5.5 seems to belong to the mathematical folklore. Note that for given r and s, part (3) of the proposition is valid for any n such that $r \leq 2^n < s$.

6. Cohomology

The goal of this section is to describe Hopf's methods for proving Theorem 5.1. The key ingredient is the notion of cohomology (technically known as singular $\mathbb{F}_2$-cohomology) from algebraic topology.

For each topological space X, there is a ring $H^*(X)$ called the **cohomology ring of X with coefficients in $\mathbb{F}_2$**. This ring is graded in the sense that it is actually a direct sum $\oplus_n H^n(X)$ of $\mathbb{F}_2$-vector spaces $H^n(X)$ for $n \geq 0$, and the multiplication respects the grading in the sense that it takes $H^n(X) \times H^m(X)$ into $H^{n+m}(X)$.

For each continuous function $f : X \to Y$, there is also a ring homomorphism $f^* : H^*(Y) \to H^*(X)$. Notice that the ring homomorphism and the continuous function go in opposite directions.

The construction respects composition. Given $f : X \to Y$ and $g : Y \to Z$, we can construct $(gf)^* : H^*(Z) \to H^*(X)$. This homomorphism is always equal to the composition $f^* \circ g^*$.

We won't need very much about cohomology, other than its value on a few concrete examples. These calculations can be found in many graduate textbooks on algebraic topology, such as [**H**].

DEFINITION 6.1. **Real projective n-space** $\mathbb{R}P^n$ is the topological space of all lines through the origin in $\mathbb{R}^{n+1}$.

It is easy to verify that the real projective line $\mathbb{R}P^1$ is isomorphic to the circle S^1. However, $\mathbb{R}P^2$ is already somewhat exotic in that it is a "non-orientable" surface. The topology of real projective spaces is a surprisingly deep subject.

THEOREM 6.2.

(1) *For all $n \geq 1$, the ring $H^*(\mathbb{R}P^n)$ is isomorphic to the truncated polynomial ring $\mathbb{F}_2[\alpha]/\alpha^{n+1}$, where α belongs to $H^1(\mathbb{R}P^n)$.*
(2) *For all $n \geq 1$ and $m \geq 1$, the ring $H^*(\mathbb{R}P^n \times \mathbb{R}P^m)$ is isomorphic to the truncated polynomial ring $\mathbb{F}_2[\alpha, \beta]/(\alpha^{n+1}, \beta^{m+1})$, where α and β both belong to $H^1(\mathbb{R}P^n \times \mathbb{R}P^m)$.*

Consider a linear subspace V of $\mathbb{R}^{n+1}$, and let $\mathbb{R}P(V)$ be the space of lines through the origin in V. If $m+1$ is the dimension of V, then V is linearly isomorphic to $\mathbb{R}^{m+1}$, which implies that $\mathbb{R}P(V)$ is homeomorphic to $\mathbb{R}P^m$.

The linear inclusion $V \to \mathbb{R}^{n+1}$ induces a continuous function $\mathbb{R}P(V) \to \mathbb{R}P^n$. Such a function is called a **standard inclusion**.

THEOREM 6.3.

(1) *In cohomology, any standard inclusion induces the homomorphism*
$$\mathbb{F}_2[\alpha]/\alpha^{n+1} \to \mathbb{F}_2[\beta]/\beta^{m+1}$$
that takes α to β.
(2) *Let x be any fixed element of $\mathbb{R}P^n$, and consider the inclusion $\{x\} \times \mathbb{R}P^m \to \mathbb{R}P^n \times \mathbb{R}P^m$. In cohomology, this function induces the homomorphism*
$$\mathbb{F}_2[\alpha, \beta]/\alpha^{n+1}, \beta^{m+1} \to \mathbb{F}_2[\beta]/\beta^{m+1}$$
that takes α to zero and takes β to β.

Symmetrically to part (2) of Theorem 6.3, the cohomology of the inclusions $\mathbb{R}P^n \times \{y\} \to \mathbb{R}P^n \times \mathbb{R}P^m$ can also be described concretely.

DEFINITION 6.4. A continuous function $f : \mathbb{R}P^n \times \mathbb{R}P^m \to \mathbb{R}P^k$ is **axial** if there exist fixed elements x in $\mathbb{R}P^n$ and y in $\mathbb{R}P^m$ such that the restrictions $\{x\} \times \mathbb{R}P^m \to \mathbb{R}P^k$ and $\mathbb{R}P^n \times \{y\} \to \mathbb{R}P^k$ are standard inclusions.

PROPOSITION 6.5. *In cohomology, an axial function $f : \mathbb{R}P^n \times \mathbb{R}P^m \to \mathbb{R}P^k$ induces the ring homomorphism $f^* : \mathbb{F}_2[\gamma]/\gamma^{k+1} \to \mathbb{F}_2[\alpha, \beta]/\alpha^{n+1}, \beta^{m+1}$ that takes γ to $\alpha + \beta$.*

PROOF. Let x and y be elements of $\mathbb{R}P^n$ and $\mathbb{R}P^m$ as in Definition 6.4. Consider the diagram

$$\begin{array}{ccccc} \{x\} \times \mathbb{R}P^m & \longrightarrow & \mathbb{R}P^n \times \mathbb{R}P^m & \longleftarrow & \mathbb{R}P^n \times \{y\} \\ & {\scriptstyle g} \searrow & \downarrow {\scriptstyle f} & \swarrow {\scriptstyle h} & \\ & & \mathbb{R}P^k, & & \end{array} \tag{6.1}$$

where the two horizontal maps are the obvious inclusions. The two triangles in the diagram are to be interpreted as composition relations. For example, g is the composition of an inclusion followed by f.

Now apply cohomology to all four spaces and all five functions. The result is another diagram

$$\begin{array}{ccccc} \mathbb{F}_2[\beta]/\beta^{m+1} & \longleftarrow & \mathbb{F}_2[\alpha,\beta]/\alpha^{n+1},\beta^{m+1} & \longrightarrow & \mathbb{F}_2[\alpha]/\alpha^{m+1} \\ & \nwarrow^{g^*} & \uparrow f^* & \nearrow^{h^*} & \\ & & \mathbb{F}_2[\gamma]/\gamma^{k+1} & & \end{array} \tag{6.2}$$

of the same shape. Notice that all of the functions now are reversed, but the two triangles still are composition relations.

The functions g and h in Diagram 6.1 are standard inclusions by assumption. Therefore, by part (1) of Theorem 6.3, the ring homomorphism g^* takes γ to β, and the ring homomorphism h^* takes γ to α.

By part (2) of Theorem 6.3, the leftward horizontal function in Diagram 6.2 takes α to 0 and takes β to β. Similarly, the rightward horizontal function takes α to α and β to 0. Now a short algebra exercise proves that f^* takes γ to $\alpha+\beta$. □

PROPOSITION 6.6. *Suppose that a sums-of-squares formula of type* $[r,s,t]$ *exists over* $\mathbb{R}$. *There exists an axial function* $f : \mathbb{R}P^{r-1} \times \mathbb{R}P^{s-1} \to \mathbb{R}P^{t-1}$.

PROOF. Suppose that a sums-of-squares formula of type $[r,s,t]$ exists over $\mathbb{R}$. The z's assemble into a bilinear function $F : \mathbb{R}^r \times \mathbb{R}^s \to \mathbb{R}^t$ such that the norm of $F(v,w)$ is the product of the norms of v and w. Since 0 is the only vector of norm zero, it follows that F restricts to a function $F : (\mathbb{R}^r - 0) \times (\mathbb{R}^s - 0) \to (\mathbb{R}^t - 0)$.

Now F is bilinear, so it induces a map $f : \mathbb{R}P^{r-1} \times \mathbb{R}P^{s-1} \to \mathbb{R}P^{t-1}$. In order to get f, you don't really need that F is bilinear. All you need is that it respects scalar multiplication.

Next, let x be any fixed non-zero element of $\mathbb{R}P^{r-1}$, and let v be any non-zero vector in $\mathbb{R}^r$ that lies on the line defined by x. The function $F(v,-) : \mathbb{R}^s \to \mathbb{R}^t$ is linear because F is bilinear. It is also an inclusion since the norm of $F(v,w)$ equals $|v| \cdot |w|$. This shows that the restriction $\{x\} \times \mathbb{R}P^{s-1} \to \mathbb{R}P^{t-1}$ of f is a standard inclusion.

A symmetric argument works for $\mathbb{R}P^{r-1} \times \{y\} \to \mathbb{R}P^{t-1}$. □

We are now ready to piece the above results together into a proof of Theorem 5.1. We remind the reader that this proof is not entirely self-contained because we take for granted the existence of cohomology and the computations in Theorems 6.2 and 6.3

SKETCH PROOF OF THEOREM 5.1. Suppose that a sums-of-squares formula of type $[r,s,t]$ exists over $\mathbb{R}$. Let $f : \mathbb{R}P^{r-1} \times \mathbb{R}P^{s-1} \to \mathbb{R}P^{t-1}$ be the continuous function constructed in Proposition 6.6. Take the cohomology of the map f to obtain a ring homomorphism

$$\mathbb{Z}/2[\gamma]/\gamma^t \to \mathbb{Z}/2[\alpha,\beta]/(\alpha^r,\beta^s).$$

By Proposition 6.5, this homomorphism takes γ to $\alpha+\beta$.

Since $\gamma^t = 0$, it follows that $(\alpha+\beta)^t = 0$. Use the binomial theorem to expand $(\alpha+\beta)^t$, where the terms are sorted by decreasing powers of α (and increasing powers of β). This yields the expression

$$\begin{aligned}&\alpha^t + \cdots + \tbinom{t}{r}\alpha^r\beta^{t-r} + \\ &\tbinom{t}{r-1}\alpha^{r-1}\beta^{t-r+1} + \cdots + \tbinom{t}{t-s+1}\alpha^{t-s+1}\beta^{s-1} + \\ &\tbinom{t}{t-s}\alpha^{t-s}\beta^s + \cdots + \beta^t.\end{aligned}$$

The first and third rows of terms are all zero because $\alpha^r = 0$ and $\beta^s = 0$. Each term in the second group must also be zero, so the coefficients $\binom{t}{i}$ must be zero in $\mathbb{F}_2$ for $t - s < i < r$. This establishes the result. $\square$

7. The Atiyah condition

Here is another theorem about non-existence of sums-of-squares formulas.

THEOREM 7.1 (Atiyah). *Let* $c = \lfloor \frac{s-1}{2} \rfloor + 1$. *If a sums-of-squares formula of type* $[r, s, t]$ *exists over* $\mathbb{R}$, *then* 2^{c-i} *divides* $\binom{t}{i}$ *for* $t - r < i < c$.

The original reference for this result is [**At**]. The proof uses K-theory, which is a generalization of the cohomology discussed in Section 6. At the time, K-theory was a relatively new technique. Theorem 7.1 demonstrated its power and gracefulness.

At face value, Theorem 7.1 is neither stronger nor weaker than Theorem 5.1. On the one hand, we are requiring binomial coefficients to be not merely even but to be divisible by lots of powers of 2. On the other hand, the range for i is smaller. The article [**K**] contains some results on the question of determining whether Theorem 5.1 or Theorem 7.1 is a stronger result, given values of r and s. This was a research project at Joe Gallian's REU program in Summer 2005.

The next two examples demonstrate that sometimes the Atiyah condition is better, and sometimes the Hopf-Stiefel condition is better.

EXAMPLE 7.2. Does a sums-of-squares formula of type $[13, 13, 16]$ exist? The Hopf-Stiefel condition tells us to inspect the binomial coefficients $\binom{16}{i}$ for $3 < i < 13$. All of these numbers are even, so the Hopf-Stiefel condition is inconclusive.

On the other hand, the Atiyah condition tells us to check whether 2^{7-i} divides $\binom{16}{i}$ for $3 < i < 7$. Note that 2^3 does not divide $\binom{16}{4}$. Therefore, a sums-of-squares formula of type $[13, 13, 16]$ does not exist.

EXAMPLE 7.3. Does a sums-of-squares formula of type $[17, 17, 25]$ exist? The Atiyah condition tells us to check whether 2^{9-i} divides $\binom{25}{i}$ for $8 < i < 9$, which is vacuously true. Therefore, the Atiyah condition is inconclusive.

On the other hand, the Hopf-Stiefel condition tells us to inspect the binomial coefficients $\binom{25}{i}$ for $8 < i < 17$. Note that $\binom{25}{9}$ is odd. Therefore, a sums-of-squares formula of type $[17, 17, 25]$ does not exist.

In fact, the Hopf-Stiefel condition implies that a formula of type $[17, 17, 31]$ does not exist, but a formula of type $[17, 17, 32]$ might exist.

8. Division algebras

The following definition is a natural generalization of the notion of a normed algebra from Section 2.

DEFINITION 8.1. An $\mathbb{R}$-**division algebra** is a finite-dimensional $\mathbb{R}$-vector space A equipped with a bilinear map $\mu : A \times A \to A$ such that

$$\mu(x, y) = 0 \text{ if and only if } x = 0 \text{ or } y = 0.$$

Compare the definition of $\mathbb{R}$-division algebras to the definition of normed $\mathbb{R}$-algebras given in Definition 2.1. It is easy to show that if A is a normed $\mathbb{R}$-algebra, then it must be an $\mathbb{R}$-division algebra. This uses the fact that a vector in $\mathbb{R}^n$ is zero if and only if its norm is zero.

The following theorem about $\mathbb{R}$-division algebras, despite its innocuous appearance, is a very deep result.

THEOREM 8.2 (Bott-Milnor, Kervaire, 1958). *If A is an $\mathbb{R}$-division algebra, then the dimension of A is 1, 2, 4, or 8.*

The original references for this theorem are [**BM**] and [**Ke**]. The following proof, relying on Theorem 7.1, is more graceful than the original arguments.

PROOF. Suppose that A is an $\mathbb{R}$-division algebra of dimension n. The multiplication in A is a bilinear map $\mathbb{R}^n \times \mathbb{R}^n \to \mathbb{R}^n$. As described in Section 6, we can use this bilinear map to construct an axial function $\mathbb{R}P^{n-1} \times \mathbb{R}P^{n-1} \to \mathbb{R}P^{n-1}$.

Although Theorem 7.1 is stated in terms of non-existence of sums-of-squares formulas, the proof of the theorem actually establishes a non-existence result for axial functions. Let $c = \lfloor \frac{n-1}{2} \rfloor + 1$. According to Theorem 7.1, 2^{c-i} divides $\binom{n}{i}$ for $0 < i < c$. In particular, with $i = 1$, we conclude that 2^{c-1} divides n. It is easy to check that this can occur only if n equals 1, 2, 4, or 8. □

The careful reader may have noticed that we have *not* claimed in Theorem 8.2 that the classical algebras $\mathbb{R}$, $\mathbb{C}$, $\mathbb{H}$, and $\mathbb{O}$ are the only $\mathbb{R}$-division algebras. In fact, there are many other examples. See, for example, the article [**D**].

9. Arbitrary fields

My interest in sums-of-squares formulas comes out of the question of formulas over arbitrary fields.

Let F be any field. A sums-of-squares formula of type $[r, s, t]$ over F is a polynomial identity

$$\left(x_1^2 + \cdots + x_r^2\right)\left(y_1^2 + \cdots + y_s^2\right) = z_1^2 + \cdots + z_t^2,$$

where the x's and y's are indeterminate variables and the z_i's are bilinear expressions in the x's and y's with coefficients in F.

Suppose that F has characteristic 2. The formula $(a+b)^2 = a^2 + b^2$ generalizes to

$$\left(x_1^2 + \cdots + x_r^2\right)\left(y_1^2 + \cdots + y_s^2\right) = \left(x_1y_1 + x_1y_2 + \cdots + x_ry_{s-1} + x_ry_s\right)^2.$$

In other words, a formula of type $[r, s, 1]$ always exists over F.

This demonstrates two important points. First, the existence of a formula of a certain type can depend on which field you're working over. Second, the problem is trivial if the characteristic of the field is 2.

The recent papers [**DI1**] and [**DI2**] generalize the Stiefel-Hopf condition and the Atiyah condition to arbitrary fields, as long as the characteristic of the field does not equal 2. The statements of the theorems are exactly the same, with $\mathbb{R}$ replaced by an arbitrary field. The proofs are also similar. Instead of using cohomology and K-theory from topology, we use motivic cohomology and algebraic K-theory from algebraic geometry.

It is an open problem to find numbers r, s and t and fields F and F' (of characteristic different from 2) such that a formula of type $[r, s, t]$ exists over F but does not exist over F'. It is possible that no such examples exist. This is probably a very hard problem.

References

[A] J. F. Adams, *Vector fields on spheres* Ann. of Math. **75** (1962) 603–632.

[Ad] J. Adem, *Construction of some normed maps*, Bol. Soc. Mat. Mexicana **20** (1975) 59–75.

[At] M. F. Atiyah, *Immersions and embeddings of manifolds*, Topology **1** (1962) 125–132.

[B] J. C. Baez, *The octonions*, Bull. Amer. Math. Soc. **39** (2002) 145–205.

[Be] F. Behrend, *Über Systeme reeller algebraischer Gleichungen*, Compositio Math. **7** (1939) 1–19.

[BM] R. Bott and J. Milnor, *On the parallelizability of the spheres*, Bull. Amer. Math. Soc. **64** (1958) 87–89.

[Br] Brahmagupta, *Brahmasphuta-siddhanta*, Chapter 18, Verses 65–66, translated in H. T. Colebrooke, *Algebra, with arithmetic and mensuration, from the Sanscrit of Brahmegupta and Bhascara*, London, 1817.

[CS] J. H. Conway and D. A. Smith, *On quaternions and octonions: their geometry, arithmetic, and symmetry*, A K Peters, 2003.

[D] E. Dieterich, *Classification, automorphism groups and categorical structure of the two-dimensional real division algebras*, J. Algebra Appl. **4** (2005) 517–538.

[DI1] D. Dugger and D. C. Isaksen, *The Hopf condition for bilinear forms over arbitrary fields*, Ann. of Math. **165** (2007) 943–964.

[DI2] D. Dugger and D. C. Isaksen, *Algebraic K-theory and sums-of-squares formulas*, Doc. Math. **10** (2005) 357–366.

[E] L. Euler, Letter CXXV to C. Goldbach, 12 April 1749, reprinted in P. H. Fuss, *Correspondance Mathématique et Physique de quelques célèbres géomètres du XVIIIeme siècle*, Tome I (1843), Johnson Reprint Corp., 1968.

[H] A. Hatcher, *Algebraic topology*, Cambridge Univ. Press, 2002.

[Ho] H. Hopf, *Ein topologishcer Beitrag zur reelen Algebra*, Comment. Math. Helv. **13** (1940/41) 219–239.

[Hu1] A. Hurwitz, *Über die Komposition der quadratischen Formen von beliebig vielen Variabeln*, Nach. V. der. Ges. der Wiss., Gottingen, Math. Phys. kl. (1898) 309–316. Reprinted in *Math. Werke*, Band 2, 565–571, Birkhäuser, 1963.

[Hu2] A. Hurwitz, *Über die Komposition der qudratischen Formen*, Math. Ann. **88** (1923) 1–25. Reprinted in *Math. Werke*, Band 2, 641–666, Birkhäuser, 1963.

[K] D. M. Kane, *Lower Bounds on the size of sums-of-squares formulas*, J. Number Theory, to appear.

[Ka] G. Károlyi, *A note on the Hopf-Stiefel function*, European J. Combin. **27** (2006) 1135–1137.

[Ke] M. Kervaire, *Non-parallelizability of the n-sphere for $n = 7$*, Proc. Nat. Acad. Sci. U.S.A. **44** (1958) 280-283.

[L] K. Y. Lam, *Topological methods for studying the composition of quadratic forms*, Quadratic and Hermitian Forms, (Hamilton, Ont., 1983), 173–192, Canadian Math. Soc. Conference Proceedings **4**, Amer. Math. Soc., 1984.

[P] A. Pfister, *Zur Darstellung von -1 als Summe von Quadraten in einem Körper*, J. London Math. Soc. **40** (1965) 159–165.

[Pr] S. Prakash, *A critical study of Brahmagupta and his works*, Indian Inst. Astronomical and Sanskrit Research, 1968.

[R] J. Radon, *Lineare Scharen orthogonaler Matrizen*, Abh. Math. Sem. Univ. Hamburg **1** (1922) 1–14. Reprinted in *Collected Works*, Volume 1, 307–320, Birkhäuser, 1987.

[S] B. J. Sanderson, *A non-immersion theorem for real projective space*, Topology **2** (1963) 209–211.

[Sh1] D. B. Shapiro, *Products of sums of squares*, Expo. Math. **2** (1984) 235–261.

[Sh2] D. B. Shapiro, *Compositions of quadratic forms*, de Gruyter Expositions in Mathematics **33**, de Gruyter, 2000.

[St] E. Stiefel, *Über Richtungsfelder in den projektiven Räumen und einen Satz aus der reelen Algebra*, Comment. Math. Helv. **13** (1940/41) 201–218.

Department of Mathematics, Wayne State University, Detroit, MI 48202
E-mail address: isaksen@math.wayne.edu

Contemporary Mathematics
Volume **479**, 2009

Product-free subsets of groups, then and now

Kiran S. Kedlaya

Dedicated to Joe Gallian on his 65th birthday and the 30th anniversary of the Duluth REU

1. Introduction

Let G be a group. A subset S of G is *product-free* if there do not exist $a, b, c \in S$ (not necessarily distinct[1]) such that $ab = c$.

One can ask about the existence of large product-free subsets for various groups, such as the groups of integers (see next section), or compact topological groups (as suggested in [**11**]). For the rest of this paper, however, I will require G to be a finite group of order $n > 1$. Let $\alpha(G)$ denote the size of the largest product-free subset of G; put $\beta(G) = \alpha(G)/n$, so that $\beta(G)$ is the density of the largest product-free subset. What can one say about $\alpha(G)$ or $\beta(G)$ as a function of G, or as a function of n? (Some of our answers will include an unspecified positive constant; I will always call this constant c.)

The purpose of this paper is threefold. I first review the history of this problem, up to and including my involvement via Joe Gallian's REU (Research Experience for Undergraduates) at the University of Minnesota, Duluth, in 1994; since I did this once already in [**11**], I will be briefer here. I then describe some very recent progress made by Gowers [**7**]. Finally, I speculate on the gap between the lower and upper bounds, and revisit my 1994 argument to show that this gap cannot be closed using Gowers's argument as given.

Note the usual convention that multiplication and inversion are permitted to act on subsets of G, i.e., for $A, B \subseteq G$,

$$AB = \{ab : a \in A, b \in B\}, \qquad A^{-1} = \{a^{-1} : a \in A\}.$$

2. Origins: the abelian case

In the abelian case, product-free subsets are more customarily called *sum-free* subsets. The first group in which such subsets were studied is the group of integers $\mathbb{Z}$; the first reference I could find for this is Abbott and Moser [**1**], who expanded

2000 *Mathematics Subject Classification.* Primary 20D60; secondary 20P05.

The author was supported by NSF CAREER grant DMS-0545904 and a Sloan Research Fellowship.

[1]In some sources, one does require $a \neq b$. For instance, as noted in [**9**], I mistakenly assumed this in [**11**, Theorem 3].

upon Schur's theorem that the set $\{1, \ldots, \lfloor n!e \rfloor\}$ cannot be partitioned into n sum-free sets. This led naturally to considering sum-free subsets of finite abelian groups, for which the following is easy.

THEOREM 2.1. *For G abelian, $\beta(G) \geq \frac{2}{7}$.*

PROOF. For $G = \mathbb{Z}/p\mathbb{Z}$ with $p > 2$, we have $\alpha(G) \geq \lfloor \frac{p+1}{3} \rfloor$ by taking

$$S = \left\{ \left\lfloor \frac{p+1}{3} \right\rfloor, \ldots, 2\left\lfloor \frac{p+1}{3} \right\rfloor - 1 \right\}.$$

Then apply the following lemma. □

LEMMA 2.2. *For G arbitrary, if H is a quotient of G, then*

$$\beta(G) \geq \beta(H).$$

PROOF. Let S' be a product-free subset of H of size $\alpha(H)$. The preimage of S' in G is product-free of size $\#S'\#G/\#H$, so $\alpha(G) \geq \alpha(H)\#G/\#H$. □

In fact, one can prove an exact formula for $\alpha(G)$ showing that this construction is essentially optimal. Many cases were established around 1970, but only in 2005 was the proof of the following result finally completed by Green and Ruzsa [**8**].

THEOREM 2.3 (Green-Ruzsa). *Suppose that G is abelian.*

(a) *If n is divisible by a prime $p \equiv 2 \pmod 3$, then for the least such p, $\alpha(G) = \frac{n}{3} + \frac{n}{3p}$.*
(b) *Otherwise, if $3|n$, then $\alpha(G) = \frac{n}{3}$.*
(c) *Otherwise, $\alpha(G) = \frac{n}{3} - \frac{n}{3m}$, for m the exponent (largest order of any element) of G.*

One possible explanation for the delay is that it took this long for this subject to migrate into the mathematical mainstream, as part of the modern subject of *additive combinatorics* [**15**]; see Section 4.

The first appearance of the problem of computing $\alpha(G)$ for nonabelian G seems to have been in a 1985 paper of Babai and Sós [**2**]. In fact, the problem appears there as an afterthought; the authors were more interested in *Sidon sets*, in which the equation $ab^{-1} = cd^{-1}$ has no solutions with a, b, c, d taking at least three distinct values. This construction can be related to embeddings of graphs as induced subgraphs of Cayley graphs; product-free subsets arise because they relate to the special case of embedding stars in Cayley graphs. Nonetheless, the Babai-Sós paper is the first to make a nontrivial assertion about $\alpha(G)$ for general G; see Theorem 3.1.

This circumstance suggests rightly that the product-free problem is only one of a broad class of problems about structured subsets of groups; this class can be considered a nonabelian version of additive combinatorics, and progress on problems in this class has been driven as much by the development of the abelian theory as by interest from applications in theoretical computer science. An example of the latter is a problem of Cohn and Umans [**5**] (see also [**6**]): to find groups G admitting large subsets S_1, S_2, S_3 such that the equation $a_1 b_1^{-1} a_2 b_2^{-1} a_3 b_3^{-1} = e$, with $a_i, b_i \in S_i$, has only solutions with $a_i = b_i$ for all i. A sufficiently good construction would resolve an ancient problem in computational algebra: to prove that two $n \times n$ matrices can be multiplied using $O(n^{2+\epsilon})$ ring operations for any $\epsilon > 0$.

3. Lower bounds: Duluth, 1994

Upon my arrival at the REU in 1994, Joe gave me the paper of Babai and Sós, perhaps hoping I would have some new insight about Sidon sets. Instead, I took the path less traveled and started thinking about product-free sets.

The construction of product-free subsets given in [**2**] is quite simple: if H is a proper subgroup of G, then any nontrivial coset of H is product-free. This is trivial to prove directly, but it occurred to me to formulate it in terms of permutation actions. Recall that specifying a transitive permutation action of the group G is the same as simply identifying a conjugacy class of subgroups: if H is one of the subgroups, the action is left multiplication on left cosets of H. (Conversely, given an action, the point stabilizers are conjugate subgroups.) The construction of Babai and Sós can then be described as follows.

THEOREM 3.1 (Babai-Sós). *For G admitting a transitive action on $\{1, \dots, m\}$ with $m > 1$, $\beta(G) \geq m^{-1}$.*

PROOF. The set of all $g \in G$ such that $g(1) = 2$ is product-free of size n/m. □

I next wondered: what if you allow g to carry 1 into a slightly larger set, say a set T of k elements? You would still get a product-free set if you forced each $x \in T$ to map to something not in T. This led to the following argument.

THEOREM 3.2. *For G admitting a transitive action on $\{1, \dots, m\}$ with $m > 1$, $\beta(G) \geq cm^{-1/2}$.*

PROOF. For a given k, we compute a lower bound for the average size of

$$S = \bigcup_{x \in T} \{g \in G : g(1) = x\} - \bigcup_{y \in T} \{g \in G : g(1), g(y) \in T\}$$

for T running over k-element subsets of $\{2, \dots, m\}$. Each set in the first union contains n/m elements, and they are all disjoint, so the first union contains kn/m elements. To compute the average of a set in the second union, note that for fixed $g \in G$ and $y \in \{2, \dots, m\}$, a k-element subset T of $\{1, \dots, m\}$ contains $g(1), y, g(y)$ with probability $\frac{k(k-1)}{m(m-1)}$ if two of the three coincide and $\frac{k(k-1)(k-2)}{m(m-1)(m-2)}$ otherwise. A bit of arithmetic then shows that the average size of S is at least

$$\frac{kn}{m} - \frac{k^3 n}{(m-2)^2}.$$

Taking $k \sim (m/3)^{1/2}$, we obtain $\alpha(G) \geq cn/m^{1/2}$. (For any fixed $\epsilon > 0$, the implied constant can be improved to $e^{-1} - \epsilon$ for m sufficiently large; see the proof of Theorem 6.2. On the other hand, the proof as given can be made constructive in case G is doubly transitive, as then there is no need to average over T.) □

This gives a lower bound depending on the parameter m, which we can view as the index of the largest proper subgroup of G. To state a bound depending only on n, one needs to know something about the dependence of m on n; by Lemma 2.2, it suffices to prove a lower bound on m in terms of n for all *simple* nonabelian groups. I knew this could be done in principle using the classification of finite simple groups (CFSG); after some asking around, I got hold of a manuscript by Liebeck and Shalev [**12**] that included the bound I wanted, leading to the following result from [**10**].

THEOREM 3.3. *Under CFSG, the group G admits a transitive action on a set of size* $1 < m \leq cn^{3/7}$. *Consequently, Theorem 3.1 implies* $\alpha(G) \geq cn^{4/7}$, *whereas Theorem 3.2 implies* $\alpha(G) \geq cn^{11/14}$.

At this point, I was pretty excited to have discovered something interesting and probably publishable. On the other hand, I was completely out of ideas! I had no hope of getting any stronger results, even for specific classes of groups, and it seemed impossible to derive any nontrivial upper bounds at all. In fact, Babai and Sós suggested in their paper that maybe $\beta(G) \geq c$ for all G; I was dubious about this, but I couldn't convince myself that one couldn't have $\beta(G) \geq cn^{-\epsilon}$ for all $\epsilon > 0$.

So I decided to write this result up by itself, as my first Duluth paper, and ask Joe for another problem (which naturally he provided). My paper ended up appearing as [**10**]; I revisited the topic when I was asked to submit a paper in connection with being named a runner-up for the Morgan Prize for undergraduate research, the result being [**11**].

I then put this problem in a mental deep freezer, figuring (hoping?) that my youthful foray into combinatorics would be ultimately forgotten, once I had made some headway with some more serious mathematics, like algebraic number theory or algebraic geometry. I was reassured by the expectation that the nonabelian product-free problem was both intractable and of no interest to anyone, certainly not to any serious mathematician.

Ten years passed.[2]

4. Interlude: back to the future

Up until several weeks before the Duluth conference, I had been planning to speak about the latest and greatest in algebraic number theory (the proof of Serre's conjecture linking modular forms and mod p Galois representations, recently completed by Khare and Wintenberger). Then I got an email that suggested that maybe I should try embracing my past instead of running from it.

A number theorist friend (Michael Schein) reported having attended an algebra seminar at Hebrew University about product-free subsets of finite groups, and hearing my name in this context. My immediate reaction was to wonder what self-respecting mathematician could possibly be interested in my work on this problem. The answer was Tim Gowers, who had recently established a nontrivial upper bound for $\alpha(G)$ using a remarkably simple argument.

It seems that in the ten years since I had moved on to ostensibly more mainstream mathematics, additive combinatorics had come into its own, thanks partly to the efforts of no fewer than three Fields medalists (Tim Gowers, Jean Bourgain, and Terry Tao); some sources date the start of this boom to Ruzsa's publication in 1994 of a simplified proof [**14**] of a theorem of Freiman on subsets of $\mathbb{Z}/p\mathbb{Z}$ having few pairwise sums. In the process, some interest had spilled over to nonabelian problems.

---

[2]If you do not recognize this reference, you may not have read the excellent novel *The Grasshopper King*, by fellow Duluth REU alumnus Jordan Ellenberg.

The introduction to Gowers's paper [**7**] cites[3] my Duluth paper as giving the best known lower bound on $\alpha(G)$ for general G. At this point, it became clear that I had to abandon my previous plan for the conference in favor of a return visit to my mathematical roots.

5. Upper bounds: bipartite Cayley graphs

In this section, I'll proceed quickly through Gowers's upper bound construction. Gowers's paper [**7**] is exquisitely detailed; I'll take that fact as license to be slightly less meticulous here.

The strategy of Gowers is to consider three sets A, B, C for which there is no true equation $ab = c$ with $a \in A, b \in B, c \in C$, and give an upper bound on $\#A\#B\#C$. To do this, he studies a certain *bipartite Cayley graph* associated to G. Consider the bipartite graph Γ with vertex set $V_1 \cup V_2$, where each V_i is a copy of G, with an edge from $x \in V_1$ to $y \in V_2$ if and only if $yx^{-1} \in A$. We are then given that there are no edges between $B \subseteq V_1$ and $C \subseteq V_2$.

A good reflex at this point would be to consider the eigenvalues of the adjacency matrix of Γ. For bipartite graphs, it is more convenient to do something slightly different using singular values; although this variant of spectral analysis of graphs is quite natural, I am only aware of the reference [**3**] from 2004 (and only thanks to Gowers for pointing it out). Let N be the *incidence matrix*, with columns indexed by V_1 and rows by V_2, with an entry in row x and column y if xy is an edge of Γ.

THEOREM 5.1. *We can factor N as a product $U\Sigma V$ of $\#G \times \#G$ matrices over $\mathbb{R}$, with U, V orthogonal and Σ diagonal with nonnegative entries. (This is called a* singular value decomposition *of N.)*

PROOF. (Compare [**7**, Theorem 2.6], or see any textbook on numerical linear algebra.) By compactness of the unit ball, there is a greatest λ such that $\|N\mathbf{v}\| = \lambda\|\mathbf{v}\|$ for some nonzero $\mathbf{v} \in \mathbb{R}^{V_1}$. If $\mathbf{v} \cdot \mathbf{w} = 0$, then $f(t) = \|N(\mathbf{v} + t\mathbf{w})\|^2$ has a local maximum at $t = 0$, so

$$0 = \frac{d}{dt}\|N(\mathbf{v} + t\mathbf{w})\|^2 = 2t(N\mathbf{v}) \cdot (N\mathbf{w}).$$

Apply the same construction to the orthogonal complement of $\mathbb{R}\mathbf{v}$ in $\mathbb{R}^{V_1}$. Repeating, we obtain an orthonormal basis of $\mathbb{R}^{V_1}$; the previous calculation shows that the image of this basis in $\mathbb{R}^{V_2}$ is also orthogonal. Using these to construct V, U yields the claim. □

The matrix $M = NN^T$ is symmetric, and has several convenient properties.

(a) The trace of M equals the number of edges of Γ.
(b) The eigenvalues of M are the squares of the diagonal entries of Σ.
(c) Since Γ is regular of degree $\#A$ and connected, the largest eigenvalue of M is $\#A$, achieved by the all-ones eigenvector $\mathbf{1}$.

LEMMA 5.2. *Let λ be the second largest diagonal entry of Σ. Then the set W of $\mathbf{v} \in \mathbb{R}^{V_1}$ with $\mathbf{v} \cdot \mathbf{1} = 0$ and $\|N\mathbf{v}\| = \lambda\|\mathbf{v}\|$ is a nonzero subspace of $\mathbb{R}^{V_1}$.*

[3] Since Joe is fond of noting "program firsts", I should point out that this appears to be the first citation of a Duluth paper by a Fields medalist. To my chagrin, I think it is also the first such citation of any of my papers.

PROOF. (Compare [**7**, Lemma 2.7].) From Theorem 5.1, we obtain an orthogonal basis $\mathbf{v}_1, \dots, \mathbf{v}_n$ of $\mathbb{R}^{V_1}$, with $\mathbf{v}_1 = \mathbf{1}$, such that $N\mathbf{v}_1, \dots, N\mathbf{v}_n$ are orthogonal, and $\|N\mathbf{v}_1\|/\|\mathbf{v}_1\|, \dots, \|N\mathbf{v}_n\|/\|\mathbf{v}_n\|$ are the diagonal entries of Σ; we may then identify W as the span of the $\mathbf{v}_i$ with $i > 1$ and $\|N\mathbf{v}_i\| = \lambda\|\mathbf{v}_i\|$.

Alternatively, one may note that W is obviously closed under scalar multiplication, then check that W is closed under addition as follows. If $\mathbf{v}_1, \mathbf{v}_2 \in W$, then $\|N(\mathbf{v}_1 \pm \mathbf{v}_2)\| \leq \lambda\|\mathbf{v}_1 \pm \mathbf{v}_2\|$, but by the parallelogram law

$$\begin{aligned}\|N\mathbf{v}_1 + N\mathbf{v}_2\|^2 + \|N\mathbf{v}_1 - N\mathbf{v}_2\|^2 &= 2\|N\mathbf{v}_1\|^2 + 2\|N\mathbf{v}_2\|^2 \\ &= 2\lambda^2\|\mathbf{v}_1\|^2 + 2\lambda^2\|\mathbf{v}_2\|^2 \\ &= \lambda^2\|\mathbf{v}_1 + \mathbf{v}_2\|^2 + \lambda^2\|\mathbf{v}_1 - \mathbf{v}_2\|^2.\end{aligned}$$

Hence $\|N(\mathbf{v}_1 \pm \mathbf{v}_2)\| = \lambda\|\mathbf{v}_1 \pm \mathbf{v}_2\|$. □

Gowers's upper bound on $\alpha(G)$ involves the parameter δ, defined as the smallest dimension of a nontrivial representation[4] of G. For instance, if $G = \mathrm{PSL}_2(q)$ with $qodd$, then then $\delta = (q-1)/2$.

LEMMA 5.3. *If $\mathbf{v} \in \mathbb{R}^{V_1}$ satisfies $\mathbf{v} \cdot \mathbf{1} = 0$, then $\|N\mathbf{v}\| \leq (n\#A/\delta)^{1/2}\|\mathbf{v}\|$.*

PROOF. Take λ, W as in Lemma 5.2. Let G act on V_1 and V_2 by right multiplication; then G also acts on Γ. In this manner, W becomes a real representation of G in which no nonzero vector is fixed. In particular, $\dim(W) \geq \delta$.

Now note that the number of edges of M, which is $n\#A$, equals the trace of M, which is at least $\dim(W)\lambda^2 \geq \delta\lambda^2$. This gives $\lambda^2 \leq n\#A/\delta$, proving the claim. □

We are now ready to prove Gowers's theorem [**7**, Theorem 3.3].

THEOREM 5.4 (Gowers). *If A, B, C are subsets of G such that there is no true equation $ab = c$ with $a \in A, b \in B, c \in C$, then $\#A\#B\#C \leq n^3/\delta$. Consequently, $\beta(G) \leq \delta^{-1/3}$.*

For example, if $G = \mathrm{PSL}_2(q)$ with q odd, then $n \sim cq^3$, so $\alpha(G) \leq cn^{8/9}$. On the lower bound side, G admits subgroups of index $m \sim cq$, so $\alpha(G) \geq cn^{5/6}$.

PROOF. Write $\#A = rn, \#B = sn, \#C = tn$. Let $\mathbf{v}$ be the characteristic function of B viewed as an element of $\mathbb{R}^{V_1}$, and put $\mathbf{w} = \mathbf{v} - s\mathbf{1}$. Then

$$\begin{aligned}&\mathbf{w} \cdot \mathbf{1} = 0 \\ &\mathbf{w} \cdot \mathbf{w} = (1-s)^2\#B + s^2(n - \#B) = s(1-s)n \leq sn,\end{aligned}$$

so by Lemma 5.3, $\|N\mathbf{w}\|^2 \leq rn^2sn/\delta$.

Since $ab = c$ has no solutions with $a \in A, b \in B, c \in C$, each element of C corresponds to a zero entry in $N\mathbf{v}$. However, $N\mathbf{v} = N\mathbf{w} + rsn\mathbf{1}$, so each zero entry in $N\mathbf{v}$ corresponds to an entry of $N\mathbf{w}$ equal to $-rsn$. Therefore,

$$(tn)(rsn)^2 \leq \|N\mathbf{w}\|^2 \leq rsn^3/\delta,$$

hence $rst\delta \leq 1$ as desired. □

As noted by Nikolov and Pyber [**13**], the extra strength in Gowers's theorem is useful for other applications in group theory, largely via the following corollary.

[4]One could just as well restrict to real representations, which would increase δ by a factor of 2 in some cases. For instance, if $G = \mathrm{PSL}_2(q)$ with $q \equiv 3 \pmod 4$, this would give $\delta = q - 1$.

COROLLARY 5.5 (Nikolov-Pyber). *If A, B, C are subsets of G such that $ABC \neq G$, then $\#A\#B\#C \leq n^3/\delta$.*

PROOF. Suppose that $\#A\#B\#C > n^3/\delta$. Put $D = G \setminus AB$, so that $\#D = n - \#(AB)$. By Theorem 5.4, we have $\#A\#B\#D \leq n^3/\delta$, so $\#C > \#D$. Then for any $g \in C$, the sets AB and gC^{-1} have total cardinality more than n, so they must intersect. This yields $ABC = G$. □

Gowers indicates that his motivation for this argument was the notion of a *quasi-random graph* introduced by Chung, Graham, and Wilson [**4**]. They show that (in a suitable quantitative sense) a graph looks random in the sense of having the right number of short cycles if and only if it also looks random from the spectral viewpoint, i.e., the second largest eigenvalue of its adjacency matrix is not too large.

6. Coda

As noted by Nikolov and Pyber [**13**], using CFSG to get a strong quantitative version of Jordan's theorem on finite linear groups, one can produce upper and lower bounds for $\alpha(G)$ that look similar. (Keep in mind that the index of a proper subgroup must be at least $\delta + 1$, since any permutation representation of degree m contains a linear representation of dimension $m - 1$.)

THEOREM 6.1. *Under CFSG, the group G has a proper subgroup of index at most $c\delta^2$. Consequently,*

$$cn/\delta \leq \alpha(G) \leq cn/\delta^{1/3}.$$

Moreover, for many natural examples (e.g., $G = A_m$ or $G = \mathrm{PSL}_2(q)$), G has a proper subgroup of index at most $c\delta$, in which case one has

$$cn/\delta^{1/2} \leq \alpha(G) \leq cn/\delta^{1/3}.$$

Since the gap now appears quite small, one might ask about closing it. However, one can adapt the argument of [**10**] to show that Gowers's argument alone will not suffice, at least for families of groups with $m \leq c\delta$. (Gowers proves some additional results about products taken more than two at a time [**7**, §5]; I have not attempted to extend this construction to that setting.)

THEOREM 6.2. *Given $\epsilon > 0$, for G admitting a transitive action on $\{1, \dots, m\}$ for m sufficiently large, there exist $A, B, C \subseteq G$ with $(\#A)(\#B)(\#C) \geq (e^{-1} - \epsilon)n^3/m$, such that the equation $ab = c$ has no solutions with $a \in A, b \in B, c \in C$. Moreover, we can force $B = C$, $C = A$, or $A = B^{-1}$ if desired.*

PROOF. We first give a quick proof of the lower bound cn^3/m. Let U, V be subsets of $\{1, \dots, m\}$ of respective sizes u, v. Put

$$\begin{aligned} A &= \{g \in G : g(U) \cap V = \emptyset\} \\ B &= \{g \in G : g(1) \in U\} \\ C &= \{g \in G : g(1) \in V\}; \end{aligned}$$

then clearly the equation $ab = c$ has no solutions with $a \in A, b \in B, c \in C$. On the other hand,

$$\#A \geq n - u\frac{vn}{m}, \qquad \#B = \frac{un}{m}, \qquad \#C = \frac{vn}{m},$$

and so

$$(\#A)(\#B)(\#C) \geq \frac{n^3}{m}\left(\frac{uv}{m}\right)\left(1 - \frac{uv}{m}\right).$$

By taking $u, v = \lfloor\sqrt{m/2}\rfloor$, we obtain $(\#A)(\#B)(\#C) \geq cn^3/m$.

To optimize the constant, we must average over choices of U, V. Take $u, v = \lfloor\sqrt{m}\rfloor$. By inclusion-exclusion, for any positive integer h, the average of $\#A$ is bounded below by

$$\sum_{i=0}^{2h-1}(-1)^i n\frac{u(u-1)\cdots(u-i+1)v(v-1)\cdots(v-i+1)}{i!m(m-1)\cdots(m-i+1)}.$$

(The i-th term counts occurrences of i-element subsets in $g(U)\cap V$. We find $\binom{v}{i}$ i-element sets inside V; on average, each one occurs inside $g(U)$ for $n\binom{u}{i}/\binom{m}{i}$ choices of g.) Rewrite this as

$$n\left(\sum_{i=0}^{2h-1}(-1)^i\frac{(m^{1/2})^i(m^{1/2})^i}{m^i i!} + o(1)\right),$$

where $o(1) \to 0$ as $m \to \infty$. For any $\epsilon > 0$, we have

$$(\#A)(\#B)(\#C) \geq n^3\frac{m}{m^2}\left(e^{-1} - \epsilon\right)$$

for h sufficiently large, and m sufficiently large depending on h. This gives the desired lower bound.

Finally, note that we may achieve $B = C$ by taking $U = V$. To achieve the other equalities, note that if the triplet A, B, C has the desired property, so do B^{-1}, A^{-1}, C^{-1} and C, B^{-1}, A. □

I have no idea whether one can sharpen Theorem 5.4 under the hypothesis $A = B = C$ (or even just $A = B$). It might be enlightening to collect some numerical evidence using examples generated by Theorem 3.2; with Xuancheng Shao, we have done this for $\mathrm{PSL}_2(q)$ for $q \leq 19$.

I should also mention again that (as suggested in [**11**]) one can also study product-free subsets of compact topological groups, which are large for Haar measure. Some such study is implicit in [**7**, §4], but we do not know what explicit bounds come out.

References

[1] H.L. Abbott and L. Moser, Sum-free sets of integers, *Acta Arith.* **11** (1966), 393–396.

[2] L. Babai and V.T. Sós, Sidon sets in groups and induced subgraphs of Cayley graphs, *Europ. J. Combin.* **6** (1985), 101–114.

[3] B. Bollobás and V. Nikiforov, Hermitian matrices and graphs: singular values and discrepancy, *Disc. Math.* **285** (2004), 17–32.

[4] F.R.K. Chung, R.L. Graham, and R.M. Wilson, Quasi-random graphs, *Combinatorica* **9** (1989), 345–362.

[5] H. Cohn and C. Umans, A group-theoretic approach to fast matrix multiplication, *Proceedings of the 44th Annual IEEE Symposium on Foundations of Computer Science (FOCS)*, 2003, 438–449.

[6] H. Cohn, R. Kleinberg, B. Szegedy, and C. Umans, Group-theoretic algorithms for matrix multiplication, *Proceedings of the 46th Annual IEEE Symposium on Foundations of Computer Science (FOCS)*, 2005, 379–388.

[7] W.T. Gowers, Quasirandom groups, arXiv preprint 0710.3877v1 (2007).

[8] B. Green and I.Z. Ruzsa, Sum-free sets in abelian groups, *Israel J. Math* **147** (2005), 157-189.

[9] M. Guiduci and S. Hart, Small maximal sum-free sets, preprint available at http://eprints.bbk.ac.uk/archive/00000439/.

[10] K.S. Kedlaya, Large product-free subsets of finite groups, *J. Combin. Theory Series A* **77** (1997), 339–343.

[11] K.S. Kedlaya, Product-free subsets of groups, *Amer. Math. Monthly* **105** (1998), 900–906.

[12] M. W. Liebeck and A. Shalev, Simple groups, probabilistic methods, and a conjecture of Kantor and Lubotzky, *J. Algebra* **184** (1996), 31–57.

[13] N. Nikolov and L. Pyber, Product decompositions of quasirandom groups and a Jordan type theorem, arXiv preprint math/0703343v3 (2007).

[14] I.Z. Ruzsa, Generalized arithmetical progressions and sumsets, *Acta Math. Hungar.* **65** (1994), 379–388.

[15] T. Tao and V. Vu, *Additive combinatorics*, Cambridge University Press, Cambridge, 2006.

Department of Mathematics, Massachusetts Institute of Technology, 77 Massachusetts Avenue, Cambridge, MA 02139

E-mail address: kedlaya@mit.edu

URL: http://math.mit.edu/~kedlaya/

Contemporary Mathematics
Volume **479**, 2009

Generalizations of product-free subsets

Kiran S. Kedlaya and Xuancheng Shao

ABSTRACT. In this paper, we present some generalizations of Gowers's result about product-free subsets of groups. For any group G of order n, a subset A of G is said to be product-free if there is no solution of the equation $ab = c$ with $a, b, c \in A$. Previous results showed that the size of any product-free subset of G is at most $n/\delta^{1/3}$, where δ is the smallest dimension of a nontrivial representation of G. However, this upper bound does not match the best lower bound. We will generalize the upper bound to the case of *product-poor* subsets A, in which the equation $ab = c$ is allowed to have a few solutions with $a, b, c \in A$. We prove that the upper bound for the size of product-poor subsets is asymptotically the same as the size of product-free subsets. We will also generalize the concept of product-free to the case in which we have many subsets of a group, and different constraints about products of the elements in the subsets.

1. Background

Let G be a group. A subset S of G is *product-free* if there do not exist $a, b, c \in S$ (not necessarily distinct) such that $ab = c$. One can ask about the existence of large product-free subsets for various groups, such as the groups of integers or compact topological groups. Assume that G is a finite group of order $n > 1$. Let $\alpha(G)$ denote the size of the largest product-free subset of G, and $\beta(G) = \alpha(G)/n$. The purpose is to find good bounds for $\alpha(G)$ and $\beta(G)$ as a function of G, or as a function of n.

If G is abelian, this problem was solved by Green and Ruzsa in 2005 [**3**]. They gave an exact value of $\alpha(G)$ as a function of some characteristics of the abelian group G. However, the problem is much harder for nonabelian G. The first appearance of the problem of computing $\alpha(G)$ for nonabelian G seems to have been in a 1985 paper [**2**]. The construction of product-free subsets given in [**2**] is quite simple: if H is a proper subgroup of G, then any nontrivial coset of H is product-free. Therefore, a lower bound for $\beta(G)$ can be derived: $\beta(G) \geq m^{-1}$, where m is the index of the largest proper subgroup of G. In 1997, Kedlaya [**5**] improved this bound to $cm^{-1/2}$ for some constant c by showing that if H has index k then one can in fact find a

2000 *Mathematics Subject Classification.* Primary 20D60; secondary 20P05.

The first author was supported by NSF CAREER grant DMS-0545904 and a Sloan Research Fellowship. The second author was supported by the Paul E. Gray (1954) Endowed Fund for the MIT Undergraduate Research Opportunities Program.

union of $ck^{1/2}$ cosets of H that is product-free. This gives the best known lower bound on $\alpha(G)$ for general G.

THEOREM 1. *$\beta(G) \geq cm^{-1/2}$ for some constant c.*

On the other hand, Gowers recently established a nontrivial upper bound for $\alpha(G)$ using a remarkably simple argument [**4**] (see also [**6**] in this volume). The strategy of Gowers is to consider three sets A, B, C for which there is no solution of the equation $ab = c$ with $a \in A$, $b \in B$, $c \in C$, and give an upper bound on $|A| \cdot |B| \cdot |C|$, where $|A|$ denote the order of the set A.

THEOREM 2. *If A, B, C are subsets of G such that there is no solution of the equation $ab = c$ with $a \in A$, $b \in B$, $c \in C$, then $|A| \cdot |B| \cdot |C| \leq n^3/\delta$, where δ is defined as the smallest dimension of a nontrivial representation of G. Consequently, $\beta(G) \leq \delta^{-1/3}$.*

To prove this, a certain *bipartite Cayley graph* Γ associated to G is constructed. The vertex set of Γ is $V_1 \cup V_2$, where each V_i is a copy of G, with an edge from $x \in V_1$ to $y \in V_2$ if and only if $yx^{-1} \in A$. Therefore, there are no edges between $B \subseteq V_1$ and $C \subseteq V_2$. Let N be the *incidence matrix* of Γ, with columns indexed by V_1 and rows by V_2, with an entry in row x and column y if xy is an edge of Γ. Define $M = NN^{\mathrm{T}}$; then the largest eigenvalue of M is $|A|$. Let λ denote the second largest eigenvalue of the matrix M. Gowers's theorem relies on the following two lemmas.

LEMMA 3. *The second largest eigenvalue λ of the matrix M is at most $n \cdot |A|/\delta$.*

LEMMA 4. $|A| \cdot |B| \cdot |C| \leq n^2 \cdot \frac{\lambda}{|A|}$.

It can be proved that the group G has a proper subgroup of index at most $c\delta^2$ for some constant c [**7**]. Therefore, we have the following bounds for $\alpha(G)$.

THEOREM 5. *$cn/\delta \leq \alpha(G) \leq cn/\delta^{1/3}$ for some constant c.*

Since the gap between the lower bound and the upper bound for $\alpha(G)$ appears quite small, one might ask about closing it. However, it has been proved in [**6**] that Gowers's argument alone is not sufficient since the upper bound in Theorem 2 cannot be improved if the three sets A, B, and C are allowed to be different. In addition, Gowers also made some generalizations to the case of many subsets. Instead of finding two elements a and b in two subsets A and B such that their product is in a third subset C, he proposed to find $x_1, \ldots, x_m$ in m subsets such that for every nonempty subset $F \subset \{1, 2, \ldots, m\}$, the product of those x_i with $i \in F$ lies in a specified subset.

In this paper, instead of trying to close the gap between the lower bound and the upper bound for $\alpha(G)$, we will show that Gowers's Theorem 2 can actually be generalized to *product-poor subsets* of a group. We will give the precise definition of product-poor subsets as well as the upper bound for the size of product-poor subsets in Section 2. In Section 3, we will examine Gowers's generalization to the case of many subsets, and we will further generalize it to the problem of finding $x_1, \ldots, x_m$ in m subsets such that for certain (*not* all) subsets $F \subset \{1, 2, \ldots, m\}$ the product of those x_i with $i \in F$ lies in a specified subset.

2. Product-poor subsets of a group

In this section, we will state and prove a generalization of Theorem 2. We will consider the size of the largest product-poor subset instead of product-free subset. In product-poor subsets, there are a few pairs of elements whose product is also in this set. It turns out that we can derive the same asymptotic upper bound for the size of a product-poor subset. Despite the fact that the best known lower bound and the upper bound for the size of the largest product-free subset do not coincide, these two bounds do coincide asymptotically for the largest product-poor subset. First of all, we give the precise definition of a product-poor subset.

DEFINITION 6. *A subset A of group G is p-product-poor iff the number of pairs $(a, b) \in A \times A$ such that $ab \in A$ is at most $p|A|^2$.*

We now give a generalization of Gowers's argument. We change the condition that there are no solutions of the equation $ab = c$ with $a \in A$, $b \in B$, $c \in C$, to the weaker condition that there are only a few solutions of that equation. In fact, Babai, Nikolov, and Pyber [**1**] recently discovered a more general result depending on probability distributions in G. However, here we want to emphasize the result that the upper bound of $|A| \cdot |B| \cdot |C|$ is asymptotically the same for both product-free subsets and product-poor subsets. Note that this theorem is similar to a lemma in [**4**], which is stated in Lemma 10 below.

THEOREM 7. *Let G be a group of order n. Let A, B, and C be subsets of G with orders rn, sn, and tn, respectively. If there are exactly $prstn^2$ solutions of the equation $ab = c$ with $a \in A$, $b \in B$, $c \in C$, then $rst(1-p)^2\delta \leq 1$.*

PROOF. Let $\mathbf{v}$ be the characteristic function of B, and put $\mathbf{w} = \mathbf{v} - s\mathbf{1}$. Then

$$\begin{aligned} \mathbf{w} \cdot \mathbf{1} &= 0, \\ \mathbf{w} \cdot \mathbf{w} &= (1-s)^2 \cdot sn + s^2(n - sn) = s(1-s)n \leq sn, \end{aligned}$$

so by Lemma 3, $\|N\mathbf{w}\|^2 \leq rn^2sn/\delta$.

Let N be the incidence matrix of G of size $n \times n$, in which there is an entry in row x and column y iff $xy^{-1} \in B$. Consider the submatrix N_1 of N containing those rows corresponding to the elements in C, and those columns corresponding to the elements in A. The matrix N_1 has size $tn \times rn$. Suppose that there are k_i ones in row i of N_1 $(1 \leq i \leq tn)$. Note that there exists a one-to-one correspondence between the solutions of the equation $ab = c$ with $a \in A$, $b \in B$, $c \in C$, and the nonzero entries in N_1. As a result, we have

$$k_1 + k_2 + \ldots + k_{tn} = prstn^2.$$

Using the above equality, we have

$$\begin{aligned}
\|N\mathbf{w}\|^2 &\geq (k_1 - rsn)^2 + (k_2 - rsn)^2 + \ldots + (k_{tn} - rsn)^2 \\
&= \sum_{i=1}^{tn} k_i^2 - 2rsn \sum_{i=1}^{tn} k_i + (rsn)^2 \cdot (tn) \\
&\geq \frac{1}{tn} (\sum_{i=1}^{tn} k_i)^2 - 2rsn \sum_{i=1}^{tn} k_i + (rsn)^2 \cdot (tn) \\
&= p^2 r^2 s^2 tn^3 - 2rsnprstn^2 + r^2 s^2 tn^3 \\
&= r^2 s^2 tn^3 (1-p)^2
\end{aligned}$$

Combining the upper bound and the lower bound for $\|N\mathbf{w}\|^2$ derived above, we get

$$r^2 s^2 tn^3 (1-p)^2 \leq \|N\mathbf{w}\|^2 \leq rn^2 sn/\delta,$$

hence $rst(1-p)^2\delta \leq 1$ as desired.

□

Applying Theorem 7 to the special case in which $A = B = C$, we get the following corollary about product-poor subsets of a group.

COROLLARY 8. *The size of any p-product-poor subset of group G is at most $c|A|/\delta^{1/3}$ for some constant $c = c(p)$ if p is at most $1/\delta^{1/3}$.*

On the other hand, we can construct a subset S of G by taking the union of k cosets of a nontrivial subgroup of smallest index m. We estimate in Theorem 9 an upper bound for the number p, such that the subset S is guaranteed to be p-product-poor. We do this by estimating the number of solutions of the equation $ab = c$ with $a, b, c \in S$. It turns out that if the value of k is chosen suitably ($k \sim m/\delta^{1/3}$), then the subset S constructed this way has order $\sim n/m^{1/3}$, and this subset S is $1/m^{1/3}$-product-poor. Therefore, for those groups with $m = \Theta(\delta)$ (such as $PSL_2(q)$), we can find a $c/\delta^{1/3}$-product-poor subset of order $\sim n/\delta^{1/3}$ for some constant c.

THEOREM 9. *For G admitting a transitive action on $\{1, 2, \ldots, m\}$ with $m > 3$, define*

$$S = \bigcup_{x \in T} \{g \in G : g(1) = x\}$$

for any subset T of $\{2, \ldots, m\}$ of order $k \geq 3$. There exists a T, such that the number of solutions of the equation $ab = c$ with $a, b, c \in S$ is at most $4n^2k^3/(m-3)^3$.

PROOF. Consider the following summation:

$$\begin{aligned}
&\sum_T |\{a,b\in G : a(1), b(1), ab(1)\in T\}|\\
=&\sum_{a,b} |\{T : |T|=k, a(1), b(1), ab(1)\in T\}|\\
\le&\sum_{a(1)=b(1)}\binom{m-3}{k-2}+\sum_{a(1)=ab(1)}\binom{m-3}{k-2}+\sum_{b(1)=ab(1)}\binom{m-3}{k-2}+\sum_{a,b}\binom{m-4}{k-3}\\
=&3\cdot n\cdot\frac{n}{m}\cdot\binom{m-3}{k-2}+n^2\cdot\binom{m-4}{k-3}\\
\le&4n^2\cdot\binom{m-4}{k-3}.
\end{aligned}$$

Since there are a total of $\binom{m-1}{k}$ choices for T, there exists some T of order k, such that the number of solutions of $ab=c$ with $a,b,c\in T$ is at most

$$4n^2\cdot\frac{\binom{m-4}{k-3}}{\binom{m-1}{k}}=4n^2\cdot\frac{k(k-1)(k-2)}{(m-1)(m-2)(m-3)}\le\frac{4n^2k^3}{(m-3)^3}$$

□

In Theorem 9, if we choose $k\sim m/\delta^{1/3}$, then the subset S has order $n/\delta^{1/3}$, and is $c/\delta^{1/3}$-product-poor for some constant c. This shows that the upper bound in Corollary 8 is asymptotically the best for families of groups with $m\sim\delta$.

3. Generalization to many subsets

In this section, we will study the problem of finding $x_1,\ldots,x_m$ in m subsets such that for subsets F in Γ, where Γ is a collection of subsets of $\{1,2,\ldots,m\}$, the product of those x_i with $i\in F$ lies in a specified subset. Gowers has already studied this problem in which Γ is the collection of *all* subsets of $\{1,2,\ldots,m\}$. We will first state and prove this result for the simple case when $m=3$, which was first proved in [**4**]. However, a constant factor is improved significantly compared to the result in [**4**], although the proofs are very similar. We will study the situation in which Γ is the collection of all two-element subsets of $\{1,2,\ldots,m\}$. After this, it will be easy to derive a result for an arbitrary collection Γ.

Before we prove any result, we first repeat the following lemma in [**4**], which is frequently used in the proofs of all the following results.

LEMMA 10. *Let G be a group of order n such that no nontrivial representation has dimension less than δ. Let A and B be two subsets of G with densities r and s, respectively and let k and t be two positive constants. Then, provided that $rst\ge(k^2\delta)^{-1}$, the number of group elements $x\in G$ for which $|A\cap xB|\le(1-k)rsn$ is at most tn.*

We now give an improved version of the corresponding theorem in [**4**] for the case $m=3$. The constant M is 16 in [**4**], which is much larger than our constant $3+2\sqrt{2}$.

THEOREM 11. *Let G be a group of order n such that no nontrivial representation has dimension less than δ. Let A_1, A_2, A_3, A_{12}, A_{13} and A_{23} be subsets of G of densities p_1, p_2, p_3, p_{12}, p_{13} and p_{23}, respectively. Then, provided that $p_1p_2p_{12}$,*

$p_1p_3p_{13}$ and $p_2p_3p_{23}p_{12}p_{13}$ are all at least M/δ, where $M > 3+2\sqrt{2}$, there exist elements $x_1 \in A_1$, $x_2 \in A_2$ and $x_3 \in A_3$ such that $x_1x_2 \in A_{12}$, $x_1x_3 \in A_{13}$ and $x_2x_3 \in A_{23}$.

PROOF. Choose two numbers $0 < \lambda < 1$ and $0 < \mu < 1/2$ such that

$$M \geq \frac{1}{\mu\lambda^2}, M > \frac{1}{(1-\lambda)^2}.$$

(In fact, since $M > 3+2\sqrt{2}$, we can choose $\lambda = 2-\sqrt{2}$ and $(3+2\sqrt{2})/(2M) < \mu < 1/2$.) If $p_1p_2p_{12} \geq M/\delta$, then we have

$$p_1p_2p_{12} \geq \frac{1}{\mu\lambda^2\delta}$$

from the choices of λ and μ. By Lemma 10, the number of x_1 such that

$$|A_2 \cap x_1^{-1}A_{12}| \leq (1-\lambda)p_2p_{12}n$$

is at most $\mu p_1 n$. Similarly, if $p_1p_3p_{13} \geq M/\delta$, then the number of x_1 such that

$$|A_3 \cap x_1^{-1}A_{13}| \leq (1-\lambda)p_3p_{13}n$$

is also at most $\mu p_1 n$. Therefore, since $\mu < 1/2$, we can choose $x_1 \in A_1$ such that, setting $B_2 = A_2 \cap x_1^{-1}A_{12}$ and $B_3 = A_3 \cap x_1^{-1}A_{13}$, $q_2 = (1-\lambda)p_2p_{12}$ and $q_3 = (1-\lambda)p_3p_{13}$, we have $|B_2| \geq q_2 n$ and $|B_3| \geq q_3 n$.

What remains is to show that there exist $y_2 \in B_2$, $y_3 \in B_3$ and $y_{23} \in A_{23}$, such that $y_2y_3 = y_{23}$. This is true from the following relation:

$$\begin{aligned} q_2q_3p_{23} &= p_2p_3p_{23}p_{12}p_{13}(1-\lambda)^2 \\ &\geq \frac{M}{\delta}\cdot(1-\lambda)^2 \\ &\geq \frac{1}{\delta}. \end{aligned}$$

□

In order to generalize Theorem 11 to arbitrary m, Gowers introduced a concept of *density condition* in his paper [**4**]. The basic idea is that the product of the densities of some subsets must be greater than a threshold. For example, in Theorem 11, the threshold is M/δ, and three products of densities must be greater than this threshold. In order to study the case that Γ is the collection of all two-element subsets for general m, we do need to define what density products are required to be larger than the threshold. We give the following definition about density product. Although we will not repeat the definition of density condition in [**4**], it is worth noticing that the definition given below is intrinsically the same as Gowers's definition of density condition. This similarity can be seen if we set the densities of subsets not considered in our problem to be 1.

DEFINITION 12. *Suppose that for every $1 \leq i \leq m$ we have a subset A_i of G with density p_i, and for every $1 \leq i < j \leq m$ we have a subset A_{ij} of G with density p_{ij}. For any $1 \leq i < j \leq m$, we define the (i,j)-density product to be*

$$P_{ij} = p_ip_jp_{ij}\prod_{k=1}^{i-1}(p_{ki}p_{kj})$$

Now we state a theorem similar to Gowers's generalization. If all the density products are larger than a threshold $f(m)/\delta$, we are able to find m elements $x_1, \ldots, x_m$ from m subsets, such that for any two-element subset F of $\{1, 2, \ldots, m\}$, the product of x_i with $i \in F$ is in a specified subset.

THEOREM 13. *Suppose $f(m)$ is a function such that $f(2) > 1$ and*

$$\left(1 - \sqrt{\frac{m}{f(m)}}\right)^2 f(m) \geq f(m-1)$$

for any $m = 3, 4, \ldots$. Let G be a group of order n such that no nontrivial representation has dimension less than δ. For every $1 \leq i \leq m$ let A_i be a subset of G with density p_i. For every $1 \leq i < j \leq m$ let A_{ij} be a subset of G with density p_{ij}. Suppose that all the (i,j)-density products are at least $f(m)/\delta$. Then there exist elements $x_1, \ldots, x_m$ of G such that $x_i x_j \in A_{ij}$ for every $1 \leq i < j \leq m$.

PROOF. We use induction on m. Since all the $(1,j)$-density products are at least $f(m)/\delta$, we have the inequality $1/m \cdot p_1 p_j p_{1j} \geq f(m)/(\delta m)$. Therefore, by Lemma 10, for each $1 < j \leq m$, the number of x_1 such that

$$|A_j \cap x_1^{-1} A_{1j}| \leq p_j p_{1j} \left(1 - \sqrt{\frac{m}{f(m)}}\right)$$

is at most $p_1 n/m$. It follows that there exists $x_1 \in A_1$ such that, if for every $1 < j \leq m$ we set $B_j = A_j \cap x_1^{-1} A_{1j}$, then every B_j has density at least

$$q_j = p_j p_{1j} \left(1 - \sqrt{\frac{m}{f(m)}}\right).$$

In order to use the induction hypothesis, it suffices to prove that for every $1 < i < j \leq m$, we have

$$Q_{ij} = q_i q_j p_{ij} \prod_{k=2}^{i-1} (p_{ki} p_{kj}) \geq \frac{f(m-1)}{\delta}.$$

In fact,

$$\begin{aligned} Q_{ij} &= \left(1 - \sqrt{\frac{m}{f(m)}}\right)^2 \cdot p_i p_j p_{1i} p_{1j} \prod_{k=2}^{i-1} (p_{ki} p_{kj}) \\ &= \left(1 - \sqrt{\frac{m}{f(m)}}\right)^2 \cdot p_i p_j \prod_{k=1}^{i-1} (p_{ki} p_{kj}) \\ &\geq \left(1 - \sqrt{\frac{m}{f(m)}}\right)^2 \cdot \frac{f(m)}{\delta} \\ &\geq \frac{f(m-1)}{\delta}. \end{aligned}$$

This proves the inductive step of the theorem. We take $m = 2$ as the base case, which is trivial since $f(2) > 1$.

□

For example, if we define $f(m)$ by $f(2) = 2$ and

$$f(m) = \frac{(m + 1 - f(m-1))^2}{4m} \qquad (m \geq 3),$$

then $0 \le f(m) \le m+2$ by induction on m, and

$$\left(1-\sqrt{\frac{m}{f(m)}}\right)^2 f(m) = f(m-1).$$

Consequently, the conclusion of Theorem 13 holds if we assume all the density products are at least $(m+2)/\delta$.

Using the same technique as above, we can derive similar results for arbitrary Γ. There are two aspects we need to consider. One of them is what density products should be involved in the condition. As mentioned before, the definition of density condition in [**4**] can be used if we set the densities of the sets we are not going to consider to be 1. Here we give a formal definition of density products (which is similar in the definition in [**4**]).

Definition 14. *Suppose Γ is a collection of subsets of $\{1,2,\ldots,m\}$. For each $F \in \Gamma$, A_F is a subset of G with density p_F. Let h be an integer less than m and let E be a subset of $\{h+1,\ldots,m\}$. Let $P_{h,E}$ be the collection of all the sets in Γ of the form $U \cup V$, where $\max U < h$ and V is either $\{h\}$, E or $\{h\} \cup E$. We say that the product $\prod_{F \in P_{h,E}} p_F$ is the (h,E)-density product.*

We also need to determine the lower bound for the density products. This lower bound is generally of the form $f_\Gamma(m)/\delta$. The constraint of the function $f_\Gamma(m)$ is related to the collection Γ. Using the same method as in the proof of Theorem 13, we get the following theorem for arbitrary collection Γ.

Theorem 15. *Let G be a group of order n such that no nontrivial representation has dimension less than δ. Let Γ be a collection of subsets of $\{1,2,\ldots,m\}$. For each $F \in \Gamma$, let A_F be a subset of G of density p_F. Let $h(m)$ be a positive integer such that there are at most $h(m)$ subsets in Γ containing k for any $1 \le k \le m$. Suppose all the (h,E)-density products are at least $f_\Gamma(m)/\delta$, where the function $f_\Gamma(m)$ satisfies the inequality*

$$\left(1-\sqrt{\frac{h(m)}{f_\Gamma(m)}}\right)^2 f_\Gamma(m) \ge f_\Gamma(m-1)$$

for any $m = 2,3,\ldots$, then there exist elements $x_1,\ldots,x_m$ of G such that $x_F \in A_F$ for every $F \in \Gamma$, where x_F stands for the product of all x_i such that $i \in F$.

Since the proof is very similar to that of Theorem 13, we will not go into details here.

References

[1] L. Babai, N. Nikolov, and L. Pyber, Product growth and mixing in finite groups, *Proceedings of the Nineteenth Annual ACM-SIAM Symposium on Discrete Algorithms (SODA '08)*, 2008.

[2] L. Babai and V. T. Sós, Sidon sets in groups and induced subgraphs of Cayley graphs, *Europ. J. Combin.* **6** (1985), 101–114.

[3] B. Green and I. Z. Ruzsa, Sum-free sets in abelian groups, *Israel J. Math* **147** (2005), 157–189.

[4] W. T. Gowers, Quasirandom groups, *eprint arXiv: 0710.3877* (2007).

[5] K. S. Kedlaya, Large product-free subsets of finite groups. *J. Combin. Theory Series A* **77** (1997), 339–343.

[6] K.S. Kedlaya, Product-free subsets of groups, then and now, this volume.

[7] M. W. Liebeck and A. Shalev. Simple groups, probabilistic methods, and a conjecture of Kantor and Lubotzky, *J. Algebra* **184** (1996), 31–57.

Department of Mathematics, Massachusetts Institute of Technology, 77 Massachusetts Avenue, Cambridge, MA 02139
E-mail address: `kedlaya@mit.edu`
URL: `http://math.mit.edu/~kedlaya/`

410 Memorial Drive Room 211A, Cambridge, MA 02139
E-mail address: `zero@mit.edu`

Contemporary Mathematics
Volume **479**, 2009

What is a superrigid subgroup?

Dave Witte Morris

To my friend and mentor Joseph A. Gallian on his 65th birthday

Contents

It is well known that a linear transformation can be defined to have any desired action on a basis. From this fact, one can show that every group homomorphism from $\mathbb{Z}^k$ to $\mathbb{R}^d$ extends to a homomorphism from $\mathbb{R}^k$ to $\mathbb{R}^d$, and we will see other examples of discrete subgroups H of connected groups G, such that the homomorphisms defined on H can ("almost") be extended to homomorphisms defined on all of G. First, let us see that this is related to a very classical topic in geometry, the study of linkages.

1. Rigidity of Linkages

Informally, a *linkage* is an object in 3-space that is constructed from some finite set of line segments (called "rods," or "edges") by attaching endpoints of some of the rods to endpoints of some of the other rods. (That is, a linkage naturally has the structure of a 1-dimensional simplicial complex.) It is assumed that the rods are rigid (they can neither stretch nor bend), but that the joints that connect the rods

2000 *Mathematics Subject Classification.* Primary 22E40; Secondary 20-02, 52C25.

This article is based on a talk given in various forms at several different universities, and at an MAA Mathfest. It was written during visits to the University of Chicago and the Tata Institute of Fundamental Research in Mumbai, India; I would like to thank both of these institutions for their generous hospitality. I would also like to thank David Fisher and an anonymous referee for their comments on a preliminary version. The writing was partially supported by a research grant from the National Science and Engineering Research Council of Canada.

are entirely flexible — they allow the rods to rotate freely, as long as the endpoints remain attached.

1.1. EXAMPLE (Hinge). Construct a linkage with four vertices (or "joints") A, B, C, D by putting together two different triangles ABC and BCD with the same base BC, as in Figure 1(a). The angle between the two triangles can be varied continuously, so the object has some flexibility — it is not *rigid*. (For example, the hinge can be opened wider, as in Figure 1(b).) This linkage can reasonably be called a "hinge."

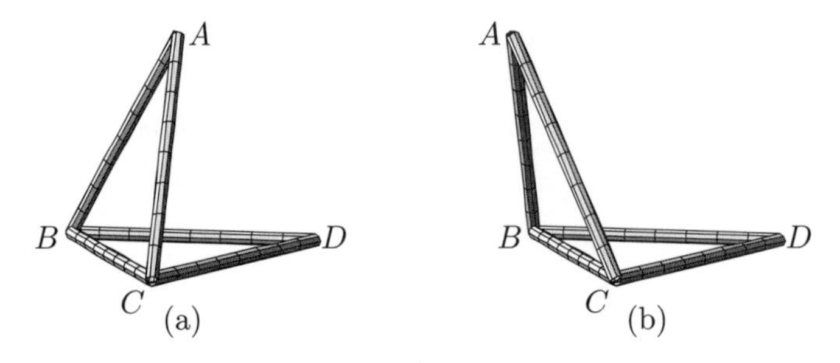

FIGURE 1. The hinge is not rigid, because the angle between the two triangles can be varied continuously, without changing the lengths of the rods in the linkage.

1.2. EXAMPLE (Tetrahedron). Construct a linkage with four vertices A, B, C, D by joining every pair of vertices with an edge, as in Figure 2. This object is rigid — it cannot be deformed.

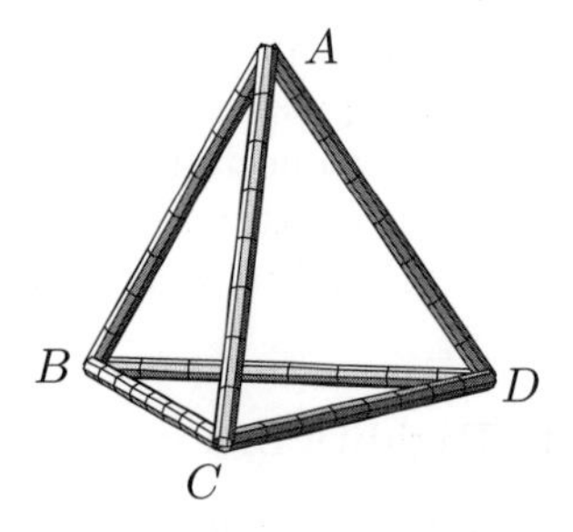

FIGURE 2. A tetrahedron cannot be deformed; it is rigid.

1.3. EXAMPLE (Double tetrahedron). Add a small tetrahedron $BCDE$ to the bottom of the tetrahedron $ABCD$, as in Figure 3. The resulting object has no deformations, so it is rigid.

However, this double tetrahedron does not have the property that is called *global rigidity*. Namely, suppose:

(1) We label each end of each rod with the name of the vertex that joins it to other rods, and then dismantle the linkage. This results in a collection of 9 rods, which are pictured in Figure 4.
(2) We then assemble these rods into a linkage, by joining together all vertices that have the same label.

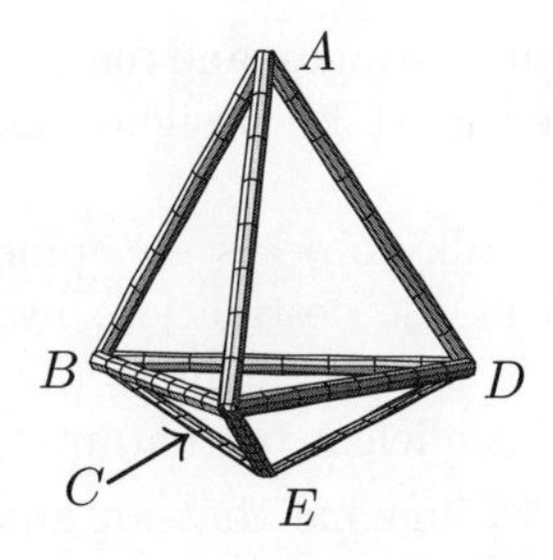

FIGURE 3. Two tetrahedra with a common face form a rigid structure.

Unfortunately, the resulting linkage may not be the one we started with; as illustrated in Figure 5, the small tetrahedron could be *inside* the larger one, instead of *outside*.

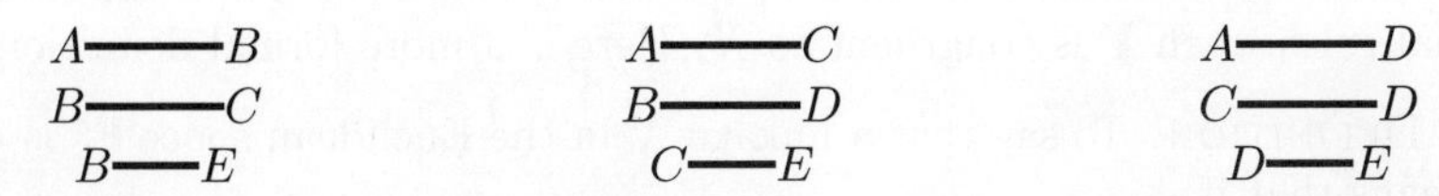

FIGURE 4. The double tetrahedron is made up of 9 rods (6 long ones and 3 short ones).

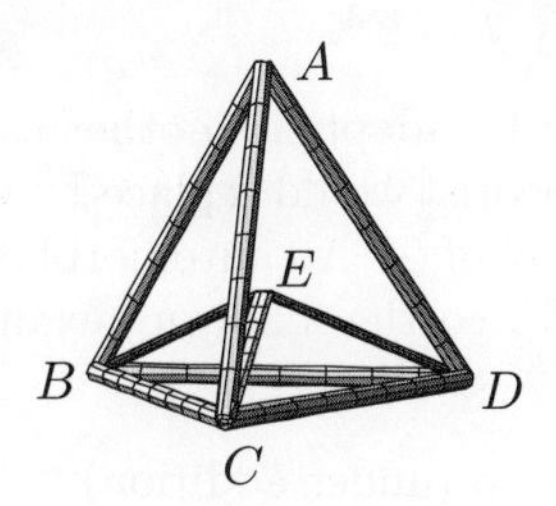

FIGURE 5. The double tetrahedron is not globally rigid: if it has been taken apart, it can be reassembled incorrectly, even if the gluing instructions are followed correctly.

In summary:

- The double tetrahedron has no small perturbations. In other words, if it is reassembled, and every rod is close to its correct position, then every rod is in *exactly* the correct position. So the object is rigid, or, more precisely, "locally rigid."
- On the other hand, the double tetrahedron is not rigid in a global sense — we say that it is not *globally rigid* — because it can be reassembled incorrectly if we do not assure that the rods are near their correct position.

1.4. EXAMPLE. A tetrahedron *is* globally rigid: its geometric structure is completely determined (up to congruence) by the combinatorial data that specify which of the rods are to be joined together.

Rigidity and global rigidity are important concepts in geometry, and also in the real world:

- Scaffolds, shelving units, bridges, and many other structures can be viewed as linkages, and they must be designed not to collapse; they must be (locally) rigid.
- Furniture and other bulky objects are sometimes shipped in pieces that are to be assembled at the destination, by following instructions of the type "insert tab A in slot B." Unless the object is globally rigid, the instructions will be insufficient to guarantee proper assembly.

Thus, it should not be hard to imagine that an analogous notion in other fields would have significant interest to researchers in that area. We will focus on the case of group theory.

2. The Analogous Notion in Group Theory

Informally, saying that a linkage X is globally rigid means that if Y is any linkage that is constructed from rods of the same lengths by using the same combinatorial rules, then Y is congruent to X. Here is a more formal definition:

2.1. Definition. To say that a linkage X in the Euclidean space $\mathbb{E}^3$ is *globally rigid* means that if

- Y is any linkage in $\mathbb{E}^3$, and
- $f\colon X \to Y$ is a combinatorial isomorphism (i.e., f is a bijection that maps each rod in X isometrically onto a rod in Y),

then f extends to an isometry $\widehat{f}$ of $\mathbb{E}^3$.

The same idea can easily be adapted to other categories of mathematical objects. For example, a group theorist would replace $\mathbb{E}^3$ with a group G, and replace X and Y with subgroups H and K of G. An automorphism of G is the group-theoretic analogue of an isometry of $\mathbb{E}^3$, so the following example shows that $\mathbb{Z}$ is globally rigid as a subgroup of $\mathbb{R}$:

2.2. Example. $\mathbb{R}$ is a group (under addition), and $\mathbb{Z}$ is a subgroup. If

- K is a subgroup of $\mathbb{R}$, and
- $\varphi\colon \mathbb{Z} \to K$ is an isomorphism,

then φ extends to an automorphism $\widehat{\varphi}$ of $\mathbb{R}$.

Proof. Let $c = \varphi(1)$ and define $\widehat{\varphi}\colon \mathbb{R} \to \mathbb{R}$ by

$$\widehat{\varphi}(x) = cx.$$

Then:

- It is obvious that $\widehat{\varphi}$ is a homomorphism.
- Since φ is injective, we know

$$c = \varphi(1) \neq \varphi(0) = 0,$$

so $\widehat{\varphi}$ is bijective.

- For any $n \in \mathbb{Z}$, we have

$$\begin{aligned} \widehat{\varphi}(n) &= cn && \text{(definition of } \widehat{\varphi}\text{)} \\ &= n \cdot \varphi(1) && \text{(definition of } c\text{)} \\ &= \varphi(n) && (\varphi \text{ is a homomorphism).} \end{aligned}$$

So $\widehat{\varphi}$ extends φ.

Thus, $\widehat{\varphi}$ is an automorphism of $\mathbb{R}$ that extends φ. □

In the above example:

(1) The group $\mathbb{R}$ is also a topological space, and the group operations of addition and negation are compatible with the topology (that is, they are continuous); thus, $\mathbb{R}$ is a *topological group*.
(2) The subgroup $\mathbb{Z}$ is discrete in $\mathbb{R}$ (i.e., has no accumulation points); so we say that $\mathbb{Z}$ is a *discrete subgroup* of $\mathbb{R}$.
(3) The homomorphism $\widehat{\varphi}$ is continuous.

Thus, the discrete subgroup $\mathbb{Z}$ is globally rigid in $\mathbb{R}$, even when we take into account the topological structure of $\mathbb{R}$:

2.3. DEFINITION. Let H be a discrete subgroup of a topological group G. Saying H is *globally rigid* in G means that if

- K is any discrete subgroup of G, and
- $\varphi\colon H \to K$ is any isomorphism,

then φ extends to a continuous automorphism $\widehat{\varphi}$ of G.

3. Definition of Superrigidity

In the definition of global rigidity (2.3), the map φ is assumed to be an isomorphism, and its image K is assumed to be contained in the same group G that contains H. "Superrigidity" is a notion that removes these restrictions. Here is a very elementary example of this that generalizes Example 2.2:

3.1. EXAMPLE. Suppose φ is any group homomorphism from $\mathbb{Z}^k$ to $\mathbb{R}^d$. (That is, φ is a function from $\mathbb{Z}$ to $\mathbb{R}^d$, and we have $\varphi(m+n) = \varphi(m) + \varphi(n)$.) Then φ extends to a continuous homomorphism $\widehat{\varphi}\colon \mathbb{R}^k \to \mathbb{R}^d$.

PROOF. Let $e_1, e_2, \ldots, e_k$ be the standard basis of $\mathbb{R}^k$, so $\{e_1, e_2, \ldots, e_k\}$ is a generating set for the subgroup $\mathbb{Z}^k$. A linear transformation can be defined to have any desired action on a basis, so there is a linear transformation $\widehat{\varphi}\colon \mathbb{R}^k \to \mathbb{R}^d$, such that

$$\widehat{\varphi}(e_i) = \varphi(e_i) \quad \text{for } i = 1, 2, \ldots, k. \tag{3.2}$$

Then:

- Since $\widehat{\varphi}$ is linear, it is continuous.
- Because $\widehat{\varphi}$ is a linear transformation, it respects addition; that is, it is a homomorphism from $\mathbb{R}^k$ to $\mathbb{R}^d$.
- From (3.2), we know that φ and $\widehat{\varphi}$ agree on $e_1, e_2, \ldots, e_k$. Thus, since $\{e_1, e_2, \ldots, e_k\}$ generates $\mathbb{Z}^k$, the two homomorphisms agree on all of $\mathbb{Z}^k$. In other words, $\widehat{\varphi}$ extends φ.

So $\widehat{\varphi}$ is a continuous automorphism that extends φ. □

In short:

Every group homomorphism from $\mathbb{Z}^k$ to $\mathbb{R}^d$ extends to a continuous homomorphism from $\mathbb{R}^k$ to $\mathbb{R}^d$. (3.3)

However, because this observation deals only with abelian groups, it is rather trivial. A superrigidity theorem is a result of similar flavor that deals with groups that are more interesting. Namely, instead of only homomorphisms into the abelian group $\mathbb{R}^d$, it is much more interesting to look at homomorphisms into matrix groups.

(Any such homomorphism is called a *group representation*, and the study of these representations is a major part of group theory.)

Let us be more precise:

3.4. NOTATION. $\mathrm{GL}(d,\mathbb{R}) = \{\, d \times d$ invertible matrices with real entries $\}$.

It is important to note that $\mathrm{GL}(d,\mathbb{R})$ is a group under multiplication. Furthermore, $\mathbb{R}^k$ is a subgroup of $\mathrm{GL}(d,\mathbb{R})$ (if $d > k$). For example,

$$\mathbb{R}^3 \cong \begin{bmatrix} 1 & 0 & 0 & * \\ 0 & 1 & 0 & * \\ 0 & 0 & 1 & * \\ 0 & 0 & 0 & 1 \end{bmatrix} \subset \mathrm{GL}(4,\mathbb{R}).$$

So any homomorphism into $\mathbb{R}^d$ can be thought of as a homomorphism into the matrix group $\mathrm{GL}(d+1,\mathbb{R})$.

Unfortunately, (3.3) does not remain valid if we replace $\mathbb{R}^d$ with $\mathrm{GL}(d,\mathbb{R})$:

3.5. EXAMPLE. Suppose φ is a group homomorphism from $\mathbb{Z}$ to $\mathrm{GL}(d,\mathbb{R})$. That is, φ is a function from $\mathbb{Z}$ to $\mathrm{GL}(d,\mathbb{R})$, and we have

$$\varphi(m+n) = \varphi(m) \cdot \varphi(n).$$

It *need not* be the case that φ extends to a continuous homomorphism from $\mathbb{R}$ to $\mathrm{GL}(d,\mathbb{R})$.

PROOF BY CONTRADICTION. Suppose there is a continuous homomorphism $\widehat{\varphi}\colon \mathbb{R} \to \mathrm{GL}(d,\mathbb{R})$, such that $\widehat{\varphi}(n) = \varphi(n)$, for all $n \in \mathbb{Z}$.

Consider the composition $\det \circ \widehat{\varphi}$. Note that:

- Since the determinant of any invertible matrix is nonzero, we see that $\det \circ \widehat{\varphi}$ is a function from $\mathbb{R}$ to $\mathbb{R}^\times$ (where $\mathbb{R}^\times$ is the set of nonzero real numbers).
- Since homomorphisms map the identity element of the domain group to the identity element of the image, we have $\varphi(0) = \mathbb{I}$ (the identity matrix). Hence,
$$\det\bigl(\varphi(0)\bigr) = \det(\mathbb{I}) = 1 > 0.$$
- Since the composition of continuous functions is continuous, and the continuous image of a connected set is continuous, we know $\det\bigl(\widehat{\varphi}(\mathbb{R})\bigr)$ is connected.

Therefore, $\det\bigl(\widehat{\varphi}(\mathbb{R})\bigr)$ is a connected subset of $\mathbb{R}^\times$ that contains the number 1. So $\det\bigl(\widehat{\varphi}(\mathbb{R})\bigr) \subset \mathbb{R}^+$. In particular, $\det\bigl(\widehat{\varphi}(1)\bigr) > 0$. Therefore

$$\det\bigl(\varphi(1)\bigr) = \det\bigl(\widehat{\varphi}(1)\bigr) > 0.$$

But φ is an arbitrary homomorphism from $\mathbb{Z}$ to $\mathrm{GL}(d,\mathbb{R})$, and it need not be the case that $\det\bigl(\varphi(1)\bigr) > 0$. (Namely, for any $A \in \mathrm{GL}(d,\mathbb{R})$, we may let $\varphi(n) = A^n$. If $\det A < 0$, then $\det\bigl(\varphi(1)\bigr) = \det A < 0$.) This is a contradiction. □

The above counterexample is based on the possibility that $\det\bigl(\varphi(1)\bigr) < 0$. However, for any n, we have

$$\det\bigl(\varphi(2n)\bigr) = \det\bigl(\varphi(n+n)\bigr) = \det\bigl(\varphi(n)\cdot\varphi(n)\bigr) = \Bigl(\det\bigl(\varphi(n)\bigr)\Bigr)^2 > 0.$$

Thus, this obstacle does not arise if we restrict our attention to even numbers. That is, in defining the extension $\widehat{\varphi}$, which interpolates a nice curve through the given

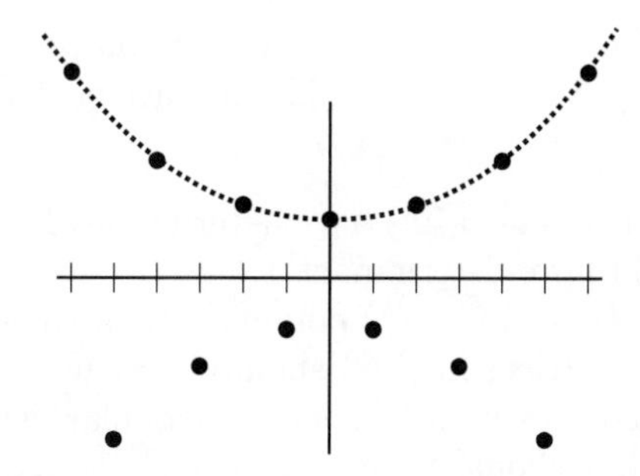

FIGURE 6. It may be necessary to ignore the values at odd numbers when interpolating.

values at points of $\mathbb{Z}$, we may have to ignore the values at odd numbers, and only match the values of φ at even numbers. An illustration of this is in Figure 6.

One can imagine that, analogously, there might be situations where it is necessary to restrict attention, not to multiples of 2, but to multiples of some other integer N. A group theorist may observe that

$$\{\text{ multiples of } N\} \text{ is a subgroup of } \mathbb{Z} \text{ that has finite index.}$$

Thus, in group-theoretic terms, the upshot of the preceding discussion is that we may need to restrict our attention to a finite-index subgroup.

The need to pass to a finite-index subgroup happens so often in the theory of infinite groups that there is a name for it: a property holds *virtually* if it becomes true when our attention is restricted to a finite-index subgroup.

3.6. EXAMPLE.

(1) To say that G is *virtually abelian* means that some finite-index subgroup of G is abelian.
(2) If G is a topological group, then, to say that G is *virtually connected* means that some finite-index subgroup of G is connected.

3.7. EXERCISE. What does it mean to say that G is *virtually finite*?

In this vein, we make the following definition:

3.8. DEFINITION. Suppose

- H is a discrete subgroup of a topological group G,
- $\varphi\colon H \to \mathrm{GL}(d,\mathbb{R})$ is a homomorphism, and
- $\widehat{\varphi}\colon G \to \mathrm{GL}(d,\mathbb{R})$ is a continuous homomorphism.

We say $\widehat{\varphi}$ *virtually extends* φ if there is a finite-index subgroup H' of H, such that $\widehat{\varphi}(h) = \varphi(h)$, for all $h \in H'$.

Although the proof is not obvious, it turns out that homomorphisms defined on $\mathbb{Z}^k$ do virtually extend to be defined on all of $\mathbb{R}^k$:

3.9. PROPOSITION. *Suppose φ is a group homomorphism from $\mathbb{Z}^k$ to* $\mathrm{GL}(d,\mathbb{R})$. *Then φ virtually extends to a continuous homomorphism $\widehat{\varphi}\colon G \to$* $\mathrm{GL}(d,\mathbb{R})$.

Unfortunately, this result is usually not useful, because it does not tell us anything about the image of $\widehat{\varphi}$ (other than that it is contained in $\mathrm{GL}(d,\mathbb{R})$). In practice, if all of the matrices in $\varphi(\mathbb{Z}^k)$ have some nice property, then it is important to know that the matrices in $\widehat{\varphi}(\mathbb{R}^k)$ also have this property. That is, if we have control on the image of φ, then we would like to have control on the image of $\widehat{\varphi}$.

3.10. EXAMPLE.

(1) If all of the matrices in $\varphi(\mathbb{Z}^k)$ have determinant 1, then all of the matrices in $\widehat{\varphi}(\mathbb{R}^k)$ should have determinant 1.
(2) If all of the matrices in $\varphi(\mathbb{Z}^k)$ commute with some particular matrix A, then all of the matrices in $\widehat{\varphi}(\mathbb{R})$ should commute with A.
(3) If all of the matrices in $\varphi(\mathbb{Z}^k)$ fix a particular vector v, then all of the matrices in $\widehat{\varphi}(\mathbb{R}^k)$ should fix v.
(4) Let $R = \begin{bmatrix} 1 & 0 & 0 & * \\ 0 & 1 & 0 & * \\ 0 & 0 & 1 & * \\ 0 & 0 & 0 & 1 \end{bmatrix} \cong \mathbb{R}^3$. If $\varphi(\mathbb{Z}^k) \subset R$, then it should be the case that $\widehat{\varphi}(\mathbb{R}^k) \subset R$. One needs to know this in order to derive Example 3.1 as a corollary of a result like Proposition 3.9.

3.11. REMARK. The problem that arises here is illustrated by the classical theory of Lagrange interpolation. This theorem states that if

$$(x_0, y_0), (x_1, y_1), \ldots, (x_n, y_n)$$

are any $n+1$ points in the plane (with $x_i \neq x_j$ whenever $i \neq j$), then there is a polynomial curve

$$y = f(x) = a_n x^n + a_{n-1}x^{n-1} + \cdots + a_0$$

of degree n that passes through all of these points. (It is easy to prove.) Unfortunately, however, even if the specified values $y_0, y_1, \ldots, y_n$ of $f(x)$ at the points $x_0, x_1, \ldots, x_n$ are well controlled (say, all are less than 1 in absolute value), it may be the case that $f(x)$ takes extremely large values at other values of x that are between x_0 and x_n, as illustrated in Figure 7.

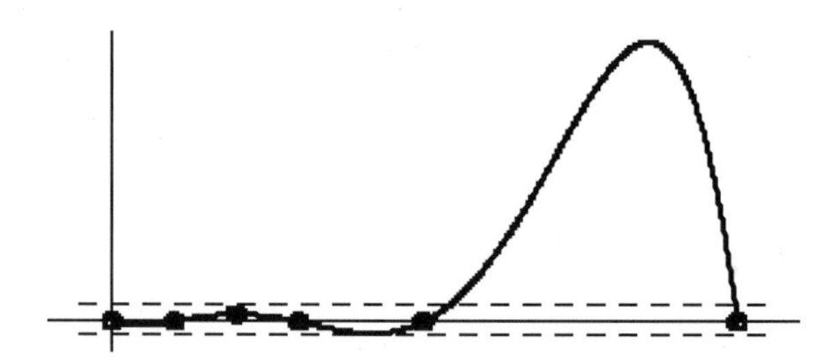

FIGURE 7. The 6 given points all lie in a small band around the x-axis, but the quintic curve that interpolates between them travels far from the x-axis.

Linear interpolation does not suffer from this defect; all of the points of the interpolating curve will lie in the convex hull of the given points.

In order to guarantee that having control on the values of φ will guarantee that we have control on the values of $\widehat{\varphi}$, we will require $\widehat{\varphi}(H)$ to be contained in a certain subgroup $\overline{\overline{H}}$ of $\mathrm{GL}(d, \mathbb{R})$ that is closely related to $\varphi(H)$. (This subgroup is called the "Zariski closure" of $\varphi(H)$.)

The formal definition of the *Zariski closure* $\overline{\overline{H}}$ of a subgroup H of $\mathrm{GL}(d,\mathbb{R})$ is not important for our purposes, if one simply accepts that it is, in a certain precise sense, the smallest natural, closed, virtually connected subgroup of $\mathrm{GL}(d,\mathbb{R})$ that contains H. It can be thought of as the group-theoretic analogue of a convex hull. (For the interested reader, a precise definition is given in §3.1 below.)

3.12. DEFINITION. Let H be a discrete subgroup of a topological group G. Saying H is *strictly superrigid* in G means, for all d, that if $\varphi\colon H \to \mathrm{GL}(d,\mathbb{R})$ is any homomorphism, then φ virtually extends to a continuous automorphism $\widehat{\varphi}\colon G \to \mathrm{GL}(d,\mathbb{R})$, such that $\widehat{\varphi}(G) \subset \overline{\overline{\varphi(H)}}$.

We have the following example:

3.13. PROPOSITION. $\mathbb{Z}^k$ *is strictly superrigid in* $\mathbb{R}^k$.

PROOF (optional). For simplicity, let us assume $k = 1$; thus, we wish to show $\mathbb{Z}$ is strictly superrigid in $\mathbb{R}$.

Given a homomorphism $\varphi\colon \mathbb{Z} \to \mathrm{GL}(d,\mathbb{R})$, let $Y = \overline{\overline{\varphi(\mathbb{Z})}}$, and let Y° be the connected component of Y that contains e (so Y° is a closed subgroup of Y). Since the Zariski closure Y has only finitely many connected components, there is some nonzero $m \in \mathbb{Z}$, such that $\varphi(m) \in Y^\circ$.

Since $\varphi(\mathbb{Z})$ is abelian, it is not difficult to see that its Zariski closure is also abelian. So Y° is a connected, abelian group of matrices; therefore, the universal cover $\widetilde{Y^\circ}$ of Y° is (isomorphic to) a simply connected, abelian group of matrices. One can show that this implies $\widetilde{Y^\circ}$ is isomorphic to $\mathbb{R}^n$, for some n. So there is no harm in assuming that $\widetilde{Y^\circ}$ is actually equal to $\mathbb{R}^n$.

- Let $\pi\colon \mathbb{R}^n \to Y^\circ$ be the covering map with $\pi(0) = e$, so π is a continuous homomorphism.
- Choose some $\overrightarrow{y} \in \mathbb{R}^n$, such that $\pi(\overrightarrow{y}) = \varphi(m)$.
- Define $\widetilde{\varphi}\colon \mathbb{R} \to \mathbb{R}^n$ by

$$\widetilde{\varphi}(x) = \frac{x}{m}\overrightarrow{y}.$$

- Let $\widehat{\varphi}\colon \mathbb{R} \to Y^\circ$ be the composition $\pi \circ \widetilde{\varphi}$.

Then:

- $\widehat{\varphi}$ is a composition of continuous homomorphisms, so it is a continuous homomorphism.
- We have

$$\widehat{\varphi}(m) = \pi\big(\widetilde{\varphi}(m)\big) = \pi\left(\frac{m}{m}\overrightarrow{y}\right) = \pi\big(\overrightarrow{y}\big) = \varphi(m),$$

  so $\widehat{\varphi}$ is equal to φ on the entire cyclic subgroup generated by m. Since $m \neq 0$, this is a finite-index subgroup of $\mathbb{Z}$.
- We have

$$\widehat{\varphi}(\mathbb{R}) = \pi\big(\widetilde{\varphi}(\mathbb{R})\big) \subset \pi(\mathbb{R}^n) = Y^\circ \subset Y = \overline{\overline{\varphi(\mathbb{Z})}}.$$

Thus, $\widehat{\varphi}$ is a continuous homomorphism that virtually extends φ, such that $\widehat{\varphi}(\mathbb{R}) \subset \overline{\overline{\varphi(\mathbb{Z})}}$. □

3.1. Definition of the Zariski closure. The concept of Zariski closure is taken from algebraic geometry. In that field, one works only with polynomials (and rational functions), not with more general continuous functions, and the notion of Zariski closure is a reflection of this. For the reader who wants details, we provide the full definition; others are welcome to skip ahead to the following section.

3.14. REMARK. In linear algebra, one works only with linear functions, and the definition of *linear span* is a reflection of this:

- A subset V of $\mathbb{R}^d$ is a *linear subspace* if it is the set of solutions of a collection of linear equations; more precisely, this means there are linear functionals $\lambda_i \colon \mathbb{R}^d \to \mathbb{R}$, such that
$$v \in V \iff \lambda_i(v) = 0, \text{ for all } i.$$
- The *linear span* $\langle S \rangle$ of a subset S of $\mathbb{R}^d$ is the unique smallest linear subspace of $\mathbb{R}^d$ that contains S.

The Zariski closure is perfectly analogous, replacing "linear functional on $\mathbb{R}^d$" with "polynomial function on $\mathrm{GL}(d,\mathbb{R})$."

3.15. DEFINITION.

- The collection $\mathrm{Mat}_{d\times d}(\mathbb{R})$ of all $d \times d$ matrices can naturally be identified with $\mathbb{R}^{d^2}$. A function $P\colon \mathrm{Mat}_{d\times d}(\mathbb{R}) \to \mathbb{R}$ is said to be a *polynomial* if becomes a polynomial (in d^2 variables) on $\mathbb{R}^{d^2}$ after making this identification.
- The group $\mathrm{GL}(d,\mathbb{R})$ is open, not closed, in $\mathrm{Mat}_{d\times d}(\mathbb{R})$. But we can think of it as a closed subgroup of the group $\mathrm{SL}(d+1,\mathbb{R})$ of $(d+1)\times(d+1)$ matrices of determinant 1, via the embedding $\rho\colon \mathrm{GL}(d,\mathbb{R}) \to \mathrm{SL}(d+1,\mathbb{R})$, defined by
$$\rho(A) = \begin{bmatrix} & A & & 0 \\ & & & \vdots \\ & & & 0 \\ 0 & \cdots & 0 & 1/\det A \end{bmatrix}.$$
A function $f\colon \mathrm{GL}(d,\mathbb{R}) \to \mathbb{R}$ is said to be a *polynomial* if there exists a polynomial $P\colon \mathrm{Mat}_{(d+1)\times(d+1)}(\mathbb{R}) \to \mathbb{R}$, such that
$$f(g) = P\big(\rho(g)\big),$$
for all $g \in \mathrm{GL}(d,\mathbb{R})$.
- A subset V of $\mathrm{GL}(d,\mathbb{R})$ is *Zariski closed* if it is the set of solutions of a collection of polynomial equations; more precisely, this means there are polynomial functions $f_i\colon \mathrm{GL}(d,\mathbb{R}) \to \mathbb{R}$, such that
$$v \in V \iff f_i(v) = 0, \text{ for all } i.$$
- The *Zariski closure* $\overline{\overline{V}}$ of a subset V of $\mathrm{GL}(d,\mathbb{R})$ is the unique smallest Zariski closed subset of $\mathrm{GL}(d,\mathbb{R})$ that contains V.

3.16. REMARK.

(1) If V is a subgroup of $\mathrm{GL}(d,\mathbb{R})$, then $\overline{\overline{V}}$ is also a subgroup of $\mathrm{GL}(d,\mathbb{R})$.
(2) $\overline{\overline{V}}$ is a closed subset of $\mathrm{GL}(d,\mathbb{R})$.
(3) $\overline{\overline{V}}$ has only finitely many connected components.

The first two of these observations are not difficult to prove. The third is rather difficult, but it is a generalization of the obvious fact that a one-variable polynomial $f(x)$ can have only finitely many zeroes.

4. Examples of Superrigid Subgroups

Proposition 3.13 tells us that $\mathbb{Z}^k$ is strictly superrigid in $\mathbb{R}^k$, and we will now see other examples of superrigid subgroups.

Let us first specify the type of group G that will be considered:

4.1. DEFINITION. We say G is a *Lie group* if it is a closed, connected subgroup of $\mathrm{GL}(d, \mathbb{R})$, for some d.

4.2. EXAMPLE. $\mathbb{R}^d$ is (isomorphic to) a Lie group.

4.3. WARNING. Other authors have a less restrictive definition of "Lie group," but this will suffice for our purposes.

Now we wish to describe the subgroups H of G that complete the analogy

$$\mathbb{Z}^k \text{ is to } \mathbb{R}^k \quad \text{as} \quad H \text{ is to } G.$$

Here are the basic properties of $\mathbb{Z}^k$:

(1) $\mathbb{Z}^k$ is a discrete subgroup of $\mathbb{R}^k$.
(2) The quotient space $\mathbb{R}^k/\mathbb{Z}^k$ is compact. (Indeed, $\mathbb{R}^k/\mathbb{Z}^k$ is the k-torus $\mathbb{T}^k$, which is well known to be compact.)

The second of these properties can be restated as the assertion that there is a compact subset of $\mathbb{R}^k$ that contains a representative of every coset of $\mathbb{Z}^k$. Thus, $\mathbb{Z}^k$ is a (cocompact) lattice, in the following sense:

4.4. DEFINITION. Suppose H is a discrete subgroup of a Lie group G. We say H is a (cocompact) *lattice* in G if there is a compact subset of G that contains a representative of every coset of H.

4.5. REMARK. Cocompact lattices suffice for most of our purposes, but we will sometimes allow H to satisfy the condition that some set of coset representatives has finite measure. Since every compact set has finite measure, but not every set of finite measure is compact, this is a more general condition.

For the moment, let us assume that G is solvable:

4.6. DEFINITION. Let G be a connected subgroup of $\mathrm{GL}(d, \mathbb{C})$. We say G is *solvable* if and only if it is upper triangular

$$G \subset \begin{bmatrix} \mathbb{C}^\times & \mathbb{C} & \mathbb{C} & \cdots & \mathbb{C} \\ & \mathbb{C}^\times & \mathbb{C} & \ddots & \vdots \\ & & \ddots & \mathbb{C} & \mathbb{C} \\ & 0 & & \mathbb{C}^\times & \mathbb{C} \\ & & & & \mathbb{C}^\times \end{bmatrix},$$

or can be made so by a change of basis.

4.7. REMARK. The following example is the base case of an inductive proof that if we restrict our attention only to connected groups, then the above definition agrees with the usual definition of solvable groups in terms of chains of normal subgroups with abelian quotient groups.

4.8. EXAMPLE. All abelian groups are solvable.

PROOF. It is well known that every matrix can be triangularized over any algebraically closed field, such as $\mathbb{C}$. (That is, there is a change of basis that makes the matrix upper triangular.) This implies that every cyclic group is solvable.

More generally, it is not difficult to show that any collection of pairwise commuting matrices can be simultaneously triangularized. (That is, there is a single change of basis that makes all of the matrices upper triangular.) This implies that every abelian group is solvable. □

4.9. EXAMPLES.

(1) Let

$$G_1 = \begin{bmatrix} 1 & 0 & 0 & \mathbb{R} \\ 0 & 1 & 0 & \mathbb{R} \\ 0 & 0 & 1 & \mathbb{R} \\ 0 & 0 & 0 & 1 \end{bmatrix} \cong \mathbb{R}^3 \text{ and } H_1 = \begin{bmatrix} 1 & 0 & 0 & \mathbb{Z} \\ 0 & 1 & 0 & \mathbb{Z} \\ 0 & 0 & 1 & \mathbb{Z} \\ 0 & 0 & 0 & 1 \end{bmatrix} \cong \mathbb{Z}^3.$$

Then:
- $H_1 \cong \mathbb{Z}^3$ and $G_1 \cong \mathbb{R}^3$, so it is clear that H_1 is a lattice in G_1.
- We have already seen that H_1 is strictly superrigid in G_1.
- G_1 is the obvious connected group containing H_1, so $\overline{\overline{H_1}} = G_1$.

(2) Let

$$G_2 = \begin{bmatrix} 1 & \mathbb{R} & \mathbb{R} & \mathbb{R} \\ 0 & 1 & \mathbb{R} & \mathbb{R} \\ 0 & 0 & 1 & \mathbb{R} \\ 0 & 0 & 0 & 1 \end{bmatrix} \text{ and } H_2 = \begin{bmatrix} 1 & \mathbb{Z} & \mathbb{Z} & \mathbb{Z} \\ 0 & 1 & \mathbb{Z} & \mathbb{Z} \\ 0 & 0 & 1 & \mathbb{Z} \\ 0 & 0 & 0 & 1 \end{bmatrix}.$$

Then:
- It is not difficult to see that H_2 is a lattice in G_2. Namely, if we let $I = [0,1]$ be the unit interval then

$$\begin{bmatrix} 1 & I & I & I \\ 0 & 1 & I & I \\ 0 & 0 & 1 & I \\ 0 & 0 & 0 & 1 \end{bmatrix}$$

  is a compact set that contains a representative of every coset.
- Our main result, to be stated below, will show that H_2 is strictly superrigid in G_2.
- G_2 is the obvious connected group containing H_2, so $\overline{\overline{H_2}} = G_2$.

(3) Let

$$G_3 = \begin{bmatrix} 1 & \mathbb{R} & \mathbb{C} \\ 0 & 1 & 0 \\ 0 & 0 & 1 \end{bmatrix} \text{ and } H_3 = \begin{bmatrix} 1 & \mathbb{Z} & \mathbb{Z}+\mathbb{Z}i \\ 0 & 1 & 0 \\ 0 & 0 & 1 \end{bmatrix}.$$

Then:
- It is not difficult to see that H_3 is a lattice in G_3. Indeed, $H_3 \cong \mathbb{Z}^3 \subset \mathbb{R}^3 \cong G_3$.

- We know that H_3 is strictly superrigid in G_3.
- G_3 is the obvious connected group containing H_3, so $\overline{\overline{H_3}} = G_3$.

4.10. EXAMPLE. Let

$$G' = \left\{ \begin{bmatrix} 1 & t & \mathbb{C} \\ 0 & 1 & 0 \\ 0 & 0 & e^{2\pi it} \end{bmatrix} \middle| \; t \in \mathbb{R} \right\} \text{ and } H' = \begin{bmatrix} 1 & \mathbb{Z} & \mathbb{Z}+\mathbb{Z}i \\ 0 & 1 & 0 \\ 0 & 0 & 1 \end{bmatrix}.$$

Unlike our previous examples, the matrix entries of elements of G' cannot be chosen independently of each other: the $(1,2)$-entry of any element of G' uniquely determines its $(3,3)$-entry. However, the relation between these entries is defined by a transcendental function, not a polynomial, so, as far as an algebraic geometer is concerned, these entries have no correlation at all. This means that in the Zariski closure of G', these entries become decoupled and can be chosen independently. Thus,

$$\overline{\overline{G'}} = \begin{bmatrix} 1 & \mathbb{R} & \mathbb{C} \\ 0 & 1 & 0 \\ 0 & 0 & \mathbb{T} \end{bmatrix}.$$

When the $(1,2)$-entry t of an element of G' is an integer, the $(3,3)$-entry $e^{2\pi it}$ is 1, so we see that $H' \subset G'$. In fact, it is not difficult to see that H' is a lattice in G'.

On the other hand, we have $H' = H_3 \subset G_3$, so H' is also a lattice G_3. Furthermore, we have

$$\overline{\overline{H'}} = \overline{\overline{H_3}} = G_3 \neq \overline{\overline{G'}}.$$

These observations can be used to show that H' is *not* strictly superrigid in G'.

4.11. PROPOSITION. *H' is not strictly superrigid in G'.*

In particular, the inclusion map $\varphi \colon H' \hookrightarrow \mathrm{GL}(3,\mathbb{C})$ does not extend to a continuous homomorphism $\widehat{\varphi} \colon G' \to \overline{\overline{H'}}$.

PROOF. Note that

$$\overline{\overline{H'}} = \overline{\overline{H_3}} = G_3$$

is abelian. Therefore, $\widehat{\varphi}$ must be trivial on the entire commutator subgroup $[G', G']$ of G'. We have

$$[G', G'] = \begin{bmatrix} 1 & 0 & \mathbb{C} \\ 0 & 1 & 0 \\ 0 & 0 & 1 \end{bmatrix} \supset \begin{bmatrix} 1 & 0 & \mathbb{Z}+\mathbb{Z}i \\ 0 & 1 & 0 \\ 0 & 0 & 1 \end{bmatrix},$$

so the supposed extension $\widehat{\varphi}$ is trivial on some nontrivial elements of H'. This contradicts the fact that φ, being an inclusion, has trivial kernel. □

4.12. REMARK.

(1) H_3 is a strictly superrigid lattice in G_3, but we constructed G' by adding some rotations to G_3 that H_3 knows nothing about. A homomorphism defined on H_3 will extend to G_3, but it need not be compatible with the additional rotations that appear in G'.
(2) One can show that the above example is typical: it is always the case that if $\overline{\overline{H}} \neq \overline{\overline{G}}$, then some of the rotations associated to elements of G do not

come from rotations associated to H. Roughly speaking, the concept of "associated rotation" can be defined by

$$\operatorname{rot}\begin{bmatrix}\alpha & * \\ 0 & \beta\end{bmatrix} = \begin{bmatrix}\alpha/|\alpha| & 0 \\ 0 & \beta/|\beta|\end{bmatrix}.$$

In general, if $\overline{\overline{H}} \neq \overline{\overline{G}}$, then the natural connected subgroup containing H is not G, but some other group; there are parts of G that have nothing to do with H. A homomorphism defined on H cannot be expected to know about the structure in this part of G, so there is no reason to expect the homomorphism to be compatible with this additional structure.

The above considerations might lead one to believe that if $\overline{\overline{H}} \neq \overline{\overline{G}}$, then H is *not* strictly superrigid in G. This conclusion is correct in spirit, but there is a technical complication[1] that leads to the fine print in the statement of the following result. The reader is invited to simply ignore this fine print.

4.13. PROPOSITION. *If H is strictly superrigid in G, then $\overline{\overline{H}} = \overline{\overline{G}} \pmod{\overline{\overline{Z(G)}}}$.*

By passing to the universal cover, let us assume that G is simply connected. Then the converse of the above proposition is true for solvable groups:

4.14. THEOREM. *A lattice H in a simply connected, solvable Lie group G is strictly superrigid if and only if $\overline{\overline{H}} = \overline{\overline{G}} \pmod{\overline{\overline{Z(G)}}}$.*

This theorem provides a complete characterization of the strictly superrigid lattices in the solvable case.

4.1. Brief discussion of groups that are not solvable. An extensive structure theory has been developed for Lie groups. Among other things, it is known that these groups can be classified into three basic types:

- solvable (e.g., $\mathbb{R}^k$),
- semisimple (e.g., $\mathrm{SL}(k,\mathbb{R})$), or
- a combination of the above (e.g., $G = \mathbb{R}^k \times \mathrm{SL}(k,\mathbb{R})$).

In the preceding section, we constructed lattices in solvable groups by taking the integer points in G. For example, $\mathbb{Z}^k$ is a lattice in $\mathbb{R}^k$. (One might note that, in the case of H_3, we used Gaussian integers, not only the ordinary integers.) It turns out that the same construction can be applied to many groups that are not solvable. For example, $\mathrm{SL}(k,\mathbb{Z})$ is a lattice in $\mathrm{SL}(k,\mathbb{R})$.

It is known that if G is a combination of a solvable group and a semisimple group, then, roughly speaking, any lattice in G also has a decomposition into a solvable part and a semisimple part. For example, $\mathbb{Z}^k \times \mathrm{SL}(k,\mathbb{Z})$ is a lattice in $\mathbb{R}^k \times \mathrm{SL}(k,\mathbb{R})$.

The following theorem shows that deciding whether or not H is superrigid reduces to the same question about its semisimple part:

4.15. THEOREM. *A lattice H in a simply connected Lie group G is superrigid if and only if*

- *the semisimple part of H is superrigid, and*

---

[1] A given group G can usually be embedded into $\mathrm{GL}(d,\mathbb{R})$ in many different ways, and $\overline{\overline{H}}$ may be equal to $\overline{\overline{G}}$ for some of these embeddings, but not others. The canonical matrix representation that can be used is the so-called "adjoint representation," which is not an embedding: its kernel is the center $Z(G)$, and the Zariski closure should be calculated modulo this kernel.

- $\overline{\overline{H}} = \overline{\overline{G}}$ (mod $\overline{\overline{Z(G)}} \cdot K$, where K is a compact, normal subgroup of $\overline{\overline{G}}$).

Although the problem for semisimple groups has not yet been settled in complete generality, a fundamental theorem of the Fields Medallist G. A. Margulis settled most cases. In particular:

4.16. THEOREM (Margulis Superrigidity Theorem). *If $k \geq 3$, then all lattices in $\mathrm{SL}(k, \mathbb{R})$ are superrigid.*

4.17. REMARK.

(1) The assumption that $k \geq 3$ is necessary: *no* lattice in $\mathrm{SL}(2, \mathbb{R})$ is superrigid. For example, if we let H be any finite-index subgroup of $\mathrm{SL}(2, \mathbb{Z})$, then H is a lattice in $\mathrm{SL}(2, \mathbb{R})$. However, it is possible to choose H to be a free group, in which case H has countless homomorphisms into $\mathrm{SL}(2, \mathbb{R})$. Some of these homomorphisms have kernels that are infinite, but the kernel of any nontrivial homomorphism defined on $\mathrm{SL}(2, \mathbb{R})$ must be finite.

(2) The work of G. A. Margulis establishes the superrigidity of lattices not only in $\mathrm{SL}(k, \mathbb{R})$, but also in any simple Lie group G satisfying the technical condition that $\mathbb{R}\text{-rank}\, G \geq 2$.

(3) The astute reader may have noticed that the modifier "strictly" is not being applied to "superrigid" in this section. The term "strictly" is used in Definition 2.3 to indicate that $\widehat{\varphi}$ is required to be exactly equal to φ (on a finite-index subgroup). Dropping this modifier means that we do not require exact equality; instead, we allow an error that is uniformly bounded (on a finite-index subgroup of H). That is, we require $\widehat{\varphi}(h) = \varphi(h)$ (mod K), where K is some compact group. Although they are always superrigid, some lattices in $\mathrm{SL}(k, \mathbb{R})$ are not strictly superrigid.

Superrigidity implies that there is a very close connection between H and G. In fact, the connection is so close that it provides quite precise information on how to obtain H from G. Namely, superrigidity implies that letting H be the integer points of G is often the only way to construct a lattice.

4.18. DEFINITION. Suppose H is a lattice in $G = \mathrm{SL}(k, \mathbb{R})$. To avoid complications, let us assume H is *not* cocompact. We say H is *arithmetic* if there is an embedding of G in $\mathrm{SL}(d, \mathbb{R})$, for some d, such that H is virtually equal to $G \cap \mathrm{SL}(d, \mathbb{Z})$.

4.19. THEOREM (Margulis Arithmeticity Theorem). *If $k \geq 3$, then every lattice in $\mathrm{SL}(k, \mathbb{R})$ is arithmetic.*

For convenience, we stated the arithmeticity theorem only for $\mathrm{SL}(k, \mathbb{R})$, but it is valid for lattices in any simple Lie group G with $\mathbb{R}\text{-rank}\, G \geq 2$. It is a truly astonishing result.

5. Why Superrigidity Implies Arithmeticity

It is not at all obvious that superrigidity has anything to do with arithmeticity, so let us give some idea of how the connection arises. We warn the reader in advance that our motivation here is pedagogical rather than logical — the main ideas in the proof of the Margulis Arithmeticity Theorem (4.19) will be presented, but there will be no attempt to be rigorous.

We are given a lattice H in $G = \mathrm{SL}(k,\mathbb{R})$, with $k \geq 3$, and we wish to show that H is arithmetic. Roughly speaking, we wish to show $H \subset \mathrm{SL}(k,\mathbb{Z})$.

Here is a loose description of the 4 steps of the proof:

(1) The Margulis Superrigidity Theorem (4.16) implies that every matrix entry of every element of H is an algebraic number.
(2) Algebraic considerations allow us to assume that these algebraic numbers are rational; that is, $H \subset \mathrm{SL}(k,\mathbb{Q})$.
(3) For every prime p, a "p-adic" version of the Margulis Superrigidity Theorem provides a natural number N_p, such that no element of H has a matrix entry whose denominator is divisible by p^{N_p}.
(4) This implies that some finite-index subgroup H' of H is contained in $\mathrm{SL}(k,\mathbb{Z})$.

Step 1. Every matrix entry of every element of H is an algebraic number. Suppose some $\gamma_{i,j}$ is transcendental. Then, for any transcendental number α, there is a field automorphism ϕ of $\mathbb{C}$ with $\phi(\gamma_{i,j}) = \alpha$. Applying ϕ to all the entries of a matrix induces an automorphism $\widetilde{\phi}$ of $\mathrm{SL}(k,\mathbb{C})$. Let

$$\varphi \text{ be the restriction of } \widetilde{\phi} \text{ to } H,$$

so φ is a homomorphism from H to $\mathrm{SL}(k,\mathbb{C})$. The Margulis Superrigidity Theorem (4.16) implies there is a continuous homomorphism $\widehat{\varphi}\colon G \to \mathrm{SL}(k,\mathbb{C})$, such that $\widehat{\varphi} = \varphi$ on a finite-index subgroup of H. (For simplicity, we have ignored the distinction between "superrigid" and "strictly superrigid.") Ignoring a finite group, let us assume $\widehat{\varphi} = \varphi$ on all of H.

Since there are uncountably many transcendental numbers α, there are uncountably many different choices of ϕ, so there must be uncountably many different k-dimensional representations $\widehat{\varphi}$ of G. However, it is well known from the the theory of "roots and weights" that G (or any connected, simple Lie group) has only finitely many non-isomorphic representations of any given dimension, so this is a contradiction.

TECHNICAL REMARK. Actually, this is not quite a contradiction, because it is possible that two different choices of φ yield the same representation of H, up to isomorphism; that is, after a change of basis. The trace of a matrix is independent of the basis, so the preceding argument really shows that the trace of $\varphi(\gamma)$ must be algebraic, for every $\gamma \in H$. Then one can use some algebraic methods to construct some other matrix representation φ' of H, such that the matrix entries of $\varphi'(\gamma)$ are algebraic, for every $\gamma \in H$.

Step 2. We have $H \subset \mathrm{SL}(k,\mathbb{Q})$. Let F be the subfield of $\mathbb{C}$ generated by the matrix entries of the elements of H, so $H \subset \mathrm{SL}(k,F)$. From Step 1, we know that this is an algebraic extension of $\mathbb{Q}$. Furthermore, because it is known that H has a finite generating set, we see that this field extension is finitely generated. Thus, F is finite-degree field extension of $\mathbb{Q}$ (in other words, F is an "algebraic number field"). This means that F is almost the same as $\mathbb{Q}$, so it is only a slight exaggeration to say that we have proved $H \subset \mathrm{SL}(k,\mathbb{Q})$.

Indeed, there is an algebraic technique, called "Restriction of Scalars" that provides a way to change F into $\mathbb{Q}$: there is a representation $\rho\colon G \to \mathrm{SL}(\ell,\mathbb{R})$, such that $\rho\big(G \cap \mathrm{SL}(k,F)\big) \subset \mathrm{SL}(\ell,\mathbb{Q})$. Thus, after changing to this new representation of G, we have the desired conclusion (without any exaggeration).

Step 3. For every prime p, there is a natural number N_p, such that no element of H has a matrix entry whose denominator is divisible by p^{N_p}. The fields $\mathbb{R}$ and $\mathbb{C}$ are complete (that is, every Cauchy sequence converges), and they obviously contain $\mathbb{Q}$. For any prime p, the p-adic numbers $\mathbb{Q}_p$ are another field that has these same properties.

The Margulis Superrigidity Theorem (4.16) deals with homomorphisms into $\mathrm{SL}(d, \mathbb{F})$, where $\mathbb{F} = \mathbb{R}$, but Margulis also proved a version of the theorem that applies when $\mathbb{F}$ is a p-adic field. Now G is connected, but p-adic fields are totally disconnected, so every continuous homomorphism from G to $\mathrm{SL}(k, \mathbb{Q}_p)$ is trivial. Thus, superrigidity tells us that φ is trivial, up to a bounded error (c.f. Remark 4.17(3)). In other words, the closure of $\varphi(H)$ is compact in $\mathrm{SL}(k, \mathbb{Q}_p)$.

This conclusion can be rephrased in more elementary terms, without any mention of the field $\mathbb{Q}_p$ of p-adic numbers. Namely, it says that there is a bound on the highest power of p that divides any matrix entry of any element of H. This is what we wanted.

Step 4. Some finite-index subgroup H' of H is contained in $\mathrm{SL}(k, \mathbb{Z})$. Let $D \subset \mathbb{N}$ be the set consisting of the denominators of the matrix entries of the elements of $\varphi(H)$.

We claim there exists $N \in \mathbb{N}$, such that every element of D is less than N. Since H is known to be finitely generated, some finite set of primes $\{p_1, \ldots, p_r\}$ contains all the prime factors of every element of D. (If p is in the denominator of some matrix entry of $\gamma_1\gamma_2$, then it must appear in a denominator somewhere in either γ_1 or γ_2.) Thus, every element of D is of the form $p_1^{m_1} \cdots p_r^{m_r}$, for some $m_1, \ldots, m_r \in \mathbb{N}$. From Step 3, we know $m_i < N_{p_i}$, for every i. Thus, every element of D is less than $p_1^{N_{p_1}} \cdots p_r^{N_{p_r}}$. This establishes the claim.

From the preceding paragraph, we see that $H \subset \frac{1}{N!} \mathrm{Mat}_{k\times k}(\mathbb{Z})$. Note that if $N = 1$, then $H \subset \mathrm{SL}(k, \mathbb{Z})$. In general, N is a finite distance from 1, so it should not be hard to believe (and it can indeed be shown) that some finite-index subgroup of H must be contained in $\mathrm{SL}(k, \mathbb{Z})$. □

Further Reading

[1] J. Graver, B. Servatius, and H. Servatius: *Combinatorial Rigidity.* American Mathematical Society, Providence, 1993. MR 1251062 (95b:52034)

A study of the rigidity of linkages in n-space, including an annotated bibliography.

[2] J. E. Humphreys: *Linear Algebraic Groups.* Springer, New York, 1975. MR 0396773 (53 #633)

A textbook on algebraic groups (including Zariski closures).

[3] G. A. Margulis: *Discrete Subgroups of Semisimple Lie Groups.* Springer, New York, 1991. MR 1090825 (92h:22021)

An encyclopedic and impressive monograph that includes proofs of the Margulis Superrigidity Theorem (4.16) and the Margulis Arithmeticity Theorem (4.19).

[4] D. W. Morris: *Introduction to Arithmetic Groups* (in preparation). http://arxiv.org/abs/math/0106063

Includes an exposition of a proof of the Margulis Superrigidity Theorem (4.16).

[5] M. S. Raghunathan: *Discrete Subgroups of Lie Groups.* Springer, New York, 1972. MR 0507234 (58 #22394a)

The place to learn basic properties of lattices.

[6] A. N. Starkov: Rigidity problem for lattices in solvable Lie groups, *Proc. Indian Acad. Sci. Math. Sci.* 104 (1994) 495–514. MR 1314393 (96d:22017)

A thorough study of global rigidity of lattices in solvable Lie groups.

[7] V. S. Varadarajan: *Lie Groups, Lie Algebras, and Their Representations.* Springer, New York, 1984. MR 0746308 (85e:22001)

A textbook on Lie groups.

[8] D. Witte: Superrigidity of lattices in solvable Lie groups, *Inventiones Math.* 122 (1995) 147–193. MR 1354957 (96k:22024)

Includes a proof of Theorem 4.15.

[9] D. Witte: Superrigid subgroups and syndetic hulls in solvable Lie groups, in: M. Burger and A. Iozzi, eds., *Rigidity in Dynamics and Geometry.* Springer, Berlin, 2002, pp. 441–457. MR 1919416 (2003g:22005)

An exposition of the proof of the superrigidity theorem for solvable groups (4.14).

[10] R. J. Zimmer: *Ergodic Theory and Semisimple Groups.* Birkhäuser, Boston, 1984. MR 0776417 (86j:22014)

Includes proofs of the Margulis Superrigidity Theorem (4.16) and the Margulis Arithmeticity Theorem (4.19). Less intimidating than [**3**], but still demanding.

Department of Mathematics and Computer Science, University of Lethbridge, Lethbridge, Alberta, T1K 3M4, Canada

Contemporary Mathematics
Volume **479**, 2009

Averaging Points Two at a Time

David Petrie Moulton

ABSTRACT. In 2006 Brendan McKay asked the following question on the newsgroup sci.math.research[**1**]: We have n points in a disk centered at the centroid of the points. We successively replace the two furthest points from each other by two copies of their average. We still have n points with the same centroid. How many such "averaging" moves are necessary to guarantee that all points lie in the concentric disk of half the radius?

This really seems to be the wrong question: it turns out that the situation is more amenable to study of we use a general Euclidean space and look at the rate of decay of the diameter of the set of points in terms of number of moves. We consider d_t, the square of the diameter after t moves (normalized by $d_0 = 1$). We compute exact values for $f(r)$, the maximum possible value of d_{rn}, for $0 \leq r < 3/2$, and we obtain some asymptotic bounds for larger values of r. In particular, we give upper and lower bounds on the number of moves necessary to guarantee that the diameter reduces by a factor of 2.

We also give a conjecture providing asymptotics for all values of r in terms of the simpler point-averaging problem in which we replace two points by just one copy of their average and ask for $g(n)$, the maximum possible distance between the last two points. We prove a large part of this conjecture, in part by studying a combinatorial problem involving "summing" moves on sets of integers.

1. Introduction

In 2006 Brendan McKay asked the following question on the Usenet newsgroup sci.math.research[**1**]: We have n points in a disk centered at the centroid of the points. We successively replace the two furthest points from each other by two copies of their average. We still have n points with the same centroid. How many such "averaging" moves are necessary to guarantee that all points lie in the concentric disk of half the radius?

This really seems to be the wrong question: it turns out that the situation is more amenable to study if we use a general Euclidean space and look at the rate of decay of the diameter of the set of points in terms of number of moves. We consider d_t, the square of the diameter after t moves (normalized by $d_0 = 1$). We compute exact values for $f(r)$, the maximum possible value of d_{rn}, for $0 \leq r < 3/2$, and we obtain some asymptotic bounds for larger values of r. In particular, we give upper and lower bounds on the number of moves necessary to guarantee that the diameter reduces by a factor of 2.

We also give a conjecture providing asymptotics for all values of r in terms of the simpler point-averaging problem in which we replace two points by just one copy of their average and ask for $g(r)$, the maximum possible distance between the last two points. We prove a large part of this conjecture, in part by studying a combinatorial problem involving "summing" moves on sets of integers. In Section 3 we give our results in this direction.

Victor Miller introduced this question to the author, Samuel Kutin, and Don Coppersmith, and together we obtained several preliminary results. First, Victor Miller showed that the sum of the squares of all distances must decrease by a factor of at least $(n-1)/(n-2)$ each move, which gives $d_{(\ln r)n+2n\ln n} \le d_0/r^2$. Instead of looking at the steady decrease of S over time, we can take a target value and see how long it takes to reach it. One can use this method to show $d_{\alpha^{-2}n} < \alpha$ for any $0 < \alpha \le 1$. This is better for smallish t and gives $d_{4n} < 1/2$. Considering the sum of squares seemed like a good idea for a while, but ended up being the wrong approach. We also had better bounds for a while when we could choose which points to average, but these were superseded by the results described in this paper.

2. Preliminaries

We begin by recalling a classical result from geometry that will be the basis of all of our calculations.

PROPOSITION 2.1 (The Parallelogram Law). *The sum of the squares of the sides of a parallelogram equals the sum of the squares of the diagonals.*

PROOF. For a parallelogram with sides $\mathbf{u}$ and $\mathbf{v}$, we have

$$(\mathbf{u}+\mathbf{v})\cdot(\mathbf{u}+\mathbf{v}) + (\mathbf{u}-\mathbf{v})\cdot(\mathbf{u}-\mathbf{v}) = 2\mathbf{u}\cdot\mathbf{u} + 2\mathbf{v}\cdot\mathbf{v}.$$

□

COROLLARY 2.2. *In a triangle with sides a, b, c, the median m to side c satisfies*

$$m^2 = \frac{a^2}{2} + \frac{b^2}{2} - \frac{c^2}{4}.$$

If $a \le b \le c$, this gives

$$m^2 \le \frac{a^2}{2} + \frac{b^2}{4} \le \frac{3b^2}{4} \le \frac{3c^2}{4}.$$

One way to think about this corollary is that adding any three squared side lengths, whether equal or not, except for three copies of the smallest squared side length, and dividing by 4 gives a bound on the squared median to the longest side.

For two points u and v in Euclidean space, we write $(uv)^2$ for the square of the distance between them. As in the Introduction, given an initial configuration of n points in Euclidean space, we let d_t be the square of the diameter of the resulting set after t moves, and we always normalize to assume $d_0 = 1$. We often write simply D to refer to the squared diameter of the set at the current point in the sequence of moves.

We start with a useful lemma that allows us to define the two functions we will study.

LEMMA 2.3. *A point resulting from some move has squared distance to any other point at most 3/4 times the squared diameter before the move.*

PROOF. Suppose two points u and v are averaged to give (one or two copies of) their midpoint w, and let x be any other point. Since u and v had maximal distance, we have $(ux)^2 \leq (uv)^2$ and $(vx)^2 \leq (uv)^2$, so by Corollary 2.2 applied to triangle uvx, we get $(wx)^2 \leq (3/4)(uv)^2$, which proves the lemma. □

COROLLARY 2.4. *No move can increase the squared diameter of a set.*

PROOF. In fact, any new squared distances after a move involve a new point and must be at most 3/4 times the previous squared diameter. □

In other words, the sequence of squared distances between the point-pairs averaged on each turn is monotonically nonincreasing.

It turns out that the number of moves needed to guarantee a certain reduction in D is proportional to n. For instance, if we have some configuration of n points that attains a certain value for d_t, then replacing each point by p copies of itself gives a configuration of pn points attaining the same value for d_{pt}. Hence we make the following definition:

DEFINITION 2.5. For $r \geq 0$, let $f(r)$ be the supremum over all $n \geq 2$, all $t \geq rn$, and all point-sets of size n with sequences of merging moves of d_t.

By Corollary 2.4, $f(r)$ is well-defined and is at most 1 for nonnegative r, and from the definition, it is monotonically nonincreasing. We will determine the exact value of $f(r)$ for $r < 3/2$ and obtain some bounds on $f(r)$ for larger values of r.

We also consider the following related problem: Given $k \geq 2$ points in Euclidean space, we make sequences of moves as before, except that instead of replacing two points by two copies of their midpoint, we replace them by only one copy, thus reducing the total number of points by 1. We refer to such a move as a *merging* move, rather than an averaging move, to minimize confusion.

DEFINITION 2.6. For $k \geq 2$, let $g(k)$ be the supremum over all point-sets of size k with sequences of merging moves of d_{k-2} (the squared diameter just before the last possible move, when exactly two points remain).

Again, g is well-defined, and it is monotonically nonincreasing. (After the first of $k-2$ moves, there are $k-3$ moves to be applied to the remaining set of squared diameter at most 1.) It seems that the values taken by f and g are the same, so we will study them together and give a precise conjecture relating them.

In order to reduce the study of f to that of g, we present one more problem: Given n numbers, all initially equal to 1, and a natural number k, we make a *summing* move by replacing each of two numbers by their sum. We ask for the longest possible length of a sequence of legal summing moves such that no number ever exceeds k.

3. Relating f to g

Throughout this section we write lg for logarithms to base 2. In Propositions 4.1 through 4.7, we will describe completely the behavior of $f(r)$ for $r < 3/2$ in terms of values of g. In this region, f is a right-continuous step function taking jumps at dyadic rationals and assuming the same values as g. We conjecture that this behavior holds in general:

CONJECTURE 3.1. *Given $r \geq 0$, take k minimal with*

$$\frac{1}{2}\lceil \lg k \rceil - 1 + \frac{k}{2^{\lceil \lg k \rceil}} > r. \tag{1}$$

Then we have

$$f(r) = g(k).$$

For convenience we write

$$h(k) = \frac{1}{2}\lceil \lg k \rceil - 1 + \frac{k}{2^{\lceil \lg k \rceil}}.$$

By induction on k we have

$$h(k) = \sum_{i=2}^{k} \frac{1}{2^{\lceil \lg i \rceil}},$$

which is the sum $1/2 + 1/4 + 1/4 + 1/8 + \dots$, with $k-1$ total terms and with up to 2^{j-1} occurrences of the term 2^{-j}. Our results from Section 4 below show that the conjecture holds for $r < 3/2$.

We can prove most of Conjecture 3.1 in the sense that we can prove the upper bound for f, and we can prove that the lower bound holds as well, provided a certain condition holds that seems always to be satisfied in practice.

To discuss the lower bound, we first need a bit more terminology. Given a configuration and a full sequence of merging moves, we can form a binary tree on the points involved. The leaves are the initial points, the root is the final point, and the two children of a point are the points that were merged to give it. We say that a binary tree is balanced if all leaves have depth one of two consecutive integers, which must be $\lceil \lg k \rceil$ and $\lceil \lg k \rceil - 1$ for a such tree with k leaves.

THEOREM 3.2. *Given $r \geq 0$ and k satisfying Inequality (1) of Conjecture 3.1, if some configuration of k points and sequence of merging moves achieving $d_{k-2} = d$ correspond to a balanced binary tree, then we have*

$$f(r) \geq d.$$

PROOF. Choose a positive integer p (to be determined below). Start with a such a configuration of k points, replace each point corresponding to a leaf at depth $\lceil \lg k \rceil$ with p copies of itself, and replace each other point (corresponding to a leaf at depth $\lceil \lg k \rceil - 1$) with $2p$ copies of itself. Now make averaging moves mimicking the original sequence of merging moves. That is, when two single points were to be merged to give one copy of their midpoint, we now average $2^i p$ copies of each (with i equal to $\lceil \lg k \rceil$ minus the height of the corresponding tree vertices) to get $2^{i+1}p$ copies of their midpoint. The total number of points in the new initial configuration is $2^{\lceil \lg k \rceil}p$, and, as the tree is balanced, the total number of averaging moves is

$$\begin{aligned}\left(\lceil \lg k \rceil 2^{\lceil \lg k \rceil - 1} - \left(2^{\lceil \lg k \rceil} - k\right)\right) p &= \left(\frac{1}{2}\lceil \lg k \rceil - 1 + \frac{k}{2^{\lceil \lg k \rceil}}\right) 2^{\lceil \lg k \rceil} p \\ &= h(k) 2^{\lceil \lg k \rceil} p.\end{aligned}$$

When one fewer than this number of moves has been made, D still is d, so we get

$$f\left(h(k) - \frac{1}{2^{\lceil \lg k \rceil} p}\right) \geq d.$$

Taking p large enough that the argument of f is at least r and applying monotonicity now yield the result. □

In particular, if some extremal configuration of k points and sequence of merging moves achieving $d_{k-2} = g(k)$ corresponds to a balanced binary tree, then we get

$$f(r) \geq g(k).$$

Now the following conjecture immediately implies the lower bound of Conjecture 3.1:

CONJECTURE 3.3. *For every value $k \geq 2$, some configuration of k points and sequence of merging moves achieving $d_{k-2} = g(k)$ correspond to a balanced binary tree.*

Since $g(k)$ is the maximum value of d_{k-2}, we know that some configuration and sequence achieve the bound. In fact, there are some such optimal sequences that do not correspond to balanced binary trees, for instance, the configuration and sequence of Figure 1, if one of the other two possible second moves is made.

Now for the upper bound, we need to study sequences of summing moves as defined at the end of the previous section. In order to do this, we also consider *doubling* moves, in which we double each of two numbers.

LEMMA 3.4. *Given n numbers, all initially equal to 1, and an upper bound k, the longest possible length of a sequence of summing and doubling moves such that the sum of the numbers never exceeds kn is at most $h(k)n$.*

PROOF. First, we claim that there exists a sequence of moves of maximum length consisting only of doubling moves. Suppose we have a sequence with at least one summing move, and let the last summing move replace each of a and b by $a+b$. We may assume $a \leq b$, and as all later moves are doubling moves, each copy of $a+b$ is doubled some number of times, and we may assume that the copy replacing a is doubled at least as many times as that replacing b.

Now change this move into a doubling move, replacing a by $2a$ and b by $2b$; this gives a new sequence whose final sum is at most the final sum of the first sequence. Thus the sum of all numbers in the new sequence also never exceeds kn. By induction on the number of summing moves, we can convert any sequence of summing and doubling moves with sum never exceeding kn into a sequence of doubling moves of the same length with sum never exceeding kn, thus proving the claim.

To simplify matters, we now consider "half-moves", in which we double just one number; a doubling move now consists of two half-moves. To get a maximum-length sequence of half-moves with sum bounded by kn, it is clear that the number of times we double each 1 can be taken to be one of two consecutive integers, namely $\lceil \lg k \rceil$ and $\lceil \lg k \rceil - 1$. If we start by doubling each 1 exactly $\lceil \lg k \rceil - 1$ times, then the sum of all numbers is

$$2^{\lceil \lg k \rceil - 1} n,$$

and since each further half-move increases the sum by $2^{\lceil \lg k \rceil - 1}$, the maximum number of further moves then allowable is at most

$$\frac{kn - 2^{\lceil \lg k \rceil - 1} n}{2^{\lceil \lg k \rceil - 1}} = \left(\frac{2k}{2^{\lceil \lg k \rceil}} - 1 \right) n.$$

Adding these to the first $(\lceil \lg k \rceil - 1)n$ half-moves gives a bound of at most

$$\left(\lceil \lg k \rceil - 2 + \frac{2k}{2^{\lceil \lg k \rceil}} \right) n = 2h(k)n$$

half-moves, which proves the desired upper bound on the maximum number of moves. □

COROLLARY 3.5. *Given n numbers, all initially equal to 1, and an upper bound k, the longest possible length of a sequence of summing moves such that no number ever exceeds k is at most $h(k)n$.*

THEOREM 3.6. *Given $r \geq 0$ and k minimal satisfying Inequality 1 of Conjecture 3.1, we have*

$$f(r) \leq g(k).$$

PROOF. Suppose we have n points and make at least $rn + 1$ averaging moves. We now create a corresponding sequence of summing moves. Label each point initially with 1, and when making an averaging move, label each copy of the midpoint with the sum of the labels of the two averaged points. Now the sequence of labels are numbers on which we perform summing moves.

Since k was chosen to be minimal satisfying Inequality 1, we have

$$r \geq h(k-1)$$

and

$$rn + 1 > h(k-1)n.$$

By Corollary 3.5, we know that some label must exceed $k - 1$, so must be at least k.

Let u be a point with label $l \geq k$. Then l is the number of initial points that are "ancestors" of u. That is, if we restrict our attention to those l initial points and one copy of each midpoint created from them, we get a sequence of merging moves producing u. (Some of these initial points may be equal, but this does not cause problems, as long as we use a full tree of merging moves producing u.)

Since the number of initial points in this sequence of merging moves is at least k, and g is nonincreasing, the two points merged to produce u must have had squared distance at most $g(k)$. This means that before the averaging move creating u in the original sequence, when at most rn averaging moves had been made, D (the current squared diameter) was at most $g(k)$. That is, once rn averaging moves have been made, D is at most $g(k)$, proving the result. □

Now Theorems 3.2 and 3.6 show that Conjecture 3.3 implies Conjecture 3.1. Since Conjecture 3.3 is true for at least for $k \leq 8$, as we will see in the next section, and seems very reasonable, this gives strong evidence for Conjecture 3.1.

4. Determining $f(r)$ for $r < 3/2$

Using Theorems 3.2 and 3.6, we can obtain exact values of $f(r)$ for $r < 3/2$. In each case, we first prove the result for g and then appeal to Theorems 3.2 and 3.6.

First, we note that at least $n/2$ moves must be made to ensure that the squared diameter decreases below 1 at all.

PROPOSITION 4.1. *For $0 \leq r < 1/2$ we have*

$$f(r) = g(2) = 1.$$

PROOF. First, $g(2) = 1$ is clear, since we make no moves. Then by Theorems 3.2 and 3.6, we have $f(r) = 1$, giving the result. □

Now we see what happens to D after $n/2$ moves.

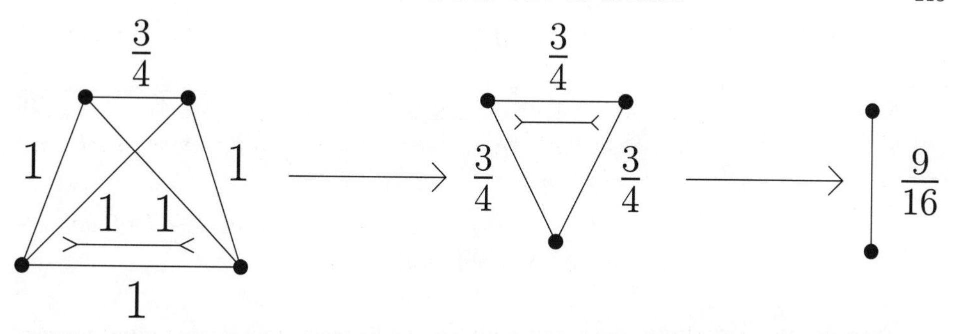

FIGURE 1. A (schematic of a 3-dimensional) configuration giving $g(4) \geq 9/16$ and $f(r) \geq 9/16$ for $r < 1$.

PROPOSITION 4.2. *For $1/2 \leq r < 3/4$ we have*

$$f(r) = g(3) = \frac{3}{4}.$$

PROOF. Lemma 2.3 shows that if we start with 3 points, then after the first merging move d_1 is at most $3/4$, and by Corollary 2.2 this is achieved if the three points are the vertices of an equilateral triangle. This proves $g(3) = 3/4$, and since the corresponding rooted binary tree is balanced, Theorems 3.2 and 3.6 give $f(r) = 3/4$. □

What happens after we make averaging $3n/4$ moves? We show that D must then be at most $9/16$.

PROPOSITION 4.3. *For $3/4 \leq r < 1$ we have*

$$f(r) = g(4) = \frac{9}{16}.$$

PROOF. Start with 4 points. After we merge two of them, then, by Lemma 2.3, two of the remaining squared distances are at most $3/4$. Now Corollary 2.2 applied to the three remaining points gives

$$g(4) \leq \frac{3}{4} \cdot \frac{3}{4} = \frac{9}{16}.$$

Taking one point at each vertex of the tetrahedron of Figure 1 and making the sequence of moves indicated there (with each inward-pointing double arrow indicating a pair of points to be merged) gives equality. Again, Theorems 3.2 and 3.6 give $f(r) = 9/16$. □

Note that if we made a different second move, the tree would not be balanced. Notice also that if we start with a regular tetrahedron of squared diameter 1, then after the first move, the squared side lengths of the remaining triangle will be 1, $3/4$, and $3/4$, and the squared diameter after the second move will be $3/4 - 1/4 = 1/2 < 9/16$. So, in order to maximize the squared distance between the last two points in this case, we must take an initial squared distance to be less than 1.

What about $f(1)$?

PROPOSITION 4.4. *For $1 \leq r < 9/8$ we have*

$$f(r) = g(5) = \frac{7}{16}.$$

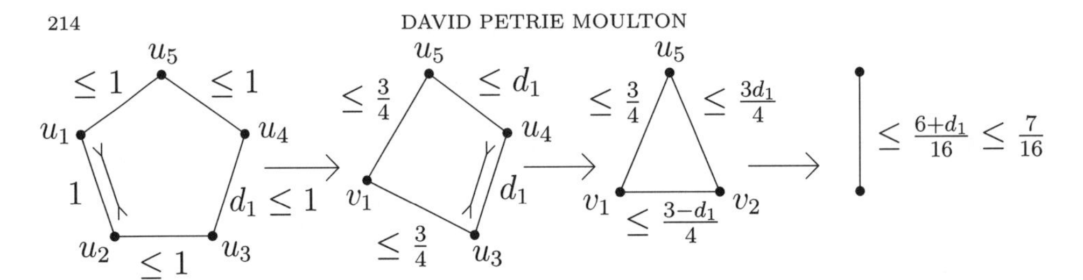

FIGURE 2. A proof of $g(5) \le 7/16$.

PROOF. Start with 5 points $u_1, \ldots, u_5$, as in Figure 2. The first move merges, say, u_1 and u_2, with $(u_1u_2)^2 = 1$, to give v_1. If the second move merges v_1, then $d_1 \le 3/4$ and $g(4) = 9/16$ give

$$\begin{aligned} d_3 &\le d_1 g(4) \\ &\le \frac{3}{4} \cdot \frac{9}{16} \\ &= \frac{27}{64} \\ &< \frac{7}{16}. \end{aligned}$$

So assume that the second move merges u_3 and u_4 to give v_2, and note that we have $d_1 = (u_3u_4)^2 \le 1$. Now Lemma 2.3 gives $(v_1u_3)^2 \le 3/4$ and $(v_1u_4)^2 \le 3/4$, as well as $(v_2u_5)^2 \le 3d_1/4$. The first formula of Corollary 2.2 gives $(v_1v_2)^2 \le (3 - d_1)/4$, and, finally, the first inequality of Corollary 2.2 (as formulated in the paragraph after the corollary) gives

$$\begin{aligned} d_3 &\le \frac{(v_1v_2)^2}{2} + \frac{(v_2u_5)^2}{4} \\ &\le \frac{1}{2}\left(\frac{3-d_1}{4}\right) + \frac{1}{4}\left(\frac{3d_1}{4}\right) \\ &= \frac{6+d_1}{16} \\ &\le \frac{7}{16} \end{aligned}$$

and $g(5) \le 7/16$. If we take all initial squared distances to be 1, then equality holds throughout, and we get $g(5) = 7/16$. Since the tree corresponding to these moves is balanced, by Theorems 3.2 and 3.6 we get $f(r) = 7/16$. □

Next we determine $g(6)$.

PROPOSITION 4.5. *For $9/8 \le r < 5/4$ we have*

$$f(r) = g(6) = \frac{25}{64}.$$

PROOF. Start with a configuration of 6 points and make the first three merging moves. We have two cases, based on the numbers of ancestors (as defined in the proof of Theorem 3.6) of each of the three remaining points among the original points.

Case 1: One of the remaining points, say u, is one of the original points (which is the same as having one such ancestor). Then the other two points, v and w, have, say, a and $b < a$ such ancestors, respectively, with $a + b = 5$. Then, since u and v have $1 + a$ ancestors between them, by Proposition 4.3 we have

$$\begin{aligned}(uv)^2 &\leq g(1+a)\\ &\leq g(4)\\ &= \frac{9}{16}\end{aligned}$$

and, from Proposition 4.4,

$$\begin{aligned}(vw)^2 &\leq g(a+b)\\ &= g(5)\\ &= \frac{7}{16}.\end{aligned}$$

Now Corollary 2.2 gives

$$\begin{aligned}d_4 &\leq \frac{(vw)^2}{2} + \frac{(uv)^2}{4}\\ &\leq \frac{1}{2}\cdot\frac{7}{16} + \frac{1}{4}\cdot\frac{9}{16}\\ &= \frac{23}{64}\\ &< \frac{25}{64}.\end{aligned}$$

Case 2: All three remaining points have 2 ancestors. Let u, v, and w be the points created by the first, second, and third moves, respectively. As in the proof of Proposition 4.4, we have

$$(uv)^2 \leq \frac{3 - d_1}{4},$$

and, as d_1 is the squared diameter of the set of ancestors of v and w, we get

$$(vw)^2 \leq d_1 g(4) = \frac{9d_1}{16}.$$

Now Corollary 2.2 gives

$$\begin{aligned}d_4 &\leq \frac{(uv)^2}{2} + \frac{(vw)^2}{4}\\ &\leq \frac{1}{2}\cdot\frac{3-d_1}{4} + \frac{1}{4}\cdot\frac{9d_1}{16}\\ &= \frac{24 + d_1}{64}\\ &\leq \frac{25}{64}.\end{aligned}$$

The larger bound, from Case 2, gives us $g(6) \leq 25/64$. If, in Case 2, we let the squared distance between the two ancestors of w be $3/4$ and all other squared distances between pairs of initial points be 1, and if we merge v and w next, then repeated applications of Corollary 2.2 give

$$\begin{aligned} d_4 &= \frac{(uv)^2}{2} + \frac{(uw)^2}{2} - \frac{(vw)^2}{4} \\ &= \frac{1}{2} \cdot \frac{1}{2} + \frac{1}{2} \cdot \frac{9}{16} - \frac{1}{4} \cdot \frac{9}{16} \\ &= \frac{25}{64}, \end{aligned}$$

so we have $g(6) = 25/64$. Again, the tree is balanced, so Theorems 3.2 and 3.6 give us $f(r) = 25/64$. □

Next, we determine $g(7)$.

PROPOSITION 4.6. *For* $5/4 \leq r < 11/8$ *we have*

$$f(r) = g(7) = \frac{21}{64}.$$

PROOF. As in the proof of Proposition 4.5, we have three cases, based on the numbers of ancestors of the final three points, which we label u, v, and w.

Case 1: The numbers of ancestors of u, v, and w are 3, 3, and 1, respectively. Then we have $(uw)^2 \leq g(4)$ and $(vw)^2 \leq g(4)$, and, since Case 2 of the proof of Proposition 4.5 cannot apply to the 6 ancestors of u and v, we get the stronger bound of Case 1, namely $(uv)^2 \leq 23/64$. Putting these together gives the bound

$$\begin{aligned} d_5 &\leq \frac{(uv)^2}{2} + \frac{(uw)^2}{4} \\ &\leq \frac{1}{2} \cdot \frac{23}{64} + \frac{1}{4} \cdot \frac{9}{16} \\ &= \frac{41}{128} \\ &< \frac{21}{64}. \end{aligned}$$

Case 2: The numbers of ancestors of u, v, and w are a, $b < a$, and 1, respectively. We have

$$\begin{aligned} d_5 &\leq \frac{(uv)^2}{2} + \frac{(uw)^2}{4} \\ &\leq \frac{g(a+b)}{2} + \frac{g(a+1)}{4} \\ &\leq \frac{g(6)}{2} + \frac{g(5)}{4} \\ &= \frac{1}{2} \cdot \frac{25}{64} + \frac{1}{4} \cdot \frac{7}{16} \\ &= \frac{39}{128} \\ &< \frac{21}{64}. \end{aligned}$$

Case 3: The numbers of ancestors of u, v, and w are 3, 2, and 2, respectively. We have

$$\begin{aligned} d_5 &\leq \frac{(uv)^2}{2} + \frac{(uw)^2}{4} \\ &\leq \frac{g(5)}{2} + \frac{g(5)}{4} \\ &= \frac{3}{4} \cdot \frac{7}{16} \\ &= \frac{21}{64}. \end{aligned}$$

The largest bound, from Case 3, gives us $g(7) \leq 21/64$. Take Case 3, with the squared distance between one ancestor of v and one ancestor of w equal to $3/4$ and all other squared distances between pairs of initial points equal to 1. Two applications of Corollary 2.2 give $(vw)^2 = 7/16$. Since the ancestors of u and v are the vertices of a regular 5-simplex, by the calculation of the lower bound of Proposition 4.4, we have $(uv)^2 = 7/16$ and, similarly, $(uw)^2 = 7/16$. Then u, v, and w are the vertices of an equilateral triangle, and we get $d_5 = 21/64$ and $g(7) = 21/64$. If move 5 merges v and w, then the tree is balanced, so Theorems 3.2 and 3.6 give us $f(r) = 21/64$. □

Finally, we determine $g(8)$.

PROPOSITION 4.7. *For* $11/8 \leq r < 3/2$ *we have*

$$f(r) = g(8) = \frac{75}{256}.$$

PROOF. Again, we have three cases, based on the numbers of ancestors of the final three points, which we label u, v, and w.

Case 1: One of the remaining points, say w, is one of the original points. Now we have, as before,

$$\begin{aligned} d_6 &\leq \frac{(uv)^2}{2} + \frac{(uw)^2}{4} \\ &\leq \frac{g(7)}{2} + \frac{g(5)}{4} \\ &= \frac{1}{2} \cdot \frac{21}{64} + \frac{1}{4} \cdot \frac{7}{16} \\ &= \frac{35}{128} \\ &< \frac{75}{256}. \end{aligned}$$

Case 2: The numbers of ancestors of u, v, and w are 3, 3, and 2, respectively. Again, as $g(6)$ is not attainable by two points with 3 ancestors each, we get

$$\begin{aligned} d_6 &\leq \frac{(uv)^2}{2} + \frac{(uw)^2}{4} \\ &\leq \frac{1}{2} \cdot \frac{23}{64} + \frac{1}{4} \cdot \frac{7}{16} \\ &= \frac{37}{128} \\ &< \frac{75}{256}. \end{aligned}$$

Case 3: The numbers of ancestors of u, v, and w are 4, 2, and 2, respectively, giving

$$\begin{aligned} d_6 &\leq \frac{(uv)^2}{2} + \frac{(uw)^2}{4} \\ &\leq \frac{g(6)}{2} + \frac{g(6)}{4} \\ &= \frac{3}{4} \cdot \frac{25}{64} \\ &= \frac{75}{256}. \end{aligned}$$

The largest bound, from Case 3, gives us $g(8) \leq 75/256$. In Case 3, let x and y be the two points merged to give w, with y formed after u, v, and x. Take all initial squared distances equal to 1, except that the squared distance between the two ancestors of y is $3/4$, and the squared distance between one ancestor of u and one ancestor of v is $9/16$. Then y will have squared distance $9/16$ to all of u, v, and x; and x will have squared distance $1/2$ to both u and v. From this we get

$$(uw)^2 = \frac{1}{2} \cdot \frac{9}{16} + \frac{1}{2} \cdot \frac{1}{2} - \frac{1}{4} \cdot \frac{9}{16} = \frac{25}{64}$$

and, by symmetry, $(vw)^2 = 25/64$. But three more applications of Corollary 2.2 show that u and v also have squared distance $25/64$, so u, v, and w form a regular triangle, giving

$$d_6 = \frac{3}{4} \cdot \frac{25}{64} = \frac{75}{256}.$$

If we merge u and v next, then the tree will be balanced, so Theorems 3.2 and 3.6 give us $f(r) = 75/256$. □

5. Reducing D to $\frac{1}{4}$

Now McKay asked about reduction of the radius by a factor $1/2$, which corresponds to reducing D to at most $1/4$, so we ask how many moves that takes. We cannot yet answer that exactly, but we bound the required number of moves to a relatively small interval.

Using the techniques from Section 4, we do not get exact bounds for $g(k)$ for $k \geq 9$, but we get the inequalities

$$\frac{1}{4} \leq f(r) \leq g(9) \leq \frac{69}{256} \quad \text{for } \frac{3}{2} \leq r < \frac{25}{16}$$

$$\frac{29}{128} \leq f(r) \leq g(10) \leq \frac{65}{256} \quad \text{for } \frac{25}{16} \leq r < \frac{13}{8}$$

$$\frac{27}{128} \le f(r) \le g(11) \le \frac{117}{512} \quad \text{for } \frac{13}{8} \le r < \frac{27}{16}.$$

These show that the smallest value of k with $g(k) \le 1/4$ is 9, 10, or 11, and the smallest value with $g(k) < 1/4$ is 10 or 11. Similarly, the smallest value of r with $f(r) \le 1/4$ is in the interval $[3/2 \,..\, 13/8]$, and the smallest value with $f(r) < 1/4$ is in the interval $[25/16 \,..\, 13/8]$. We can reformulate these results as follows:

PROPOSITION 5.1. *The minimum value of r such that for all $n \ge 2$, any sequence of at least rn averaging moves on a set of n points is guaranteed to reduce the diameter by at least a factor of 2 is in the interval $[3/2 \,..\, 13/8]$. Similarly, the minimum value of r such that the diameter is guaranteed to reduce by more than a factor of 2 is in the interval $[25/16 \,..\, 13/8]$.*

Let us consider the asymptotics of $f(r)$ as r goes to ∞.

LEMMA 5.2. *For $r, s \ge 0$ with at least one of r and s rational we have*

$$f(r+s) \le f(r)f(s).$$

PROOF. We suppose that r is rational, with denominator q. Take any configuration of n points with squared diameter 1 and any sequence of $m \ge (r+s)n$ averaging moves, and let d be d_m, the squared diameter after this sequence. If we replace each point and each move by q copies of itself, we get a set of qn points and a sequence of qm averaging moves with $d_{qm} = d$.

Now rqn is an integer, and after the first rqn moves, we have a configuration of qn points with squared diameter at most $f(r)$. Then after the remaining $qm - rqn \ge sqn$ moves, the squared diameter d is at most $f(r)f(s)$. This means that after making the m moves on the original configuration, the squared diameter d is at most $f(r)f(s)$, which proves the lemma. □

By Lemma 5.2, $\log f(r)$ is a subadditive function of r, so by Fekete's Lemma [**2**] (applied to this logarithm), the limit

$$\lim_{r\to\infty} f(r)^{\frac{1}{r}}$$

exists and is equal to the infimum of $f(r)^{1/r}$, and we would like to obtain bounds on its value.

If we start with a regular 2^p-simplex of squared diameter 1, after 2^{p-1} merging moves, we get a regular 2^{p-1}-simplex of squared diameter $1/2$, as can be seen by considering the case $p = 2$ discussed after the proof of Proposition 4.3. By induction we get

$$g(2^p) \ge \frac{1}{2^{p-1}},$$

and since the corresponding tree is balanced, Theorem 3.2 gives

$$f(r) \ge \frac{1}{2^{p-1}}$$

for $r < p/2$ with p a natural number. By starting with a regular $3 \cdot 2^p$-simplex, we similarly obtain

$$g(3 \cdot 2^p) \ge \frac{3}{2^{p+2}}$$

and (subtracting 2 from p)

$$f(r) \ge \frac{3}{2^p}$$

for $r < p/2 - 1/4$ with $p \geq 2$ an integer.

Before I had discovered the proof for Theorem 3.6, I proved the upper bounds for Propositions 4.3 and 4.4 by viewing them as linear-programming problems. In order to get an upper bound for $f(r)$ as r goes to ∞, Victor Miller and I produced a general linear program and obtained its optimum, which gave the upper bound

$$f(r) \leq \left(\frac{3}{4}\right)^p,$$

with r the closest multiple of 2^{-p} to $\frac{p}{3} + \frac{1}{9}$ and p a natural number. In particular, this gives the sharp upper bounds $f(0) \leq 1$ and $f(1/2) \leq 3/4$ and $f(3/4) \leq 9/16$, as well as the bounds $f(9/8) \leq 27/64$ and $f(57/32) \leq 243/1024 < 1/4$.

These results from the previous two paragraphs give the bounds

$$\left(\frac{1}{4}\right)^r \leq f(r) < \left(\frac{27}{64}\right)^{r-\frac{1}{2}}$$

for all $r \geq 0$ and bound the limit discussed above by

$$\frac{1}{4} \leq \lim_{r\to\infty} f(r)^{\frac{1}{r}} \leq \frac{27}{64}.$$

In fact, the upper bound $f(13/8) \leq 117/512$ from the beginning of the section, together with Fekete's Lemma, gives the better asymptotic upper bound

PROPOSITION 5.3.

$$\lim_{r\to\infty} f(r)^{\frac{1}{r}} \leq \left(\frac{117}{512}\right)^{\frac{8}{13}} < .40316861.$$

I think the upper bound probably can be improved by getting asymptotics for g in the manner of the upper-bound proofs of Propositions 4.5, 4.6, and 4.7, although I feel that the lower bound should be closer to the actual value of the limit.

6. Open Questions

We conclude with some open questions.

- What is $f(3/2) = g(9)$? I originally suspected that it equaled $1/4$, but now I am not so sure, since I might have a construction that does better, and it might not even be a dyadic rational.
- Does f or g take non-dyadic-rational values?
- Does some optimal merging configuration of 9 points have a balanced tree? Again, I thought the answer was "yes", but if my new construction works and is optimal, then the answer is "no". If so, then for every value of k is there an optimal k-point merging configuration with a balanced tree?
- If not, does $f(r)$ decrease at a value of r not predicted by Conjecture 3.1?
- Is g strictly monotonically decreasing?
- Does g satisfy the inequality

$$g(2k) \geq \frac{g(k)}{2}?$$

- Does f satisfy the corresponding inequality

$$f\left(r + \frac{1}{2}\right) \geq \frac{f(r)}{2}?$$

- What is the value of the limit

$$\lim_{r\to\infty} f(r)^{\frac{1}{r}}?$$

7. Acknowledgments

I would like to thank Brendan McKay for the original problem; Victor Miller, Samuel Kutin, Don Coppersmith, and Amit Khetan for helpful conversations; the referee for useful suggestions; and Joseph Gallian for running an excellent REU program.

References

[1] Brendan McKay, posting to sci.math.research, November 1, 2006.
[2] Michael Fekete, "Uber die Verteilung der Wurzeln bei gewissen algebraischen Gleichungen mit. ganzzahligen Koeffizienten," *Mathematische Zeitschrift* **17** (1923), pp. 228–249.

CENTER FOR COMMUNICATIONS RESEARCH, PRINCETON, NJ

Contemporary Mathematics
Volume **479**, 2009

Vertex Algebras as twisted Bialgebras: On a theorem of Borcherds

Manish M. Patnaik

ABSTRACT. Following Borcherds, we show a certain class of vertex algebras can be uniquely constructed from a bialgebra together with a twisted multiplication by a bicharacter. We illustrate this construction in the case of Heisenberg and lattice vertex algebras. As a consequence, we see that these vertex algebras can be recovered from their 2-point correlation functions and their underlying bialgebra structure.

To Joe Gallian on his 65th birthday

CONTENTS

1. Introduction

(1.1) Today, vertex algebras have become rather ubiquitous in mathematics. Among the subjects on which they bear direct influence, we can list representation theory, finite group theory, number theory (both through the classical theory of elliptic functions and the modern geometric Langlands theory), combinatorics, and algebraic geometry. Despite their firm entrenchment within the world of pure mathematics, vertex algebras (or rather their constituent elements *vertex operators*) actually first arose in the 1970s within the early string theory literature concerning dual resonance models [**Sch73**]. Independently of this, [**LW78**] Lepowsky and Wilson were interested in constructing representations of a twisted form of the affine Lie algebra $\widehat{sl}_2$ using differential operators acting on the ring of infinite polynomials

2000 *Mathematics Subject Classification.* Primary 17B69.

$\mathbb{C}[x_1, x_2, \ldots]$. To do so, the main difficulty they faced was in representing a certain infinite dimensional Heisenberg algebra. This they achieved using rather complicated looking formulas. Howard Garland then observed that essentially the same formulas occurred within the physics literature, and he thus imported the "vertex operator" into mathematics. Shortly thereafter, Igor Frenkel and Victor Kac **[FK81]** (and G. Segal **[Seg81]** independently) constructed the *basic representation* of untwisted affine Lie algebras, again using the newly christened vertex operators. Within the world of representation theory of affine Lie algebras, vertex operators were thus seen to play a central role.

Another significant impetus to the development of vertex operators came from the theory of finite groups. In the classification of finite groups, the Fischer-Greiss Monster (aka, "friendly giant") is the largest of the 26 sporadic simple groups. A number of very mysterious empirical phenomena concerning this group began to surface (see **[Bor98b]** for a review). These results collectively went under the title *Monstrous moonshine,* and one particular facet of moonshine was the prediction was that most natural representation of this finite simple group was actually of infinite dimension! The question then arose as to explicitly construct such a representation. Again, vertex operators proved to be essential here, and in **[FLM84]**, I. Frenkel, J. Lepowsky, and A. Meurman constructed the sought-after Moonshine module with the aid of vertex operators. Richard Borcherds **[Bor86]** then proposed an axiomatic framework to deal systematically with all of these occurrences of vertex operators, and define what we today refer to as a *vertex algebra.* Then in **[FLM88]**, the authors extended Borcherds' ideas and introduced the important notion of a *vertex operator algebra.* Moreover, they showed that their Moonshine module had a natural vertex operator algebra structure and that the Monster group could be realized as the group of automorphisms of this vertex operator algebra. Motivated by these results, it was widely believed by the mid 1980s that similar constructions should exist for other sporadic simple groups. However, it was not until the work of John Duncan **[Dun06, Dun05]**, nearly two decades after the initial work of **[FLM88]**, that this was to come to fruition.

Though there is no geometry in this paper, it was motivated by a connection between vertex algebras and algebraic geometry. H. Nakajima **[Nak97]** and I. Grojnowski **[Gro96]** have constructed the *Fock space* representation of an infinite dimensional Heisenberg algebras on the cohomology of the Hilbert scheme of points on an algebraic surface. This Fock space is none other than the ring of symmetric functions in infinitely many variables, the working ground for many combinatorialists. The fact that this same Fock space also possesses a vertex operator algebra structure has facilitated the migration of results back and forth between combinatorics and the theory of vertex algebras. For example, the Boson-Fermion correspondence from vertex algebras has been shown by I. Frenkel to encode (or be encoded by) the combinatorics of Littlewood-Richardson coefficients **[FJ88]**. Returning to our geometric setup, certain "halves" of vertex operators have been shown to act on the cohomology of Hilbert schemes by means of cup-product with the Chern classes of certain tautological bundles **[Gro96]**. Essentially this observation was enough to unravel the cup-product structure on Hilbert schemes, a problem that had otherwise seemed intractable **[LS01]**. I. Frenkel has then posed the question of understanding the full vertex algebra structure geometrically. Despite interesting progress **[Leh99]**, the answer to this question still remains quite mysterious. Our approach was a synthetic one which attempted to first pare down the axioms of a vertex algebra so that each part may have a clear geometric meaning. This we do by means of Borcherds' theorem.

(1.2) *What exactly is a vertex algebra?* Unfortunately, this question does not have a very short answer. For example, the statement that a vertex algebra is a particular type of algebra is false! Rather one may think of a vertex algebra as a vector space V (usually infinite dimensional) endowed with an infinite family of multiplications $\circ_n$,

$$\begin{array}{rcl} \circ_n : V \times V & \to & V \\ a \times b & \mapsto & a \circ_n b \end{array}$$

These products are neither associative nor commutative, but satisfy conditions somewhere in between, the so called Borcherds identities [**Bor86**]. An alternative starting point for vertex algebras is the axiomatic treatment of Frenkel-Huang-Lepowsky [**FHL93**]. Here one begins with a vector space V equipped with a "state-field correspondence," i.e. a linear map,

$$\begin{array}{rcl} Y(\cdot, z) : V & \to & \mathrm{End}(V)[[z, z^{-1}]] \\ a & \mapsto & Y(a, z) \end{array}$$

The $Y(\cdot, z)$ may be viewed as a generating function encoding all the infinite products $\circ_n$ into one object: indeed, applying $Y(a, z)$ to b, we obtain

$$Y(a, z)b = \sum_{n \in \mathbb{Z}} (a \circ_n b) z^n.$$

One then develops a formal calculus to deal with these series and phrases the axioms of a vertex algebra in terms of properties of $Y(\cdot, z)$. This viewpoint is reviewed in section 3.

There is yet another related viewpoint put forth by Borcherds which purports to "make the theory of vertex algebras trivial" [**Bor01**]. In other words, a fairly elaborate categorical framework is constructed in which one can view vertex algebras as commutative ring objects. Borcherds' motivation for such a description seems to have come from the important question of quantizing vertex algebras, but in our exposition we will not be concerned with any such quantum deformation. Neither will we deal with the categorial framework which Borcherds has constructed. Rather, our goal will be to present in as simple-minded a way as possible the new construction of vertex algebras contained in [**Bor01**]. The reader is warned here of the obvious danger that such a simplification necessarily introduces and is referred to the original works [**Bor98a, Bor01**] for further motivations.

In our formulation, the essence of Borcherds' construction is that starting from some rather innocuous algebraic elements– a commutative, cocommutative bialgebra V equipped with a derivation $T : V \to V$ and a symmetric bicharacter $r : V \otimes V \to \mathbb{C}[(z-w)^{-1}]$– one can show how to construct the seemingly richer structure of a vertex algebra on V. In this light, a natural predecessor to Borcherds' result is the simple fact (due also to Borcherds) that a commutative vertex algebra structure on V, i.e., one in which the fields $Y(a, z)v \in End(V)[[z]]$ are just power series, is equivalent to the structure of a commutative algebra with derivation on V. Briefly, this equivalence is constructed as follows: given a commutative vertex algebra, we may define a product on V by

$$a \cdot b := [Y(a,z)b]_{z^0}$$

where by $[A(z)]_{z^0}$ denotes the coefficient of the constant term of $A(z) \in V((z))$. The axioms of a vertex algebra then provide V with a natural derivation compatible with this algebra structure. Conversely, given a commutative unital algebra with derivation T, we can construct a state-field correspondence by the formula

$$Y(a,z)b = (e^{zT}a) \cdot b.$$

The axioms for a vertex algebra then follow readily from the ring properties V.

Since the most interesting vertex algebras are not commutative, we might wonder how to modify the above construction by somehow introducing singularities (i.e., negative powers of z) into the picture. Borcherds answer to this question is as follows: given a *bi*algebra V, a derivation T compatible with the bialgebra structure, and bicharacter $r : V \otimes V \to \mathbb{C}[(z-w)^{-1}]$ he defines a state-field correspondence through the the following elegant formula which combines all the data which we are given:

$$Y(a,z)b = \sum_{(a)} T^{(k)}(a') \cdot b' \cdot r(a'', b''),$$

where the coproduct on V is written as $\Delta(a) = \sum_{(a)} a' \otimes a''$ and where $T^{(k)} := \frac{T^k}{k!}$ are the *divided powers* of T. Our first goal will be to directly prove that this formula does in fact define a vertex algebra on V.

The next question we address is what types vertex algebras can be constructed using the new approach. Along these lines, we define a certain class, called r-vertex algebras, which contain the vertex algebras constructed as above from commutative, cocommutative bialgebras V. For example, Heisenberg and lattice vertex algebras are contained in this class. The second main result in this note is a uniqueness theorem which shows that if the underlying bialgebra of an r-vertex algebra V is *primitively generated*, then it is *the* vertex algebra as constructed above from a bialgebra V, derivation T, and bicharacter r determined by the 2-point functions. There are many other natural examples of vertex algebras in representation theory, among them the vacuum modules for affine Kac-Moody and Virasoro algebras. These vertex algebras do not have a commutative algebra structure and thus do not fall under the twisted bicharacter construction presented in this paper. It is an interesting question as to how this construction can be modified to account for such examples.

We would also like to add that since this paper was written, Anguelova and Bergvelt have extended and clarified many aspects of Borcherds proposal for quantizing vertex algebras in their interesting paper [**AB07**].

(1.3) This paper is organized as follows: In section 2, we set up some preliminary algebraic machinery. In section 3, we review some basics from the theory of vertex algebras, and in section 4, we state and prove Borcherds' theorem. In section 5, we state a converse, and we conclude with some examples in section 6.

Acknowledgements: This paper grew out of discussions with Igor Frenkel during the Spring 2003 semester. The observation that Borcherds"s bicharacter $r(\cdot,\cdot)$ was merely a 2-point function, which informs much of this note, is due to him. I would like to also thank him for his encouragement and many helpful suggestions on this manuscript. Also, I would like to thank Minxian Zhu and John Duncan for many tutorials on vertex algebras. This research was conducted while the author was supported by an NSF graduate research fellowship.

Finally, I would like to dedicate this paper to Joe Gallian on the occasion of his 65th birthday. During the summer of 2000, I had the wonderful opportunity to participate in his REU program and to be so pleasantly initiated into the world of mathematical research. His unrivaled enthusiasm and dedication have been an inspiration ever since.

2. Algebraic Preliminaries

We fix an algebraically closed field k of characteristic zero.

(2.1) For us, an **algebra** A will be a commutative $\mathbb{Z}_+$-graded k-algebra with unit, $A = \oplus_{n=0}^{\infty} A_n$, where each A_n is a finite dimensional k-vector space.

Dually, a **coalgebra** will be a cocommutative $\mathbb{Z}_+$-graded k-coalgebra (C, Δ, ϵ) where $\Delta : C \to C \otimes C$ is the comultiplication. We adopt Sweedler notation to write the coproduct, $\Delta(x) = \sum_{(x)} x' \otimes x''$. Generalizing, we will also write for example

$$\Delta_2(x) = (\Delta \otimes 1)\Delta(x) = (1 \otimes \Delta)\Delta(x) = \sum_{(x)} x^{(1)} \otimes x^{(2)} \otimes x^{(3)}.$$

Writing $C = \oplus_{n=0}^{\infty} C_n$, then we assume C_n is a finite k-vector space, $\epsilon(C_n) = 0$ for $n \neq 0$, and $\Delta C_n \subset \sum_{i=0}^{n} C_i \otimes C_{n-i}$. We say that a coalgebra is **connected** if $C_0 = k$.

We define a **bialgebra** to be an algebra A with a coalgebra structure (A, Δ, ϵ) such that the gradings on A as an algebra and coalgebra agree and Δ and ϵ are algebra morphisms. We say that an element of a bialgebra is **primitive** if $\Delta(x) = x \otimes 1 + 1 \otimes x$ and **group-like** if $\Delta(x) = x \otimes x$.

(2.2) By a **bialgebra with derivation**, we shall mean bialgebra V with a linear map $T : V \to V$ of degree 1 satisfying

(1) $T(1) = 0$
(2) $T(a \cdot b) = T(a) \cdot b + a \cdot T(b)$
(3) $T(\Delta(a)) = \sum_{(a)} T(a') \otimes a'' + \sum_{(a)} a' \otimes T(a'')$

We say that a set $\{a_\alpha | a_\alpha \in V\}_{\alpha \in S}$ is a T**-generating** set of an algebra V if $\{T^{(k)} a_\alpha\}_{\alpha \in S}$ generates V as an algebra. Here, as usual, $T^{(k)} := \frac{T^k}{k!}$ denotes the divided powers of the operators T.

(2.3) We next describe the additional piece of information on a bialgebra with derivation which allows us to construct interesting vertex algebras.

DEFINITION 2.3.1. Let V be a bialgebra with derivation T. Then a **bicharacter** is a bilinear map $r_{z,w} : V \times V \to k[(z-w)^{\pm 1}]$ satisfying

(1) $r(a, b) = r(b, a)$;
(2) $r(a, 1) = \epsilon(a)$;
(3) $r(a \cdot b, c) = \sum_c r(a, c') r(b, c'')$
(4) $r(Ta, b) = -\frac{d}{dz} r(a, b)$
(5) $r(a, Tb) = \frac{d}{dw} r(a, b)$.
(6) For $a \in V_m$ and $b \in V_n$, $r(a, b) = \frac{c(a,b)}{(z-w)^{-m-n}}$, where $c(c, b)$ is some scalar.

Suppose now (V, T, r) is a bialgebra with derivation and bicharacter. In what follows, we will also make use of the following multivariable generalization of r.

DEFINITION 2.3.2. Let $a_1, \ldots, a_n \in V$. Then we recursively define

$$r_{z_1,\ldots,z_n} : V \times V \cdots V \to k[(z_i - z_j)^{\pm 1}]_{1 \le i \ne j \le n},$$

to be

$$r_{z_1,\ldots,z_n}(a_1,\ldots,a_n) = \sum_{(a_1),(a_2),\ldots,(a_n)} r_{z_2,\ldots,z_n}(a_2',\ldots,a_n')\Pi_{j=2}^{n} r_{z_1,\ldots,\hat{z}_j,\ldots z_n}(a_1^{(j-1)}, a_2'',\ldots,\widehat{a_j''},\ldots a_n'')$$

where a hat over a variable signifies that this variable is to omitted.

EXAMPLE 2.3.3. To make things more transparent, we unravel the above formula to get,

$$r_{x,y,z}(a,b,c) = \sum_{(a),(b),(c)} r_{y,z}(b',c')r_{x,y}(a',b'')r_{x,z}(a'',c'')$$

and similarly

$$r_{x,y,z,w}(a,b,c,d) = \sum_{(a),(b),(c),(d)} r_{x,y}(a^{(1)},b^{(1)})r_{x,z}(a^{(2)},c^{(1)})r_{x,w}(a^{(3)},d^{(1)})r_{y,z}(b^{(2)},c^{(2)})r_{y,w}(b^{(3)},d^{(2)})r_{z,w}(c^{(3)},d^{(3)})$$

Using the properties of r and the (co)commutativity of V, we then have,

LEMMA 2.3.4. *Let $a_1, \ldots, a_n \in V$ and $\sigma \in S_n$ a permutation. Then*

$$r_{z_1,\ldots,z_n}(a_1,\ldots,a_n) = r_{z_{\sigma(1)},\ldots,z_{\sigma(n)}}(a_{\sigma(1)},\ldots,a_{\sigma(n)}).$$

(2.4) Given a bialgebra V, we can form its graded dual $V^\vee = \oplus_{n=0}^{\infty} V_n^*$. By a **bilinear form** on V we will mean a non-degenerate, symmetric, bilinear form $(\cdot,\cdot)$ on the vector space V satisfying

(1) $(V_m, V_n) = 0$ unless $m = n$ (identifying V with $V^\vee$ naturally)
(2) $(1, v) = \epsilon(v)$

3. Vertex Algebras

(3.1) We must first introduce the concept of a *field.*

DEFINITION 3.1.1. Let V be a vector space over k. A formal power series

$$A(z) = \sum_{j\in\mathbb{Z}} A_{(j)} z^{-j} \in \mathrm{End}\, V[[z^{\pm 1}]]$$

is called a **field** on V if for any $v \in V$ we have $A_{(j)} \cdot v = 0$ for large enough j.
In case V is $\mathbb{Z}_+$ graded, we can define a field of **conformal dimension** $d \in \mathbb{Z}$ to be a field where each $A_{(j)}$ is a homogeneous linear operator of degree $-j + d$

A useful procedure for dealing with fields will be normal ordering, defined as follows.

DEFINITION 3.1.2. Let Let $A(z), B(w)$ be fields, and denote by $A(z)_+$ and $A(z)_-$ the non-negative and negative parts (in powers of z) of $A(z)$. Then define $: A(z)B(w) := A(z)_+B(w) + B(w)A(z)_-$ to be **normally ordered product** of $A(z)$ and $B(w)$.

(3.2) We are now in a position to define the main object of study,

DEFINITION 3.2.1. **[FBZ01**, p.20] A **vertex algebra** $(V, |0\rangle, T, Y)$ is a collection of data:

- (state space) a $\mathbb{Z}_+$ graded k-vector space $V = \oplus_{m=0}^{\infty} V_m$, with each $\dim V_m < \infty$
- (vacuum vector) $|0\rangle \in V_0$
- (translation operator) $T : V \to V$ of degree 1.
- (vertex operators) $Y(\cdot, z) : V \to \mathrm{End}(V)[[z^{\pm 1}]]$ taking each $A \in V_m$ to a field
$$Y(A, z) = \sum_{n \in \mathbb{Z}} A_{(n)} z^{-n-1}$$
of conformal dimension m (i.e., deg $A_{(n)} = -n + m - 1$).

satisfying the following axioms:

- (vacuum axiom) $Y(|0\rangle, z) = \mathrm{Id}_V$. Furthermore, for any $A \in V$, $Y(A, z)|0\rangle \in V[[z]]$ and $Y(A, z)|0\rangle|_{z=0} = A$.
- (translation axiom) For every $A \in V$, $[T, Y(A, z)] = \frac{d}{dz} Y(A, z)$ and $T|0\rangle = 0$
- (locality) For every $A, B \in V$, there exists a positive integer N such that $(z-w)^N [Y(A, z), Y(B, w)] = 0$ as formal power series in $\mathrm{End}(V)[[z^{\pm 1}, w^{\pm 1}]]$.

(3.3) *Example: Heisenberg Vertex Algebra:* Let $\mathfrak{h}$ be an l-dimensional complex vector space equipped with a non-degenerate, symmetric, bilinear form $(\cdot, \cdot)$. The *Heisenberg Lie algebra* is defined to be

$$\hat{\mathfrak{h}} = \mathfrak{h} \otimes \mathbb{C}[t, t^{-1}] \oplus \mathbb{C}K$$

with relations

$$[K, \mathfrak{h}] = 0 \quad \text{and} \quad [h_1 \otimes t^m, h_2 \otimes t^n] = (h_1, h_2) m \delta_{m,-n} K.$$

We will denote

$$h(n) := h \otimes t^n, h \in \mathfrak{h}.$$

For each λ in the dual space $\mathfrak{h}^*$, we may construct the Fock space by the following induction:

$$\Pi^\lambda = U(\hat{\mathfrak{h}}) \otimes_{U(\mathfrak{h} \otimes \mathbb{C}[t] \oplus \mathbb{C}K)} \mathbb{C}$$

where $\mathbb{C}$ is the one dimensional space on which $\mathfrak{h} \otimes t\mathbb{C}[t]$ acts as zero, K acts as the identity, and $\mathfrak{h} \otimes t^0$ acts via λ. As vector spaces, one can show that

$$\Pi^\lambda = \mathrm{Sym}(\mathfrak{h} \otimes t^{-1} \mathbb{C}[t^{-1}]),$$

where $\mathrm{Sym}(V)$ is the symmetric algebra of a vector space V.

The space $\Pi := \Pi^0$ has a vertex algebra structure with vertex operators are described as follows: for $h \in \mathfrak{h}$, set

$$h(z) = \sum_{n \in \mathbb{Z}} h(n) z^{-n-1}$$

where $h(n)$ acts as the operator $h \otimes t^n$ on the $\hat{\mathfrak{h}}$-module Π. Then explicitly, K acts as the identity, $h(n)$ acts by multiplication for $n < 0$, the derivation defined by $h(n) h_0(-s) = n \delta_{n,s} (h, h_0)$ for $n > 0$, $h_0 \in \mathfrak{h}$, and as the operator 0 for $n = 0$. Set

$$Y(h_1(j_1) \cdots h_k(j_k), z) = \frac{1}{(-j_1 - 1)! \cdots (-j_k - 1)!} : \partial_z^{j_1 - 1} h_1(z) \cdots \partial_z^{-j_k - 1} h_k(z) :,$$

for $h_1, \ldots, h_k \in \mathfrak{h}$. The verification of the vertex algebra axioms is left as an exercise, and will also follow from our theorem below.

(3.4) Let $(V, |0\rangle, T, Y)$ be a vertex algebra, and assume that it is equipped with a non-degenerate, symmetric, bilinear form $(\cdot,\cdot)$ which identifies V with its restricted dual $V^\vee$. With this assumption, denote $\langle 0| \in V$ the dual to the element $|0\rangle \in V$. Then, we introduce the

DEFINITION 3.4.1. Given $a_1, \ldots, a_n \in V$. We define the n-**point functions** to be

$$(\langle 0|, Y(a_1, z_1) \cdots Y(a_n, z_n)|0\rangle) \in k[[z_1, z_2, \ldots, z_n]][(z_i - z_j)^{\pm 1}]_{1 \le i < j \le n}$$

4. Borcherds' theorem

(4.1) In this section, we state and prove Borcherds's theorem [**Bor01**]. Let V be a bialgebra with derivation T and bicharacter r. We aim to construct a vertex algebra on V with translation operator given by V and 2-point functions specified by r. We first define the following maps which will allow us to construct vertex operators,

DEFINITION 4.1.1. Let

$$\Phi^n(z_1, \ldots, z_n) : V \times V \times \cdots \times V \to V[[z_1, z_2, \cdots, z_n]][(z_i - z_j)^{-1}]$$

send

$$(a_1, a_2, \ldots, a_n) \mapsto \sum_{(a_1)\cdots(a_n)} \sum_{j_1, \ldots j_k \ge 0} T^{(j_1)}(a_1')\cdot T^{(j_2)}(a_2') \cdots T^{(j_n)}(a_n') r_{z_1, \ldots, z_n}(a_1'', a_2'', \ldots, a_n'').$$

So, for example, we have that

$$\Phi^2(z, w)(a, b) = \sum D^{(j)}(a')D^{(k)}(b')r_{z-w}(a'', b'').$$

The following is easy to see from (co)commutativity of V and the corresponding facts for r,

LEMMA 4.1.2. *Let $a_1, \ldots, a_n \in V$ and $\sigma \in S_n$ a permutation. Then*

$$\Phi^n(z_1, \ldots, z_n)(a_1, \ldots, a_n) = \Phi^n(z_{\sigma(1)}, \ldots, z_{\sigma(n)})(a_{\sigma(1)}, \ldots, a_{\sigma(n)})$$

as elements of $V[[z_1, \ldots, z_n]][(z_i - z_j)^{-1}]_{1 \le i < j \le n}$.

(4.2) Following Borcherds, we make the following fundamental definition:

DEFINITION 4.2.1. Let $a, b \in V$. Then define the field $Y(a, z)$ by

$$Y(a, z_1)Y(b, z_2)|0\rangle = \Phi(z_1, z_2)(a, b).$$

In particular, we have $Y(a, z_1)b = \Phi(z_1, 0)(a, b)$.

With the above definition, we can verify the following formula for multiplication.

LEMMA 4.2.2. *Let $a_1, \ldots, a_n \in V$. Then $Y(a_1, z_1) \cdots Y(a_n, z_n)|0\rangle = \Phi(a_1, \ldots, a_n)$*

PROOF. The proof is a straightforward computation. For ease of notation, we just sketch the case $n = 3$ which already illustrates all the main ideas. Let $a, b, c \in V$. Then from the above definition, we see that

$$Y(a, z_1)Y(b, z_2)Y(c, z_3)|0\rangle = Y(a, z_1) \sum_{(b),(c)} \sum_{k,l \geq 0} z_2^k z_3^l T^{(k)}(b')T^{(l)}(c')r_{z_2,z_3}(b'', c'').$$

Expanding

$$\sum_{(b),(c)} r_{z_2,z_3}(b'', c'') \sum_{k,l \geq 0} Y(a, z_1)[T^{(k)}(b')T^{(l)}(c')]z_2^k z_3^l,$$

we get

$$\sum_{(b),(c)} r_{z_2,z_3}(b'', c'') \sum_{k,l \geq 0} z_2^k z_3^l \sum_{(a)} r_{z_1}(a'', [T^{(k)}(b')T^{(l)}(c')]'') \sum_{j \geq 0} z_1^j T^{(j)}(a')[T^{(j)}(b')T^{(k)}(c')]'',$$

which using the fact that $\Delta(a \cdot b) = \Delta(a) \cdot \Delta(b)$ and property (3) of 2.3.1 becomes,

$$\sum_{(b),(c)} r_{z_2,z_3}(b'', c'') \sum_{k,l \geq 0} z_2^k z_3^l \sum_{(a)} r_{z_1}(a^{(2)}, [T^{(j)}(b')]'')r_{z_1}(a^{(3)}, [T^{(k)}(c')]'') \sum_{j \geq 0} z_1^j T^{(j)}(a')[T^{(j)}(b')]'[T^{(k)}(c')]'.$$

Observing that,

$$\Delta T^{(k)}(a) = \sum_{j=0}^{k} \sum_{(a)} D^{(k-j)}(a') \otimes D^{(j)}a'',$$

we collect terms in the above coefficients $T^{(j)}(a^{(1)})T^{(k)}(b^{(1)})T^{(l)}(c^{(1)})$ to get

$$\sum_{j,k,l \geq 0} \sum_{(a),(b),(c)} T^{(j)}(a^{(1)})T^{(k)}(b^{(1)})T^{(l)}(c^{(1)}) \sum_{p,q \geq 0} z_1^j z_2^{k+p} z_3^{l+q} r_{z_1}(a^{(2)}, T^{(p)}(b^{(2)})r_{z_1}(a^{(2)}, T^{(q)}(c^{(2)}).$$

But Taylor's formula and the property (5) of 2.3.1 give that

$$r_{z,w}(a, b) = \sum_{n \geq 0} r_z(a, T^{(n)}(b))w^n.$$

Applying this fact to the above sum, we conclude that the terms with coefficient $T^{(j)}(a^{(1)})T^{(k)}(b^{(1)})T^{(l)}(c^{(1)})$ is just $r_{z_1,z_2,z_3}(a'', b'', c'')$ which concludes the proof. □

(4.3) We are now ready to state the first theorem of this note,

THEOREM 4.3.1. *Let V be a bialgebra with derivation T and bicharacter r. Define fields $Y(a, z)$ as in Definition 4.2.1. Then $(V, |0\rangle, T, Y)$ is a vertex algebra.*

PROOF. It is obvious that the map $Y(\cdot, z) : V \to V[[z]][z^{-1}]$ takes $a \in V_m$ to a field of conformal dimension m. So, let us verify the vacuum, translation, and locality axioms.

Vacuum: For $a \in V$, the above definition gives that $Y(a, z)|0\rangle = \sum_{k=0}^{\infty} T^{(k)}(a)z^k$, from which the vacuum axiom follows easily.

Translation: The translation axiom requires that

$$[T, Y(a, z)]b = \frac{d}{dz} Y(a, z)b.$$

Expanding the left hand side, we get

$$TY(a, z)b = T[\sum_{(a),(b),k \geq 0} T^{(k)}(a')b'r_z(a'', b'')z^k]$$

and

$$Y(a,z)Tb = \sum_{(a),(b),k\geq 0} T^{(k)}(a')T(b')r_z(a'',b'')z^k + \sum_{(a),(b),k\geq 0} T^{(k)}(a')b'r_z(a'',Tb'')z^k$$

Since

$$T[T^{(k)}(a')b'] = (k+1)T^{(k+1)}(a')b' + T^{(k)}(a')T(b'),$$

this difference equals

$$\sum_{(a),(b),k\geq 0} (k+1)T^{(k+1)}(a')b'r_z(a'',b'') - T^{(k)}(a')b'r_z(a'',Tb'')z^k$$

which is easily seen to be equal to

$$\frac{d}{dz}Y(a,z)b = \sum_{(a),(b),k\geq 0} T^{(k)}(a')\cdot b'\frac{d}{dz}[r_z(a'',b'')]z^k + T^{(k)}(a')\cdot b'r_z(a'',b'')\frac{d}{dz}[z^k].$$

Locality: Locality essentially follows from the Lemma 4.2.2 above. Indeed, let $a, b \in V$. Then

$$\Phi^3_{(z_1,z_2,0)}(a,b,c) = Y(a,z_1)Y(b,z_2)c$$

and similarly

$$\Phi^3_{(z_2,z_1,0)}(b,a,c) = Y(b,z_2)Y(a,z_1)c.$$

So, by Lemma 4.2.2, we see that both

$$\Phi^3_{(z_1,z_2,0)}(a,b,c) = \Phi^3_{(z_2,z_1,0)}(b,a,c) \in End(V)[[z_1^{\pm 1}, z_2^{\pm 1}]][(z_1-z_2)^{-1}]$$

which means that there is some positive integer N such that

$$(z-w)^N[Y(a,z_1), Y(b,z_2)]c = 0.$$

Furthermore, this number N does not depend on c as is clear from looking at the explicit formula for Φ in Definition 4.1.1.

□

In case our bialgebra V also has a bilinear form, we have the following formula for the n-point functions.

COROLLARY 4.3.2. *Let $(V, |0\rangle, T, Y)$ be the vertex algebra constructed above and let $(\cdot,\cdot)$ be a bilinear form on V. Then*

$$(\langle 0|, Y(a_1,z_1)\cdots Y(a_n,z_n)|0\rangle) = r_{z_1,z_2,\ldots,z_n}(a_1,a_2,\ldots,a_n)$$

5. Converse to Borcherds's theorem

(5.1) In this previous section, we showed how starting from a bialgebra V together with a derivation T and a bicharacter r we can construct a vertex algebra. Now, given a vertex algebra $(V, |0\rangle, T, Y)$ whose underlying space V also has a bialgebra structure with derivation T, we might wonder to what extent does this completely determine the vertex algebra structure. By itself, V and T cannot know about the singularities (hence, 2-point functions) of $Y(\cdot, z)$, so in order to get a meaningful answer to our question, we have to also feed in the information of r. If the bialgebra structure on V is not reflected in the vertex algebra structure, we will have no hope of recovering the vertex algebra structure. So, in addition to r, we need to assume some compatibility conditions between the vertex and bialgebra structures. These considerations are formalized in the following definition.

DEFINITION 5.1.1. An r**-vertex algebra** consists of the following data:

- a vertex algebra $(V, |0\rangle, T, Y)$
- a bialgebra structure with derivation (V, T) with compatible gradings.
- a bilinear form $(\cdot, \cdot)$ on V

satisfying the following additional axioms:

- for $a \in V$ primitive, $a_{(-1)}b = a \cdot b$
- for $a, b \in V$ primitive, $a_{(0)}b = 0$

(5.2) With the above Definition 5.1.1, we can state a partial converse to Borcherds's theorem. This is not the strongest statement that can be made, but the following theorem (or a slight variation of it) suffices to deal with all of the r-vertex algebras which we know.

THEOREM 5.2.1. *Let V be a connected bialgebra with derivation T, bicharacter r and bilinear form $(\cdot, \cdot)$. Assume that V is T-generated by a set of primitive elements $\{a_\alpha\}_{\alpha \in S}$. Then there exists a unique r-vertex algebra structure on V such that the 2-point functions of V are given by r.*

PROOF. In the previous section, we showed how to construct a vertex algebra on V with two-point functions specified by r. It is clear that this vertex algebra is actually a r-vertex algebra.

To prove the uniqueness, we shall make us of the following.

THEOREM 5.2.2. [**FBZ01**, p.70] *Let $V = \oplus_{n=0}^{\infty} V_n$ be a $\mathbb{Z}_+$-graded vector space, $|0\rangle \in V_0$ a non-zero vector, and T a degree 1 endomorphism of V. Let S be a (countable) set and $\{a^\alpha\}_{\alpha \in S}$ a collection of homogenous vectors in V. Suppose we are given fields*

$$a^\alpha(z) = \sum_{n \in \mathbb{Z}} a^\alpha_{(n)} z^{-n-1}$$

such that the following conditions hold:

1. *For all α, $a^\alpha(z)|0\rangle = a^\alpha + z(\cdots)$*
2. *$T|0\rangle = 0$ and $[T, a^\alpha(z)] = \partial_z a^\alpha(z)$ for all $\alpha \in S$.*
3. *The fields $a^\alpha(z)$ are mutually local*
4. *V is spanned by the vectors*

$$a^{\alpha_1}_{(j_1)} \cdots a^{\alpha_m}_{(j_m)}|0\rangle, j_i < 0$$

Then these structures together with the vertex operation

$$Y(a^{\alpha_1}_{(j_1)} \cdots a^{\alpha_m}_{(j_m)}|0\rangle, z) = \frac{1}{(-j_1-1)!} \cdots \frac{1}{(-j_m-1)!} : \partial_z^{-j_1-1} a^{\alpha_1}(z) \cdots \partial_z^{-j_m-1} a^{\alpha_m}(z) :$$

give rise to a unique vertex algebra structure on V satisfying (1)-(4) above and such that $Y(a^\alpha, z) = a^\alpha(z)$.

Assume that we have a r-vertex algebra satisfying the hypothesis of Theorem 5.2.1. First we contend that the vectors $a^{\alpha_1}_{(j_1)} \cdots a^{\alpha_m}_{(j_m)}|0\rangle$, $j_i < 0$ generate V as a vector space. Indeed,

LEMMA 5.2.3. *Let $a \in V$ be primitive and $b \in V$. For $k \geq 0$, we have that*

$$T^{(k)}(a) \cdot b = a_{(-k-1)}b.$$

PROOF. For $k = 0$, we have that $a \cdot b = a_{(-1)}b$ since a is primitive and V is a r-vertex algebra. The general case follows by induction using the following two facts: first, if a is primitive, then so is $T^{(k)}(a)$; and second, for any vertex algebra $Y(Ta, z) = \frac{d}{dz}Y(a, z)$. □

This shows that any r-vertex algebra satisfying the hypothesis of Theorem 5.2.1, it will also satisfy the hypothesis of the Reconstruction theorem. Therefore, in order to show uniqueness, we just need to verify that the fields $Y(a_\alpha, z)$ are uniquely specified by the given information. So, let us suppose that $(V, |0\rangle, T, Y)$ and $(V, |0\rangle, T, \widetilde{Y})$ are two r-vertex algebra structures satisfying the hypotheses of the theorem. We may as well assume that $(V, |0\rangle, T, Y)$ is the vertex algebra constructed in Theorem 5.2.1.

Write $\widetilde{Y}(a, z) = \sum_{n\in\mathbb{Z}} \widetilde{a}_{(n)}z^{-n-1}$ and $Y(a, z) = \sum_{n\in\mathbb{Z}} a_{(n)}z^{-n-1}$. Then the lemma above allows us to conclude that $\widetilde{Y}(a, z)_+ = Y(a, z)_+$.

To show that $\widetilde{Y}(a, z)_- v = Y(a, z)_- v$, we first check the result for v a primitive element of minimal positive degree d. Note than since V is primitively generated, the minimal degree of a primitive element is also the minimal positive degree of any element of V. By assumption, $a_{(0)}v = 0$, so let us focus on $a_{(n)}v$ for $n > 0$. Each $a_{(n)}$ has strictly negative degree as an operator on V, so by our assumption on v we conclude that $a_{(n)}v = 0$ unless $n = d$, in which case $a_{(n)}v = \lambda|0\rangle$. Furthermore, we can determine this value of λ from the 2-point functions:

$$r(a, b) = (\langle 0|, Y(a, z)b) = (\epsilon, \lambda|0\rangle)z^{-d-1} = \lambda z^{-d-1}.$$

Hence for a primitive and v primitive of minimal degree $Y(a, z)v = \widetilde{Y}(a, z)v$.

Now, we observe the following simple,

LEMMA 5.2.4. *Suppose* $Y(a, z)b = \widetilde{Y}(a, z)b$, *then* $Y(Ta, z)b = \widetilde{Y}(a, z)b$ *and* $Y(b, z)a = \widetilde{Y}(b, z)a$

Since V is T-generated from a set of (minimal degree) primitives, we see that $Y(a, z)b$ is determined for all $a, b \in V$ primitive. So in particular, $a_{(n)}b$ is determined for all $a, b \in V$ primitive. It is easy to compute these to be:

- $a_{(0)}b = 0$.
- for $n > 0$, $a_{(n)}b = 0$ if $n \neq \deg(b)$
- for $n > 0$, $a_{(n)}b = \mathsf{Coeff}(r_{z,0}(a, b))$, where Coeff is the coefficient of the power of z.

To finish the proof of the theorem, we need to understand the action of $Y(a, z)_-$ on products of primitive elements. This is achieved through the following,

PROPOSITION 5.2.5. *Let* $\widetilde{Y}$ *be a* r*-vertex algebra, and let* a, v *be a primitive. Then*

$$\widetilde{Y}(a, z)_-(v \cdot w) = \widetilde{Y}(a, z)_- v \cdot w + v \cdot \widetilde{Y}(a, z)_- w$$

PROOF OF PROPOSITION. It is enough to show that for $n \geq 0$

$$\widetilde{a}_{(n)}[v \cdot w] = \widetilde{a}_{(n)}[v] \cdot w + w \cdot \widetilde{a}_{(n)}[w].$$

This will follow essentially from the well known Borcherds commutator formulas, a special case of which is,

$$[a_{(m)}, v_{(-1)}] = \sum_{j=0}^{n} \binom{m}{j} (a_{(j)}v)_{(m-1-j)}.$$

But since for a and v primitive, $a_{(j)}v$ is always a scalar, so the only way $(a_{(j)}v)_{(m-1-j)}$ can be a nonzero operator is when $m = j$. So, the formula reduces to

$$[a_{(m)}, v_{(-1)}]w = (a_{(m)}b)_{(-1)}w.$$

Expanding this formula out and recalling that for $x \in V$ primitive $x_{(-1)}y = x \cdot y$, we see that the claim is verified. □

From Proposition 5.2.5 and the fact that V is T-generated by the set $\{a_\alpha\}$, the theorem is proven.

□

6. Examples

In this section, we show how the Heisenberg and lattice Vertex algebras may be given the structure of a r-vertex algebra.

(6.1) *Heisenberg Vertex Algebras:* We can describe an r-vertex algebra structure on the space $\Pi := \Pi^0$ of Example 3.3. First, we will need to define a grading on Π, which is achieved by setting

$$\deg 1 = 0 \quad \text{and} \quad \deg h_1(-j_1) \cdot h_2(-j_2) \cdots h_k(-j_k) \cdot 1 = j_1 + \cdots j_k$$

and taking the grading by degree. Now Π has an algebra structure by construction, and it is also equipped with a derivation $T : \Pi \to \Pi$ defined by $T \cdot 1 = 0$ and $T(h(-j) \cdot 1) = jh(-j-1)$. Additionally, Π has the structure of a bialgebra with derivation if we take $h(-1), h \in \mathfrak{h}$ to be primitive elements and $\epsilon(1) = 1$. Also, we note that Π has a bilinear form $(\cdot, \cdot)$ defined by $(1,1) = 1$ and such that $h(n)$ and $h(-n)$ are adjoint.

With these preliminaries, we can use Theorem 4.3.1 to construct an r-vertex algebra structure on Π with

$$r_{z,w}(h(-1), h'(-1)) = \frac{(h, h')}{(z-w)^2}.$$

Using Theorem 5.2.1, we easily conclude that this vertex algebra structure coincides with the one introduced in Example 3.3.

(6.2) *Lattice Vertex Algebras:* Next, we consider the vertex algebra associated to a positive-definite, even lattice L. Suppose L is a free abelian group of rank l together with a positive, definite, integral bilinear form $(\cdot, \cdot) : L \otimes L \to \mathbb{Z}$ such that (α, α) is even for $\alpha \in L$. Let $\mathfrak{h} = \mathbb{C} \otimes_{\mathbb{Z}} L$ equipped with the natural $\mathbb{C}$-linear extension of the form $(\cdot, \cdot)$ from L. Construct Π as above associated to $\mathfrak{h}$, and set $V_L = \Pi \otimes \mathbb{C}[L]$ where $\mathbb{C}[L]$ is the group algebra of L with basis $e^\alpha, \alpha \in L$. We will often write e^α for $1 \otimes e^\alpha$ and $e^0 = 1$.

Since Π and $\mathbb{C}[L]$ are bialgebras, V_L also has a bialgebra structure with gradation given by

$$\deg(e^\alpha) = \frac{1}{2}(\alpha, \alpha) \quad \text{and} \quad \deg(h \otimes t^n) = n, \ h \in \mathfrak{h}.$$

There is also a derivation T acting on V_L by $T \cdot 1 = 0$, $T(h(-n) \otimes 1) = nh(-n-1) \otimes 1$, and $T(1 \otimes e^\alpha) = \alpha(-1) \otimes e^\alpha$. With this data V_L is a bialgebra with derivation. V_L is also equipped with a bilinear form defined by $(1 \otimes 1, 1 \otimes 1) = 1$ and $(h \otimes e^\alpha, g \otimes e^\beta) = (h, h')(\alpha, \beta)$.

We proceed to give two descriptions of r-vertex algebra structures on V_L, and then show they are isomorphic. This common vertex algebra structure is none other than the **lattice vertex algebra** (of an even lattice). First, there is the usual vertex algebra constructed with vertex operators specified as follows: Let π_1 be the representation of $\hat{\mathfrak{h}}$ on Π defined above in Example 3.3. Also define a representation π_2 of $\hat{\mathfrak{h}}$ on $\mathbb{C}[L]$ (the action of scaling by the abelian subalgebra of constants) by setting

$$\pi_2(K) = 0 \text{ and } \pi_2(h(n))e^{\alpha} = \delta_{n,0}(\alpha, h)e^{\alpha} \text{ for } h \in \mathfrak{h}, \alpha \in L, n \in \mathbb{Z}.$$

Then V_L becomes a $\hat{\mathfrak{h}}$ module by $\pi = \pi_1 \otimes 1 + 1 \otimes \pi_2$. Then for $h \in \mathfrak{h}$, set $h(z) = \sum_{n\in\mathbb{Z}} h(n)z^{-n-1}$ where $h(n)$ acts as the operator $h \otimes t^n$ on the $\hat{\mathfrak{h}}$-module V_L. Then, set

$$Y(h_1(j_1)\cdots h_k(j_k)\otimes e^{\alpha}, z) = \frac{1}{(-j_1-1)!\cdots(-j_k-1)!} : \partial_z^{j_1-1}h_1(z)\cdots\partial_z^{-j_k-1}h_k(z)\Gamma_{\alpha}(z) :$$

where

$$\Gamma_{\alpha}(z) = e^{\alpha}z^{\alpha}e^{-\sum_{j<0}\frac{z^{-j}}{j}\alpha(-j)}e^{-\sum_{j>0}\frac{z^{-j}}{j}\alpha(-j)}$$

where $z^{\alpha}(a\otimes e^{\beta}) = z^{(\alpha,\beta)}a\otimes e^{\beta}$ and $e^{\alpha}(a\otimes e^{\beta}) = \epsilon(\alpha,\beta)a\otimes e^{\alpha+\beta}$, and $\epsilon : L\times L \to \pm 1$ is a 2-cocycle satisfying

$$\epsilon(\alpha,\beta)\epsilon(\beta,\alpha) = (-1)^{(\alpha,\beta)}.$$

Secondly, we take take the r-vertex algebra defined by Theorem 4.3.1 using the bicharacter

$$r : V_L \otimes V_L \to \mathbb{C}[(z-w)^{-1}]$$

uniquely specified by $r(e^{\alpha}, e^{\beta}) = \frac{\epsilon(\alpha,\beta)}{(z-w)^{(\alpha,\beta)}}$.

In order to show that the two descriptions of V_L coincide, we would like to use Theorem 5.2.1, noting that they have the same two-point functions. Unfortunately, that theorem is not applicable since V_L is not primitively generated (the elements e^{α} are group like, not primitive). But, actually the proof of Theorem 5.2.1 can be modified in this case as follows. Denote by $(V_L, 1, T, \widetilde{Y})$ and $(V_L, 1, T, Y)$ the first and second vertex algebra descriptions above. The vertex algebra V_L is T-generated from $\mathbb{C}[L]$, so by the Reconstruction Theorem, we just need to see that the fields $Y(e^{\alpha}, z)$ and $\widetilde{Y}(e^{\alpha}, z)$. are equal.

Step 1: Given $h \in \mathfrak{h}$, let us show that the fields $Y(h(-1), z) = \widetilde{Y}(h(-1), z)$. The proof of Theorem 5.2.1 shows that $Y(h(-1), z)_+ = \widetilde{Y}(h(-1), z)_+$, so what remains is to show that $Y(h(-1), z)_- = \widetilde{Y}(h(-1), z)_-$. On Π these two agree by the same reasoning as in the proof of Theorem 5.2.1. Furthermore, on $\mathbb{C}[L]$, we contend that

$$Y(h(-1), z)_- e^{\alpha} = \frac{(h,\alpha)e^{\alpha}}{z}.$$

Indeed, since the 2-point functions are

$$(\epsilon, \widetilde{Y}(h(-1), z)e^{\alpha}) = r(h(-1), e^{\alpha}) = \frac{(h,\alpha)e^{\alpha}}{z},$$

we know that

$$\widetilde{Y}(h(-1), z)e^{\alpha} = \frac{h(-1)_{(0)}e^{\alpha}}{z} = \frac{\lambda e^{\alpha}}{z},$$

since $\widetilde{a}_{(0)}$ is a scalar on group-like elements for $a \in V_L$ primitive. Clearly λ is specified by r and so we have shown that $\widetilde{Y} = Y$ on $\mathbb{C}[L]$ as well. Now, use Proposition 5.2.5 to conclude that $\widetilde{Y} = Y$ on all of V_L.

Step 2: $[h(-1)_{(n)}, Y(e^\alpha, z)] = (\alpha, h)z^n Y(e^\alpha, z)$ and $[h(-1)_{(n)}, \widetilde{Y}(e^\alpha, z)] = (\alpha, h)z^n \widetilde{Y}(e^\alpha, z)$. This follows easily from Borcherds's commutator formula and the fact that $Y(h(-1), z) = \widetilde{Y}(h(-1), z)$. Note that in particular $[h(0), Y(e^\alpha, z)] = (\alpha, h)Y(e^\alpha, z)$ and similarly $[h(0), Y(e^\alpha, z)] = (\alpha, h)\widetilde{Y}(e^\alpha, z)$. This means that the operator $Y(e^\alpha, z)$ maps elements of the form $x \otimes e^\beta$ where $x \in \Pi$ to elements of the form $y \otimes e^{\beta+\alpha}$ for $y \in \Pi$. The operator $\widetilde{Y}(e^\alpha, z)$ also behaves similarly.

Step 3: Since e^α has degree $\frac{1}{2}(\alpha, \alpha)$, we may write

$$Y(e^\alpha, z) = \sum_{n \in \mathbb{Z}} e^\alpha[n] z^{-n-\frac{(\alpha,\alpha)}{2}}$$

and

$$Y(e^\alpha, z) = \sum_{n \in \mathbb{Z}} \tilde{e}^\alpha[n] z^{-n-\frac{(\alpha,\alpha)}{2}}$$

where $e^\alpha[n]$ and $\tilde{e}^\alpha[n]$ have degree $-n$.

Step 4: We show that $Y(e^\alpha, z)e^\beta = \widetilde{Y}(e^\alpha, z)e^\beta$ for $\alpha, \beta \in L$. Indeed, by degree considerations, we have that

$$e^\alpha[s]e^\beta = 0 \text{ for } s > -\frac{(\alpha, \alpha)}{2} - (\beta, \alpha)$$

and

$$e^\alpha[s]e^\beta = c_{\alpha,\beta} e^{\alpha+\beta} \text{ for } s = -\frac{(\alpha, \alpha)}{2} - (\beta, \alpha).$$

Similarly for $\tilde{e}^\alpha[s]$, and examining 2-point functions we see that the $c_{\alpha,\beta}$ are completely determined. Now, using induction and the fact that

$$\frac{d}{dz} Y(e^\alpha, z) = Y(Te^\alpha, z) =: Y(\alpha(-1), z)Y(e^\alpha, z)$$

and

$$\frac{d}{dz} \widetilde{Y}(e^\alpha, z) = \widetilde{Y}(Te^\alpha, z) =: \widetilde{Y}(\alpha(-1), z)\widetilde{Y}(e^\alpha, z)$$

we see that $Y(e^\alpha, z)e^\beta = \widetilde{Y}(e^\alpha, z)e^\beta$.

Step 5: Using Step 2, we may compute $Y(e^\alpha, z)h_1(-n_1)\cdots h_k(-n_k)e^\beta$ and $\widetilde{Y}(e^\alpha, z)h_1(-n_1)\cdots h_k(-n_k)e^\beta$ and see that they are equal by Step 1 and 4.

References

[AB07] I. Anguelova and M. Bergvelt. H_D-Quantum Vertex Algebras and Bicharacters. *arXiv: 0707.1528*, 2007.

[Bor86] R. Borcherds. Vertex Algebras, Kac-Moody Algebras, and the Monster. *Proc. Nat. Acad. Sci. U.S.A.*, 83(10):3068–3071, 1986.

[Bor98a] R. Borcherds. Vertex Algebras. In *Topological Feld Theory, Primitive Forms, and Related Topics*, number 160 in Progr. Math., pages 35–77. Birkhauser, 1998.

[Bor98b] R. Borcherds. What is Moonshine? In *Proceedings of the International Congress of Mathematicians*, volume 1, pages 607–615, Berlin, 1998.

[Bor01] R. Borcherds. Quantum Vertex Algebras. In *Adv. Stud. Pure Math.*, number 31. Math Soc. Japan, 2001.

[Dun05] J. Duncan. Super-Moonshine for Conway's Largest Sporadic Group. *arXiv: math RT/0502267*, 2005.

[Dun06] J. Duncan. Moonshine for Rudvalis's Sporadic Group I, II. *arXiv: math.RT/0609449, math.RT/0611355*, 2006.

[FBZ01] E. Frenkel and D. Ben-Zvi. *Vertex Algebras and Algebraic Curves.* American Math. Society, 2001.

[FHL93] I. Frenkel, Y-Z. Huang, and J. Lepowsky. On Axiomatic Approaches to Vertex Operator Algebras and Modules. *Mem. Amer. Math. Soc.*, 104(494), 1993.

[FJ88] I. Frenkel and N. Jing. Vertex Representations of Quantum Affine Algebras. *Proc. Nat. Acad. Sci. U.S.A.*, 85(24):9373–9377, 1988.

[FK81] I. Frenkel and V. Kac. Basic Representations of Affine Lie Algebras and Dual Resonance Models. *Invent. Math.*, 62(1):23–66, 1980/81.

[FLM84] I. Frenkel, J. Lepowsky, and A. Meurman. A Natural Representation of the Fischer-Griess Monster with the Modular Function J as Character. *Proc. Nat. Acad. Sci. U.S.A.*, 81(10):3256–3260, 1984.

[FLM88] I. Frenkel, J. Lepowsky, and A. Meurman. *Vertex Operator Algebras and the Monster.* Academic Press, 1988.

[Gro96] I. Grojnowski. Instantons and Affine Lie Algebras. I. the Hilbert Scheme and Vertex Operators. *Math Res. Lett.*, (2):275–291, 1996.

[Leh99] M. Lehn. Chern Classes of Tautological Sheaves on Hilbert Schemes of Points on Surfaces. *Invent. Math.*, 136(1):157–207, 1999.

[LS01] M. Lehn and C. Sorger. Symmetric Groups and the Cup Product on the Cohomology of Hilbert Schemes. *Duke Math. J.*, 110(2):345–357, 2001.

[LW78] J. Lepowsky and R. Wilson. Construction of the Affine Lie Algebra $A_1^{(1)}$. *Comm. Math. Phys.*, 62(1):43–53, 1978.

[Nak97] H. Nakajima. Heisenberg Algebra and Hilbert Schemes of Points on Projective Surfaces. *Annals of Math (2)*, 145(2):379–388, 1997.

[Sch73] J. Schwartz. Dual Resonance Theory. *Phys. Rept.*, 8C:269–335, 1973.

[Seg81] G. Segal. Unitary Representations of some Infinite-Dimensional Groups. *Comm. Math. Phys.*, 80(3):301–342, 1981.

Titles in This Series